Environmental Ethics, Sustainability and Decisions

Fabio Zagonari

Environmental Ethics, Sustainability and Decisions

Literature Problems and Suggested Solutions

Springer

Fabio Zagonari
University of Bologna
Rimini, Italy

ISBN 978-3-031-21181-2 ISBN 978-3-031-21182-9 (eBook)
https://doi.org/10.1007/978-3-031-21182-9

This Springer imprint is published by the registered company Springer Nature Switzerland AG
The registered company address is: Gewerbestrasse 11, 6330 Cham, Switzerland

To Elena for her clever and tireless support.

Foreword

This book provides a summary of the main concepts involved in environmental ethics, sustainability and the related decisions. It can be read to discover the main cases and references for these three concepts, or as a consistent sequence of environmental ethics, sustainability and decisions that reveals the tight linkages among these three concepts. In particular, I focus on feasibility (whether realistic parameter values exist that would let a decision achieve its goal) rather than on reliability (whether a tight statistical relationship exists between a decision and its goal). Moreover, I refer to real policies (taxes, standards, subsidies, permits, national laws and regulations, bilateral and multilateral agreements) and projects (e.g., case studies) rather than to experimental or hypothetical decisions. Finally, I use a quantitative approach that analytically formalizes ethics, sustainability and the related decisions and demonstrates the approach using numerical exercises. The approach can be implemented either at an individual level or a country level if we assume rationality (a consistent set of informed decisions) within a normative analysis (which decision *should be* taken) rather than within a positive analysis (which decision *has been* taken). Since environmental sustainability implies seeking a compromise between economic and ecological criteria (environmental sustainability is an opportunity cost) based on specified ethical principles, the suggested decisions will fall along a spectrum between a more economic and a more ecological decision. In particular, I identify the main approaches applied to sustainable decisions (alternative objectives and contexts for policies, alternative methodologies and contexts for projects) by referring to more than 200 *theoretical* papers within the English language sustainability literature from the late 1980s to 2020 in the Scopus database. In addition, I highlight the main mistakes (missed objectives) and concerns (inadequate policies and methodologies) related to the application of these policies and methodologies, by performing statistical analyses of more than 800 *empirical* studies. Moreover, I suggest several approaches to avoid: these include applying marginal and monetary assessments for social and ecological interdependencies; disregarding equity in the case of environmental interactions; disregarding equity in the case of economic inefficiency; and using monetary or non-monetary flow assessments to take decisions about environmental stocks. Finally, I suggest what to do *in principle*: set priorities, constraints, assumptions, objectives, scales, metrics and indicators, *then* choose policy instruments or decision-making tools to measure and implement sustainability. I also suggest what to do *in practice*: in the context of inefficiency or interactions for policies, promote equity; in the context of environmental stocks for projects, choose dynamic models. This analysis is based on my 25 English empirical articles in sustainability science. Thus, this book supports efforts to achieve *consistency* between the chosen ethics, the adopted paradigm of sustainability, and the suggested decisions; these include market-based policies and agreements in cost-benefit analysis for weak sustainability and command-and-control policies and agreements in multi-criteria analysis for strong sustainability. I also consider the *transparency* of the chosen ethics behind the adopted paradigm of sustainability and the suggested decisions, such as substitution between forms of capital for weak sustainability or efforts to maintain the environmental status quo for strong sustainability.

Fabio Zagonari
University of Bologna
Bologna, Italy

Preface

Environmental sustainability is an ethical issue (Zagonari, 2020[1]), and secular and religious ethics offer complementary strategies to achieve sustainability in the short and long run, respectively (Zagonari, 2021[2]).

To reduce the complexity of the analysis, in Chap. 2 of this book, I reasonably exclude religious ethics and one's perceived duty to nature, and instead highlight two main environmental ethics based on one's duty to future generations. First, the goal will be to maximize total welfare from a representative individual perspective; second, the goal will be to minimize resource inequality from a per-capita individual perspective. Tables 2.1 and 2.2 summarize the main teleological environmental ethics (i.e., actions have a goal) and deontological environmental ethics (i.e., actions are performed for their own sake rather than based on their consequences), respectively. The world population size is relevant for both the specified ethical approaches, although its relative importance for the total welfare maximization and the resource inequality minimization depends on the share of Earth resources among developed vs. developing countries.

Chapter 3 identifies two main sustainability paradigms that lie on the spectrum between a pure ecological model and a pure economic model, together with intermediate paradigms based on the maximization of social-ecological continuity or the minimization of social-ecological impacts (Table 3.1). This spectrum ranges from weak sustainability as a pure economic model subject to considerations of intergenerational equity, and strong sustainability as a pure ecological model subject to a resilient reference status. Figures 3.1 and 3.2 depict ecological and social equilibria, respectively. Metrics are crucial for both sustainability paradigms, although indicators consistent with either weak or strong sustainability paradigms should include

[1] Zagonari, F. (2020) Environmental sustainability is not worth pursuing unless it is achieved for ethical reasons, Nature—Palgrave Communications 6: Art. No. 108.

[2] Zagonari, F. (2021) Religious and secular ethics offer complementary strategies to achieve environmental sustainability, Nature—Humanities and Social Sciences Communications 8: Art. No. 124.

both biophysical and socioeconomic dimensions within an explicit ethical framework to quantify progress towards environmental sustainability as an opportunity cost.

For weak sustainability, the objective is assumed to be efficiency; for strong sustainability, the objective is assumed to be equity.

In Chap. 4, I realistically exclude whether sustainability is achieved in terms of resource use and pollution production for the *current* status of weak or strong sustainability paradigms. Instead, I discuss policies and projects that are intended to achieve future sustainability based on the efficiency and equity objectives.

For the relationship between efficiency and policies, I rely on the simplest models to facilitate understanding by students and show that policies (taxes, standards, subsidies, permits for pollution; taxes, conservation areas, subsidies, harvest rights for resources) are not equivalent if some assumptions behind the pure economic model do not hold. For example, an economic general equilibrium model may not apply due to information asymmetry, uncertainty and imperfect competition. Disregarding equity in these situations is the first mistake. Similarly, if either stocks (short-run affects long-run) or interactions (some humans depend on other humans and vice versa) or both are relevant, shadow prices (social economic values of the environment) differ from marginal utility, both for pollution and for resources. Disregarding equity in these situations is the second mistake. Tables 4.1 and 4.3 summarize the main references to complicated models for pollution and resources (renewable or non-renewable), respectively.

For the relationship between equity and policies, I rely on the simplest models to help students understand, and show that alternative measures of inequality (Gini, Theil, and Shannon indexes) can be applied to national and international policies and agreements for individuals or groups. Similarly, alternative equity criteria (Kalai-Smorodinsky, Nash bargaining, Rawls and Harsanyi) can be applied to international agreements for individual countries or groups of countries. Tables 4.6 and 4.7 summarize the main references to the real impacts of policies and agreements for individuals and countries, respectively.

For the relationship between efficiency and projects, I discuss how cost-benefit analysis can properly deal with complicated contexts (time, space, uncertainty, linkages, inequalities) as well as to what extent this methodology can cope with its weaknesses (monetary assessment of the social and environmental features). I show that cost-benefit analysis is suitable for considering market distortions, but is inadequate for assessing environmental issues if stocks or interactions are crucial; the only exceptions are when opportunity costs and preventive costs within production approaches are applied to dynamic models with sensitivity analyses on monetary assessments.

In other words, a short-run marginal evaluation based on real or simulated market equilibrium prices is misleading: applying some production approaches (i.e., dose responses, replacement costs) or utility approaches (hedonic prices, travel costs, contingent valuation, choice experiments) in these situations is a third mistake.

For the relationship between equity and projects, I discuss how multi-criteria analysis can properly deal with complicated contexts (time, space, uncertainty, linkages, inequalities) as well as to what extent this methodology can cope with its weaknesses (assessment of relative weights). I show that multi-criteria analysis is inadequate for assessing environmental issues if stocks are crucial, unless subjective methods (the analytical hierarchy process, revised Simos' procedure, linear regression, factor analysis) are applied to dynamic models with sensitivity analyses for the relative weights of economic, social and environmental factors.

In other words, the relative weights based on the short-run perceptions could be misleading: improper use of subjective methods for assessment of relative weights within multi-criteria analysis in these situations is a fourth mistake. Participation by the people who will be affected by policy decisions is high in the context of efficiency and policies, but low in the context of equity and policies or agreements, and is medium in projects with average decisions (decisions by a representative individual) within cost-benefit analysis to depict weak sustainability and majority decisions (decisions by a majority of stakeholders) within multi-criteria analysis to depict strong sustainability.

Note that the improper application of methodologies in complicated contexts (time, space, uncertainty, linkages, inequalities) within cost-benefit analysis or multi-criteria analysis are concerns because they can lead to incorrect decisions regardless of the efficiency or the equity objectives.

By referring to the four mistakes highlighted in Chap. 4, within an education outlook, in Chap. 5, within a research outlook, I identified the 15 most important journals that *empirically* focus on sustainability and I sorted out these journals (with only a few exceptions, the estimated dummy variables used to represent the frequency of these mistakes with 1 = yes and 0 = no over the period from 1985 to 2020 confirmed the estimated variables used to represent the dynamics of these mistakes as changes over time in their frequency over time) into environmental economics journals (e.g., Environmental and Resource Economics, Journal of Environmental Management, Environment, Development and Sustainability, Environmental Science and Policy, Journal of Cleaner Production, Climate Policy), ecological economics journals (e.g., Ecological Economics, Environmental Management, Land Use Policy) and other sustainability journals (Water Resources Management, Environmental Modelling and Assessment, Science of the Total Environment, Sustainability, Water, Environmental Policy) by performing panel data probit estimations on a representative dataset of 835 *empirical* articles.

Remarkably, combining these overall results with the specific results obtained from Environmental and Resource Economics (which is concerned with information asymmetry, uncertainty, and imperfect competition, but does not apply multi-criteria analysis), the specific results obtained from Ecological Economics (which is not concerned with assumptions behind the general equilibrium model but applies multi-criteria analysis), together with the specific results obtained from both Environmental and Resource Economics and Ecological Economics (they both make the mistakes of disregarding equity if either stocks or interactions or both are relevant; they both make the mistakes of inappropriately applying production or utility approaches in short-run marginal evaluations) confirms the insights suggested in the methodological literature (Spash, 2020[3]): there is a lack of transparency of environmental economics to the general population (it does not reveal the assumptions behind the consistent suggested policies and adopted methodologies that are intended to achieve more economic sustainable decisions); environmental economics and ecological economics show more similarities than differences (Environmental and Resource Economics and Ecological Economics are close to the incorrectness frontier based on a Data Envelopment Analysis at 100% and 94%, respectively); and there is a lack of consistency in ecological economics (the suggested policies and adopted methodologies are inconsistent with the stated goal of a more ecologically sustainable solution).

Note that all four mistakes described earlier are statistically significant in my representative dataset of empirical articles, although the mistake of disregarding equity in the case of inefficiency is more relevant for solid waste management, the mistake of disregarding equity if either stocks or interactions or both are relevant is more important for climate change and groundwater management, the mistake of inappropriately applying production or utility approaches for short-run marginal evaluations within cost-benefit analysis is more relevant for biodiversity loss, and the mistake of improper use of subjective methods for assessment of relative weights within multi-criteria analysis is more important for solid waste management and groundwater management.

Next, I provide the solutions to the methodological problems highlighted in Chap. 4.

In particular,

- In a context of information asymmetry, uncertainty or imperfect competition, we should focus on equity in terms of welfare and resources for weak and strong sustainability, respectively;
- In a context of environmental interactions, we should focus on equity in terms of welfare and resources for weak and strong sustainability, respectively;
- In a context of environmental stocks within cost-benefit analysis, we should apply production approaches (opportunity costs or preventive costs) and dynamic models with sensitivity analyses for monetary assessment;
- In a context of environmental interactions within cost-benefit analysis, we should apply production approaches (opportunity costs or preventive costs) and game models with sensitivity analyses for monetary assessment;
- In a context of environmental stocks and interactions within cost-benefit analysis, we should apply production approaches (opportunity costs or preventive costs) and dynamic game models with sensitivity analyses for monetary assessment;
- And, in a context of environmental stocks within multi-criteria analysis, we should apply subjective methods and dynamic models with sensitivity analyses for relative weights.

[3] Spash, C.L. (2020) A tale of three paradigms: Realising the revolutionary potential of ecological economics, Ecological Economics 169: art. no. 106518.

Note that these suggestions are supported by examples from my 25 *empirical* articles (60% were published in the top 15 and 20 most important journals), in which I applied a wide range of policies and methodologies described in Chap. 4 (from policies, based on both efficiency and equity, to projects, using both cost-benefit analysis and multi-criteria analysis) to many topics, independently of labels such as environmental economics, ecological economics, environmental science or environmental studies (Levrel and Martinet, 2021[4]). These results suggest that sustainability is an interdisciplinary issue and that it requires an interdisciplinary research.

Thus, since consistency and transparency are ethical matters, students and the general population could become aware of possible mistakes and concerns in the literature, whereas researchers and sustainability scientists could have an incentive to be clearer and more consistent.

Therefore, environmental sustainability is an ethical issue for three key reasons: First, for the general population, sustainability is not worth pursuing unless it is achieved for ethical reasons (institutions are crucial for shaping behaviours, which I discuss in Chap. 3). Second, for decision-makers, environmental sustainability requires ethical decisions to achieve this goal (institutions are essential for suggesting values, which I discuss in Chap. 4). Third, for sustainability scientists, sustainability must be analysed using a consistent and transparent research strategy, regardless of the ethical approach to sustainability that is adopted (researchers must declare the assumptions behind weak sustainability, and apply consistent policies and methodologies to strong sustainability, which I discuss in Chap. 5).

Rimini, Italy
January 2023

Fabio Zagonari

[4] Levrel, H., Martinet, V. (2021) Ecological Economists: The Good, The Bad, And The Ugly? Ecological Economics 179: art. no. 106694.

Variables, Functions and Parameters

X_t	Stock of resource at time t
H_t	Flow of resource use at time t
Y_t	Flow of pollution production at time t
S_t	Stock of pollution at time t
P_t	Price of a generic good at time t
Q_t	Quantity of a generic good at time t
Z_{phyt}	Stocks of physical capital at time t
Z_{soct}	Stocks of social capital at time t
Z_{envt}	Stocks of environmental capital at time t
W	Wage rate
L	Labour factor
C	Unit cost
K	Overall capital
U	Utility or welfare
V	Value function
M	Monetary income
Q_1	Quantity of a good or service 1
Q_2	Quantity of a good or service 2
P_1	Price of a good or service 1
P_2	Price of a good or service 2
F	Production function for goods
G	Growth function of renewable resources
$P*$	Equilibrium price
a	Coefficients of X in G
b	Coefficients of X^2 in G
d	Good production per unit of pollution
E	Use of Earth resources
l	Coefficients of L in F
m	Coefficients of K in F
r	Interest rate from capital markets
i	Generic and local index over the book

j	Generic and local index over the book
n	Generic and local total index over the book
x	Generic outcome
p	Generic probability
α	Preference for consumption goods
β	Concern for environmental status
γ	Concern for future generations
δ	Decay rate of overall capital
ε	Intra-generational inequality aversion
ζ	Inter-generational inequality aversion
η	Sustainable average or per-capita use of resources
θ	Good production per unit of resource
λ	Shadow price for static variables
μ	Shadow price for dynamic variables
ν	Growth rate of population
ξ	Natural decay rate of pollution
π	Profit function
ρ	Inter-temporal preferences for consumption
σ	Social discount rate
τ	Concern for developing countries
υ	Concern for poorer current generation
φ	Generic and local coefficient over the book
χ	Generic and local coefficient over the book
ψ	Generic and local coefficient over the book
ω	Generic and local coefficient over the book

Notes. Latin capital letters refer to flow or stock variables, some Latin small letters refer to coefficients of variables, some Latin small letters per-capita variables, small Greek letters refer to preferences or perceptions, capital Greek letters refer to the same preferences or perceptions if two decision-makers are involved; * is used for optimal or equilibrium level of variables

Acronyms for Topics, Policies and Methodologies

AHP	Analytical Hierarchy Process
AIR	Air
ASS	Assessment
ASY	Asymmetric Information
CAP	Capability
CAPM	Capital Assessment Pricing Model
CBA	Cost Benefit Analysis
CE	Choice Experiment
CGEM	Computable General Equilibrium Model
CV	Contingent Valuation
DNS	Debt For Nature Swap
DR	Dose Response
DUT	Responsibility
DYN	Dynamic Model
EFF	Efficiency
EGEM	Economic General Equilibrium Model
EQU	Equity
EUM	Expected Utility Model
EV	Expected Value
FA	Factor Analysis
FLO	Flow
FS	Fuzzy Set
GIS	Geographic Information System
GLO	Global Scale
GRO	Protected Areas
HP	Hedonic Prices
IM	Inequality Measure
IMP	Imperfect Competition
IND	Decision without Interaction
INT	International Scale
IOM	Input-Output Model

IW	Inequality Weights
KAL	Kalai-Smorodinsky Solution
KYP	Kyoto Protocol
LAN	Land
LCA	Life Cycle Assessment
LOC	Local Scale
LR	Linear Regression
MCA	Multi-Criteria Analysis
MET	Metrics
MLCA	Monetary LCA
NAS	Nash Bargaining Solution
NAT	National Scale
OC	Opportunity Cost
PA	Production Approaches
PAA	Paris Agreement
PAR	Participation
PC	Preventive Cost
PER	Permit
PLA	Policies
POL	Pollution
POP	Population
PRO	Projects
RAW	Rawls Solution
RC	Replacement Cost
RED	Reduce Emissions from Deforestation
REG	Regional Scale
REL	Decision Interaction
RES	Resource
RIG	Rights
RIS	Risk Context
RSP	Revised Simos Procedure
SCE	Scenario
SD	Space Discount
SM	Subjective Methods
SOV	Sovereignty
SPA	Spatial Model
STA	Standard
STO	Stock
SUB	Subsidy
SWF	Social Welfare Function
TAX	Tax
TC	Travel Costs
TD	Time Discount
TEC	Technological Supports
TIS	Temporal Information System

UA	Utility Approaches
UNC	Uncertainty
WAT	Water
WLCA	Weighted LCA

Acronyms for Journals

AE	Applied Ecology
AEM	Aquaculture Economics and Management
AgEc	Agricultural Economics
AJWR	Australian Jou of Water Resources
AMF	Applied Mathematics and Finance
APFM	Asian-Pacific Financial Market
ARPS	Annual Review of Political Science
ARRE	Annual Review of Resource Economics
AtEn	Atmospheric Environment
BB	Biomass and Bioenergy
BC	Biodiversity and Conservation
CC	Climatic Change
CCE	Climate Change Economics
CD	Climate and Development
CH	Critical Horizon
CP	Climate Policy
EC	Env Conservation
EcMo	Ecological Modelling
EDS	Environment, Development and Sustainability
EE	Ecological Economics
EI	Ecological Indicators
EJ	Energy Jou
EL	Economics Letters
ELM	Env Law and Management
EM	Env Management
EMA	Env Modelling and Assessment
EMS	Env Management and Software
EnEc	Energy Economics
EnEt	Environmental Ethics
EnIn	Env Int
EnPo	Energy Policy

ENS	Energies
Env	Environmental
ENY	Energy
EP	Env Policy
EPE	Ethics, Policy and Environment
EPG	Env Policy and Governance
ERE	Env and Resource Economics
ERL	Env Research Letters
ES	Env Studies
ESD	Earth System Dynamics
ESP	Env Science and Policy
ESS	Ecosystem Services
FESE	Frontiers in Env Science and Engineering
GEC	Global Env Change
GW	Global Warming
IEA: PLE	International Environmental Agreements: Politics, Law and Economics
IEEP	Int Economics and Economic Policy
JCP	Jou of Cleaner Production
JCR	Jou of Coastal Research
JE	Journal of Economics
JEDC	Jou of Economic Dynamics and Control
JEEM	Jou of Env Economics and Management
JEEP	Jou of Env Economics and Policy
JEM	Jou of Env Management
JEPM	Jou of Env Planning and Management
JES	Journal of Economic Surveys
JESS	Jou of Env Science and Studies
JFE	Jou of Forest Economics
JH	Jou of Hydrology
Jou	Journal
JRE	Jou of Regulatory Economics
LaEc	Landscape Ecology
LE	Land Economics
LoEn	Local Environment
LUP	Land Use Policy
MASGC	Mitigation and Adaptation Strategies for Global Change
MP	Marine Policy
NaCu	Nature and Culture
NC	Nature Communications
NCC	Nature Climate Change
NH	Natural Hazards
NHSS	Nature Humanities and Social Sciences Communications
NP	Nature Palgrave Communications
NRM	Natural Resources Modelling
OCM	Ocean and Coastal Management

RCR	Resource, Conservation, Recycling
REE	Resource and Energy Economics
RP	Resources Policy
RSER	Renewable and Sustainable Energy Reviews
SD	Sustainable Development
SoSc	Social Sciences
SS	Sustainability Science
STE	Science of the Total Environment
SUS	Sustainability
TWQ	Third World Quarterly
WAT	Water
WM	Waste Management
WP	Water Policy
WR	Water Research
WRM	Water Resources Management
WRR	Water Resources Research
WST	Water Science and Technology

Contents

About the Author

Fabio Zagonari obtained Diploma in Violin at Bologna Conservatoire (Italy), Laurea in Economics at Bologna University, M.Phil. in Economics of Developing Countries at Cambridge University (UK), and Ph.D. in Economics at Ancona University (Italy). He has published on many topics (net per-capita per-year H index at 2.5), from Industrial Organization to Hospitality Management, from Decision Economics to Science Metrics, although his main focus is on Ethics, Environment and Environmental Ethics. He has held academic positions at Bologna University since 1995, with lectures mainly on Micro- and Environmental Economics. He had consultancy experiences for international organizations and national governments, agencies and foundations.

List of Figures

List of Tables

Chapter 1
Introduction

Environmental sustainability is an ethical issue,[1] and secular and religious ethics offer complementary strategies to achieve sustainability in the short and long run, respectively.[2] Thus, sustainable decisions must be developed in the context of environmental ethics. In this book, I present the main alternative environmental ethics, paradigms of sustainability, and the associated decisions, by showing how to solve problems in the literature that relate to combining environmental ethics, sustainability, and these decisions both consistently and transparently.

Note that I will show that many steps in developing consistent combinations of sustainability paradigms and solutions are permeated by environmental ethics. Moreover, I provide an original systematization rather than a comprehensive summary of the literature, although I reviewed more than 1000 references to support this book. Finally, I suggest that the decision steps should be transparently linked to the ethical steps.

Chapters 2–4 of the book include tables that summarize the *theoretical* literature (the seminal articles and more than 200 of the 1000 articles since the late 1980s), provide insights into the applied simplifications, describe two representative approaches, and discuss the approaches I did not consider. In particular, Chap. 2 focuses on environmental ethics, and I identify maximization of average or total welfare (based on a representative individual within an average perspective) and minimization of resource inequality (based on each single individual within a per-capita perspective) as the two main ethical rules. In Chap. 3, I focus on environmental sustainability, and particularly the weak and strong sustainability paradigms; the former aims to maximize welfare of the current generation, constrained by the future generation's welfare, and the latter aims at minimizing inequalities between current and future generations in terms of their access to environmental resources. In Chap. 4, I focus on decisions, and distinguish policies (plans) in Sect. 4.1 and projects (investments) in Sect. 4.2.

[1] Zagonari (2020).

[2] Zagonari (2021).

F. Zagonari, *Environmental Ethics, Sustainability and Decisions*,
https://doi.org/10.1007/978-3-031-21182-9_1

Section 4.1.1 links policies to efficiency in the context of weak sustainability by referring to pollution using two alternative approaches: reducing the quantity and improving pollution-treatment technologies. To do so, I develop simple static models to present the four main policies (taxes, standards, subsidies, permits) and develop simple dynamic models to show the impact of interactions in quantity decisions *and* assessments when the pollution stock is relevant (alternatively, think of local air pollution). Policies in Sect. 4.1.1 that support efficiency *also* refer to both renewable and non-renewable resources. I develop simple dynamic models to present the main economic and ecological policies (taxes, creation of conservation areas, subsidies, distribution of exploitation rights) and to characterize assessments and decisions when the resource stock is crucial (alternatively, think of river water flow). Section 4.1.2 discusses policies to achieve equity linked to strong sustainability, and refers to national laws and regulations that are intended to achieve equity. In that Section, I discuss inequality indicators that can be used to estimate equity at a national level (Gini, Theil, and Shannon indexes). Policies to achieve equity *also* refer to bilateral and multilateral agreements, which I accomplish by discussing static models to estimate equity at an international level (Nash bargaining, Kalai-Smorodinski and Rawls equilibria).

Section 4.2.1 of Chap. 4 discusses projects that are intended to achieve efficiency using cost–benefit analysis. I discuss how to solve the main weakness to this methodology (how to assess monetary values) as well as how this methodology can deal with alternative contexts (relevant time, space, uncertainty, linkages, inequalities) by suggesting that monetary life-cycle assessment is a promising methodology. Section 4.2.2 of Chap. 4 assesses projects intended to achieve equity from the perspective of multi-criteria analysis. I discuss how to solve the main weakness to this methodology (how to assess relative weights) as well as how this methodology can deal with alternative contexts (relevant time, space, uncertainty, linkages, inequalities) by suggesting weighted life-cycle assessment as a promising methodology.

Note that Chaps. 2–4 discuss the population, metric, and participation issues, respectively, whereas each of these Chapters presents some conceptual and numeric exercises.

This theoretical discussion lets me highlight four main potential methodological problems in *each Section of* Chap. 4. In particular, pollution policies lose efficiency to some extent in the case of information asymmetry, uncertainty, and imperfect competition. In other words, policies intended to increase efficiency for both pollution and resources are *not* equivalent if some of the assumptions behind the economic general equilibrium model do not hold. Second, pollution and resource policies might be inefficient in the case of interactions between decision-makers at international, national, regional, or local levels. In other words, the shadow prices of pollution and resources *differ from* the equilibrium market prices, both with pollution and with resource flows and stocks. Third, the equilibrium market prices identified by cost–benefit analysis in all utility approaches and some production approaches differ from shadow prices if pollution or resource stocks, with or without interactions, are involved in project decisions. Forth, the relative weights estimated by multi-criteria analysis with subjective methods might be misleading if pollution or resource stocks are involved in project decisions.

Chapter 5 summarizes the references to the *empirical* literature for a representative sample of more than 800 articles from the beginning of the literature in the late 1980s–2020, classified within a tight framework based on the theoretical concepts developed in previous Chapters of the book. In particular, I distinguish papers in terms of pollution versus resources, stocks versus flows, and air versus land versus water, by stressing the risk context and the two policies to improve efficiency for renewable resources (nature conservation) and non-renewable resources (green technologies). I distinguish each paper in terms of the problematic context (information asymmetry, uncertainty, imperfect competition, decision interactions) and the analytical scale (international, national, regional, local). I have characterised each article in terms of the policies (taxes, standards, subsidies, permits, exploitation rights) that it recommends to improve efficiency for both pollution and resources. I have also characterised policies to achieve equity for both individuals and countries (Nash bargaining, Kalai-Smorodinsky, Rawls, solidarity or capability, historical responsibility, sovereignty or grandfathering, Kyoto Protocol, Paris Agreement, Debt-for-Nature Swaps, Reduce Emissions from Deforestation). I have characterised projects to improve efficiency in the context of cost–benefit analysis in easy and problematic contexts. For monetary assessment in easy contexts, I discuss the following methods: dose responses, replacement costs, opportunity costs, and preventive costs within the production approaches; hedonic prices, travel costs, contingent valuation, and choice experiments with the utility approaches. For problematic contexts, I discuss time-based discounting, spatial discounting, expected utility, capital asset pricing models, computable general equilibrium models, and social welfare functions. I also discuss projects intended to promote equity within multi-criteria analysis in easy and problematic contexts. To assess relative weights in easy contexts, I discuss the analytical hierarchy process, revised Simos' procedure, linear regression, and factor analysis. For problematic contexts, I discuss temporal and geographical information systems, expected values, fuzzy sets, social accounting matrices, and inequality weights.

Note that I disregarded empirical articles on policies in which authors assumed that all assumptions behind the economic general equilibrium model were met, since all policies are equivalently efficient in this context. Similarly, I disregarded empirical articles on projects in which assessments of monetary values in cost–benefit analysis and of relative weights in multi-criteria analysis were not based on the stakeholder involvement, since participation is crucial for sustainable decisions.

In Sect. 5.1 of Chap. 5, this dataset let me show the statistical relevance of mistakes for objectives and concerns about policies and projects in the empirical literature over time and across journals as well as the relative importance and dynamics of policies and projects with respect to different environmental issues.

Note that I highlighted the statistical relevance of mistakes over time and across journals using panel data probit estimation, but showed the statistical relevance of concerns and environmental issues using numerical analysis and graphical analysis based on charts linked to each other by referring to high-level categories (pollution, resources, stocks, flows) and middle-level categories (efficiency, equity, cost–benefit analysis, multi-criteria analysis).

In Sect. 5.2 of Chap. 5, I discuss possible solutions to the methodological problems for both policies and projects. These solutions are supported by examples from my 25 *empirical* articles, and focus on both policies and projects within the efficiency or equity approaches.

Note that I refer to environmental equity more than justice. Justice includes both distributional equity and procedural equity based on access to adequate information and participation in decision processes, although distributional issues are crucial both for renewable and non-renewable resources and for pollution whenever there is a maximum pollution flow or stock that ecosystems can tolerate. However, I provide methodologies to account for justice where justice is more likely to be implemented, namely in projects at local or regional levels rather than in policies at national or international levels. Moreover, I develop the simplest quantitative models to show the possible methodological problems from an education perspective. However, I also mention the main advanced contributions from the literature to provide a research outlook. Finally, I refer to the two main objectives (efficiency and equity). However, I show how policies intended to improve efficiency, such as taxes, standards, subsidies, and permits, can be evaluated in terms of equity. In addition, I show how policies intended to promote equity, such as national laws and regulations and bilateral and multilateral agreements, can account for efficiency. I also show how decision methodologies used to improve efficiency can be fine-tuned to account for other features, including equity (e.g., cost–benefit analysis can account for inequalities) and decision methodologies intended to promote equity can be fine-tuned to account for other features, including efficiency (e.g., multi-criteria analysis can account for welfare).

Thus, students could skip Chap. 5, whereas researchers could skip Chap. 4. In particular, calculations of the shadow prices or implications of the economic general equilibrium assumptions based on the simple models in Chap. 4 are remarkable to students but obvious to researchers. However, Chap. 2 on environmental ethics and Chap. 3 on environmental sustainability can be useful for researchers, since these Chapters consistently summarize alternative approaches as well as the pros and cons of alternative paradigms. In other words, some Chapters are more useful to students, and others are more useful to researchers. Moreover, the models I have presented for students are simplified by referring to only two times and spaces, and use quadratic or logarithmic functions rather than more complex mathematics. This does not jeopardize the identification of possible methodological problems by researchers, since these problems are represented as positive statements in which a problem is related to a methodology but not to the formulation of the problem. In other words, the analyses are more suitable to students, but are also useful for researchers. Finally, the compact presentation of the models I have adopted for researchers does not jeopardize comprehension of these models by students, since each model starts with definitions and is coupled with illustrative examples. In other words, the discussions are more useful for researchers, but are also useful for students.

Therefore, the conceptual map for students and teachers of undergraduate students could be as follows:

- For a course on sustainable development, start with Chap. 3, skip Sect. 4.1 of Chapt. 4, and end with Sect. 4.2 of Chap. 4. This skips computable general equilibrium models and social accounting matrices, capital asset pricing models, and fuzzy sets.
- For a course on sustainability science, start with Chaps. 2 and 3, and end with Sect. 4.2 of Chap. 4. This skips the Remarks in Chaps. 2–4.

The conceptual map for researchers (and Ph.D. teachers) could be as follows:

- For a course on sustainable development, start with Chaps. 2 and 3, skip Sect. 4.1 of Chap. 4, and end with Chap. 5, thereby emphasizing the remarks on participation.
- For a course on sustainability science, start with Chaps. 2 and 3, skip Sect. 4.2 of Chap. 4, and end with Chap. 5, thereby emphasizing the remarks on population and metrics.

Note that it is possible to read only Chap. 4, since all main concepts that I introduced in Chaps. 2 and 3 are summarized, although this option will make it more difficult to achieve the main purpose of the present book.

References

Zagonari, F. (2020). Environmental sustainability is not worth pursuing unless it is achieved for ethical reasons. *Nature—Palgrave Communications, 6*(108).

Zagonari, F. (2021). Religious and secular ethics offer complementary strategies to achieve environmental sustainability. *Nature—Humanities and Social Sciences Communications, 8*(124).

Chapter 2
Environmental Ethics

This chapter discusses the main religious and secular environmental ethics. In particular, it provides two *Tables* that summarise the theological and deontological approaches in terms of their ethical rules and ethical reasons. I will show that the main *religious* ethics can be characterised as bio-centric and focused on all non-human beings (Hindus), zoo-centric and focused on animals (Buddhism), speciesist and focused on all species (Christianity), or as eco-centric based on trusteeship (Islam) and stewardship (Judaism). Moreover, I will show that within *secular* ethics, some teleological ethics (i.e., ethics based on the assumption that actions have a goal) are based on duty to nature (e.g., Attfield (1987) for bio-centrism, Dawkins (1988) for zoo-centrism, Singer (1975) for scientism, Cohen (1986) for speciesism, Hill (1983) for eco-centrism), some deontological ethics (i.e., ethics based on the assumption that actions are performed for their own sake rather than based on their consequences) are based on duty to nature (e.g., Taylor (1986) for bio-centrism, Schmidtz (1998) for zoo-centrism, Rolston (1979) for speciesism, Leopold (1949) for eco-centrism); some teleological ethics are based on duty to future generations (e.g., Aristotle and Harsanyi); some deontological ethics are based on duty to future generations (e.g., Kant and Rawls); some teleological secular ethics are based on the rights of nature (e.g., Varner (1998) for bio-centrism, Regan (1983) for zoo-centrism, Naess (1973) for eco-centrism); and some deontological secular ethics are based on the rights of nature (e.g., Webster (2003) for zoo-centrism, Palmer (2010) for scientism, Wilson (2016) for eco-centrism). Finally, I will provide some remarks on population issues.

Environmental ethics is the study of ethical questions raised by human relationships with the nonhuman environment (Palmer et al., 2014). Note that the purpose of this Chapter is not to provide a complete list of references, but rather to present a summary of environmental ethics for global and local sustainability (see Zagonari, 2020, for a recent review of religious environmental ethics; and Zagonari, 2019; for a recent review of secular environmental ethics).

In particular, Tables 2.1 and 2.2 summarize teleological and deontological ethics, respectively, by providing representative references rather than a comprehensive literature review.

F. Zagonari, *Environmental Ethics, Sustainability and Decisions*,
https://doi.org/10.1007/978-3-031-21182-9_2

Table 2.1 Relationships between ethical rules and ethical reasons in teleological approaches

		Ethical reasons			
		Duty to NH	Rights of NH	Duty to FG	Rights of FG
Ethical rules		Hedonic NH versus *eudemonic* versus **pragmatic**	NH welfare versus *no injuries*	Utilitarian FG versus *eudemonic*	FG welfare versus *no damages*
Maximize welfare	HB	INC	INC	WS, AG, DG, Carson (1962), Becker (1974)	UNF
	HB > NH	Cohen (1986) (Spe), Singer (1975) (Sci), Attfield (1987) (Bio), *Hill (*1983*) (Eco)*	Varner (1998) (Bio)	Harsanyi (1977), *Aristotle (*2014*)*	BPP, *PPP*, *ESP*
	HB & NH	Dawkins (1988) (Zoo), Johnson (1991) (Eco)	Naess (1973) (Eco)	Parfit (1984)	UNR
Minimize inequality	Resources	UNR	UNR	Dworkin (1981)	Lomasky (1987), *Nozick (*1974*)*
	Actions	UNR	UNR	Arneson (1989), *Sen (*1999*)*	UNR
	Outcomes	**Norton** (1991) **(Eco)**	*Regan (*1983*) (Zoo)*	*Barry (*1977*)*, Sher (1979)	UNF

Table 2.1 describes teleological ethics (i.e., in which actions have a goal) related to non-humans (NH) and future generations (FG). Teleological duties refer to consequences for NH and FG. Interests and rights (columns) can refer to individual entities, such as anthropocentrism (unspecified), scientism (Sci), zoocentrism (Zoo), and biocentrism (Bio), or collective entities, such as speciesism (Spe) or ecocentrism (Eco). Here, rights protect interests; Zoo includes both wild and domestic animals, and Bio is based on evolutionary theory. Ethical rules (rows) refer to social groups (i.e., all human beings). Here, anthropocentric approaches focus on interests and rights of current generation (CG) or FG; HB refers to strong anthropocentric approaches (i.e., intrinsic value to HB alone), HB > NH refers to weak anthropocentric approaches (i.e., a greater amount of intrinsic value to HB than to NH), HB & NH refers to non-anthropocentric approaches; hedonic and utilitarian approaches focus on pleasure or pain (NH) and welfare (CG and FG); a eudemonic approach (italics) of environmental virtue ethics focuses on prosperity and flourishing; a pragmatic approach (bold) focuses on practical achievements. *Other abbreviations HB* human beings; *WS* weak sustainability; *AG* a-growth; *DG* de-growth; *BPP* the beneficiary pays principle; *PPP* the polluter pays principle; *ESP* the equal sacrifice principle; *INC* incompatible; *UNF* unfeasible (i.e., theoretically unsuccessful) due to uncertainty about future technologies or preferences; *UNR* feasible but unreliable (i.e., untrustworthy in practice)

Table 2.2 Relationships between ethical rules and ethical reasons in deontological approaches

		Ethical reasons			
		Duty to NH	Rights of NH	Duty to FG	Rights of FG
Ethical rules		NH intrinsic value versus **pragmatic**	NH *freedom* versus no obstacles	FG freedom versus CG *virtue*	FG *freedom* versus no obstacles
Maximize health	HB	INC	INC	UNF	UNF
	HB > NH	INC	INC	UNF	UNF
	HB & NH	Leopold (1949) (Eco)	Palmer (2010) (Sci)	Sandler (2007) *(Eco)*	*PP, RP*
Minimize inequality	Resources	Taylor (1986) (Bio)	Wilson (2016) (Eco)	SS, Rawls (1971)	DRD (1986)
	Actions	Schmidtz (1998) (Zoo)	Cochrane (2012) (Zoo)	Kant (1996)	UDHR (1948)
	Outcomes	Callicott (1985) (Eco), Rolston (1979) (Spe), Schweitzer (1987) (Bio) **Minteer** (2012) **(Eco)**	*Webster* (2003) *(Zoo)*	Reuther (1975) (Eco)	Hayward (2004)
Equilibrium		Buddhism (Sci), Hinduism (Bio)			
Harmony		Confucianism (Eco), Taoism (Eco)			

(continued)

Table 2.2 (continued)

		Ethical reasons			
		Duty to NH	Rights of NH	Duty to FG	Rights of FG
Stewardship				Judaism (Eco)	
Trusteeship				Islam (Eco)	

Table 2.2 describes deontological ethics (i.e., actions performed for their own sake rather than based on their consequences) with respect to non-humans (NH) and future generations (FG). Deontological duties refer to actions towards NH and FG. Duties and rights (columns) can refer to individuals, such as anthropocentrism (unspecified), scientism (Sci), zoocentrism (Zoo), or biocentrism (Bio), or to collective entities such as speciesism (Spe) or ecocentrism (Eco). Here, rights protect actions; Zoo includes both wild and domestic animals; Bio is based on evolutionary theory. Ethical rules (rows) can refer to single individuals (underlined), as in the example of religious rules for each human being, or to social groups (no format), as in the case of secular rules for all human beings. Here, anthropocentric approaches focus on duties and rights to the current generation (CG) or FG; ecocentric approaches refer environmental health (e.g., integrity and stability); HB refers to strong anthropocentric approaches (i.e., intrinsic value to HB alone), HB > NH refers to weak anthropocentric approaches (i.e., a greater amount of intrinsic value to HB than to NH), HB & NH refers to non-anthropocentric approaches; a non-eudemonic approach (italics) of environmental virtue ethics is based on respect, benevolence, moderation, humility, compassion, courage, and simplicity; a pragmatic approach (boldface) focuses on practical achievements. *Other abbreviations HB* human beings; *SS* strong sustainability; *UDHR* United Nations Universal Declaration of Human Rights; *DRD* United Nations Declaration on the Right to Development; *PP* precautionary principle; *RP* reversibility principle; *INC* incompatible; *UNF* unfeasible (i.e., theoretically unsuccessful) due to uncertainty about future technologies or preferences

Each cell in Table 2.1 tries to answer the following question: Which duty to nature or future generations (specified in the table's columns) enables achievement of the environmental goal specified in the table's rows? In other words, start with the goal specified in a given row, then read across to answer the question. Note that Table 2.1 is based on the assumption that the measure of environmental sustainability cannot be provided by human beings, although alternative natural indicators can be applied (i.e., natural measures such as ecosystem resilience and ecological footprint cannot be replaced by human measures such as equal capabilities and genuine investments in case the former measures come into conflict with the latter measures) (Dicks, 2017).

Each cell in Table 2.2 tries to answer the following question: Which environmental consequences (specified in its row) are implied by the duty to nature or future generations (specified in its column)? In other words, start with the duties specified in the columns, then read down to answer the question. Note that knowledge about ecological systems, the state of the world, human psychology, and social institutions is crucial to support good ethical reasoning. Within the teleological approach, this helps to identify ethical actions, practices, and laws that can achieve a specified environmental goal; within the deontological approach, this helps to consider the environmental consequences of given ethical values, principles, and rules.

These tables include *also* two additional environmental ethics. First, environmental virtue ethics (i.e., a description of the effects of virtues such as justice, moderation, compassion, truthfulness, and hope, on an overall benevolent disposition to accomplish goals such as environmental sustainability) is context-dependent (Abbate, 2014): it is closer to a teleological approach in its eudemonic version (Hill, 1983) and to a deontological approach in its non-eudemonic version (Sandler, 2007). Second, environmental pragmatism (i.e., a focus on what to do to achieve goals such as environmental sustainability, irrespective of any theoretical foundations) is inadequate whenever principles come into conflict (Lowe, 2019): it is consistent with both approaches because it seeks actions, practices, and laws that can achieve a specified environmental goal in the teleological approach (Norton, 1991), and can examine the environmental consequences of given ethical values, principles, and rules in the deontological approach (e.g., Minteer, 2012). Note that beauty ethics (i.e., an aesthetic judgment about the sublimity of nature resulting in an outflowing of moral forces) can be disregarded because it depends on space (culture) and time (history) (Jaffe, 2015). Similarly, ethical psychologism (i.e., emotional responses to nature) can be disregarded since these responses are based on subjective relational values (Kasperbauer, 2015). In other words, these two environmental ethics are inadequate to cope with an urgent and global issue such as environmental sustainability.

Three main definitions are used in Tables 2.1 and 2.2.

1. Intrinsic value (Table 2.2) should be interpreted as contrasting with the instrumental value (i.e., the value as a means to some end) and extrinsic value (i.e., the value based on its relation to another valuable thing) within secular ethics, and as consisting of intrinsic properties (i.e., its metaphysical status) within religious

ethics. Anthropocentrism (strong) claims that only human beings have intrinsic value.

2. Moral status (i.e., the factor should be taken into account in decision-making) requires intrinsic value to be good, which is a property that can be based on alternative sources. Anthropocentrism (strong) claims that only human beings have moral status or (weak) that human beings are much more morally significant than any other living organisms (scientism, zoocentrism, biocentrism) or collective things (speciesism, ecocentrism). This is depicted by human beings HB > NH non-human beings in both Tables 2.1 and 2.2. Note that Vinnari et al. (2017), by focusing on farmed and wild animals, present utilitarianism as the main consequentialist ethical theory, but distinguish egalitarianism from the main deontologist ethical theory, since they do not jointly consider both humans and non-humans.
3. Moral rights are applied to humans and some animals only to enforce some moral status. Anthropocentrism claims that protection of human actions (Table 2.2) and interests (Table 2.1) is more important than the protection of actions and interests of other living things.

Note that distributive justice (i.e., the allocation of burdens and benefits in societies) and participative justice (i.e., the involvement of those affected in making the decisions) are more important in strong sustainability (which is more concerned about inter-generational justice; Table 2.2) than in weak sustainability (i.e., which is more concerned about intra-generational justice; Table 2.1). Most non-anthropocentric ethics claims that nonhuman organisms and collectives are not worthy of justice, although they have a moral status, whereas some non-anthropocentric ethics claims that distributive justice should include nonhuman beings (Caney, 2018).

Alternatively, Tables 2.1 and 2.2 can be read as follows. Responsibility to nature and non-human beings can be direct (intrinsic views in Keitsch, 2018) (I column in Table 2.1: teleological speciesism to zoocentrism based on consequences rather than actions if non-human beings are believed to have desires and hopes; I column in Table 2.2: deontological biocentrism to eco-centrism based on actions rather than consequences if nature has an intrinsic value) (Coyne, 2017; Diehm, 2014). Relational values (i.e., pertaining non-substitutable relationships between people and nature), within the Life Framework, are associated with Living in and as, whereas intrinsic and instrumental values are associated with Living from and with, respectively (O'Connor & Kenter, 2019). Responsibility to nature and non-human beings can be anthropocentric and indirect (extrinsic views in Keitsch, 2018) (III column in Table 2.1: teleological based on utilitarian or eudemonistic approaches; III column in Table 2.2: deontological based on freedom or virtue) (Gansmo Jakobsen, 2017; Taylor, 1986). Intrinsic value of nature or non-human beings could come from God (but then God will cope with nature) or from biotic community (i.e., the human and nature interdependence and mutual causality) up to biospherical egualitarism (i.e., all organisms have the same rights to live and flourish). Rights of nature or non-human beings enforce direct duties to nature or non-human beings (II column in Table 2.1: teleological protection of interests, by introducing positive rights of non-human beings or negative rights of the current generation; II column in Table 2.2: deontological protection of actions, by introducing positive rights of non-human beings or negative rights of the current generation). Non-human beings have no rights (i.e., no II columns in Tables 2.1 and 2.2), since a formal justice with egalitarian (i.e., the same respect to each non-human individual) and non-perfectionist (i.e., there is no a better non-human individual deserving a higher level of moral regard) perspectives should be applied not only to animals characterized by consciousness, intentionality and sentience as subjects-of-a-life, but to all organisms, with human beings having duty to help non-human victims of injustice and to never harm non-human individuals, and non-human rights overriding human benefits. Maximization of welfare (I–III rows in Tables 2.1 and 2.2) and minimization of outcome inequality (VI row in Tables 2.1 and 2.2) refer to humans and non-human beings as individual entities (scientism, zoocentrism). Minimization of resource inequality (IV row in Tables 2.1 and 2.2) refers to humans and non-humans as social entities (ecocentrism), whereas minimization of action inequality (V row in Tables 2.1 and 2.2) refers to humans and non-humans as individual entities (scientism, zoocentrism). Individual ethical rules come from religions (e.g., respect, compassion, not-harming, non-violence, not-stealing, non-possession) and biology (e.g., nature recycles everything, runs on sunlight, uses only the energy it needs). Descartes is not mentioned, since non-human beings are things to be used. Kant is included in Table 2.2, although only non-human beings characterized by moral agency (i.e., an autonomous agent with the ability to control decisions through free will) should *not* be treated as a means to an end. Weak sustainability resembles Judaism; a-growth is close to Islam; de-growth evokes the approach in Christianity; and strong sustainability resembles the precepts of Hinduism or Buddhism (Zagonari, 2020).

However, Tables 2.1 and 2.2 can be criticized as follows. People born as a result of current generation's actions and policies would not have been born at all so they are not harmed by those actions or policies (contingency of future generations upon decisions of the current generation: Parfit, 1984) (i.e., no II and IV columns in Tables 2.1 and 2.2). All moments are actual (time in intergenerational relations should be regarded as cosmic rather than simply historical or psychological: Griffith, 2017) (i.e., no future). All pains are actual (non-identity: Perret, 2003) (i.e., no generations). Date and space of one's birth is a matter of chance (but then I am not responsible for others' misfortune) or chosen by God (but then God will cope with future generations). Intrinsic value of future generations could come from God or from reciprocity (but then achieving human continuity might clash with reducing human pain) (Jonas, 1979). Future generations is an abstraction, since there is a continuous overlapping of next generations (i.e., I and II columns only in Tables 2.1 and 2.2). Future generations cannot have rights or even claims in the present (i.e., no IV columns in Table 2.1), since they do not exist (non-existence: Maklin, 1981): but concessional view (Elliot, 1989) (i.e., rights can exist presently without a holder because they correlate with duties). Future generations cannot have a right to resources that do not exist at the time of their existence (i.e., no IV columns in Table 2.2), because such a right cannot be satisfied (no satisfaction: De George, 1981): but constitutive view (Sterba, 1980) (i.e., a present action that may be a cause for a legitimate complaint represents a future right). Subjects of rights could be social roles or status-functions rather than actual persons (e.g., human collectives or human beings). Within comparative standards, apart from the egalitarian standard (I column in Table 2.1) and the commutative standard represented by strong sustainability (III column in Table 2.1), one could mention an aggregative standard based on total or average basket maximization in a transgenerational community (the most disadvantaged generations make the greatest sacrifices so justice as effectiveness), whereas within non-comparative standards, one could quote the sufficientarian standard (enough resources to pursue the aims and aspirations people affirm) (Frankfurt, 1987) and the communitarian standard (preservation of the cultural identity of communities) (Thompson, 2009). The correct application of a discount factor to market evaluation of future generations' welfare as surplus within a welfare maximization framework (II row In Table 2.1) means attaching a smaller value to it (i.e., Hardin, 1974: social justice is more important than intergenerational equity in life boat ethics). The incorrect application of a discount factor to resources could lead to a debt towards the environment and towards future generations if it is inconsistent with natural dynamics (i.e., Azar and Holmberg, 1995: social values of ecosystems are underestimated by marginal market evaluations). The precautionary and reversibility principles (III row In Table 2.2) assume a current generation risk aversion and amount to MaxMin applied to future generation benefits arising from current (technological) risk (see Sect. 4.2.1.3 in Chap. 4). Individual ethical rules about next generation rather than to future generations come from religions (e.g., the Golden rule, love, respect, honor, and cherished attitudes) and philosophy (e.g., sensitivity and humility; rationality, consistency, and universality).

Note that weak sustainability (here defined as maximization of the current and future generations's welfare so intra and inter-generational participatory and distributive justice is excluded) will be better characterized in Chap. 3 as maximization of the current generation's welfare subject to future generations's welfare being equal or larger than the current generation's welfare. Next, strong sustainability (here defined as minimization of inequality of resources to current and future generations so intra and inter-generational participatory and distributive justice is included) will be better characterized in Chap. 3 as resources to future generations being equal or larger than resources to the current generation.

In particular, some issues are unsolved by Tables 2.1 and 2.2.

- Within a western approach: humans (subject) are different from non-humans (object), where nature has its own dynamic laws (anyway, laws refer to ecocentrism and speciesism rather than to scientism, zoocentrism, biocentrism) so it has intrinsic value (although a moral standing is not intrinsic, since nature is one possible long-run dynamic status, in Darwinian terms). Should humans leave nature untouched? Yes, humans should maintain naturally disproportionate species (although huge social impacts could arise) and should not preserve naturally extinction species (although duties to future generations would suggest the opposite).

- Within an eastern approach: humans are included in nature (including pain, killing, death) so humans can use nature (like other animals), but each human has no intrinsic value larger than non-humans. Should humans leave each single pieces of nature untouched? Yes, humans should preserve each single non-human beings (although future generations will not be affected) and should foster wildness for its biodiversity (although wild nature is worse without humans).

In other words, the western approach needs the eastern approach and vice versa for them to be *realistic* towards environmental sustainability.

Teleological duties to non-human beings in terms of individual entities (scientism, zoocentrism, biocentrism) are *redundant* for weak sustainability (i.e., welfare of future animals is independent on welfare of current animals), in terms of social entities (speciesism, ecocentrism) are *unfounded* (i.e., there is no welfare of species or ecosystems). Teleological duties to future generations are founded (on reciprocity or psychological concerns of the current generation for next generation combined with an infinite chain of subsequent generations), although *potentially redundant* (due to future technologies and preference) for weak sustainability. In summary, teleological (secular) duties can be reliable for weak sustainability if based on duties to future generations.

Deontological duties to non-human beings in terms of individual entities (scientism, zoocentrism, biocentrism) are *redundant* for strong sustainability in secular ethics (although these duties can be instrumentally assumed if properly shared, but they exclude justice between human beings and nature), but they are founded in religious ethics (Buddhism, Hinduism); duties to non-human beings in terms of social entities (speciesism, ecocentrism) are *unfounded* (intrinsic values do not imply moral standing) in secular ethics (although these duties can be instrumentally assumed if properly shared, but they exclude justice between human beings and nature), but they are founded in religious ethics (Taoism, Confucianism). Duties to future generations are founded (on reciprocity or psychological concerns of the current generation for next generation combined with an infinite chain of subsequent generations) in secular ethics and founded (on stewardship in Judaism, trusteeship in Islam) in religious ethics. In summary, deontological (secular) duties are reliable for strong sustainability if based on duties to future generations.

Note that religious ethics fix individual rules which are independent of goals (e.g., human and non-human pain alleviation is crucial in *many* religious ethics but disregarded in *most* secular ethics; intra-generational equity is crucial in *all* religious ethics and disregarded in *all* secular ethics), although it can have beneficial impacts on nature preservation. Moreover, virtue ethics is useless, since virtue sources are *unfounded*. Finally, environmental pragmatism is inadequate, since it is not based on ethical principles.

2.1 Main Insights of Chap. 2

In this Chapter, I stressed teleological duties to future generations as the (secular) ethics required to achieve sustainability (e.g., Aristotle in eudemonic terms based on flourishing and Harsanyi in utilitarian terms based on welfare, but both with human beings more important than non-human beings from an average person's perspective). This excludes distributive and participative justice: ecosystems have *no* intrinsic value and humans are not considered to be part of nature. Weak sustainability is an example of teleological duty to future generations. Next, I stressed deontological duties to future generations as the (secular) ethics required to achieve sustainability (e.g., Kant in terms of freedom and Rawls in terms of resources, both focused on human beings from a per-capita perspective). This includes distributive and participative justice: ecosystems have intrinsic value but *not* moral standing, which is indirectly due to the current generation's duty towards future generations, but humans are considered to be outside nature, otherwise each single human would have no intrinsic value. Strong sustainability is an example of deontological duty to future generations. Thus, I showed that environmental ethics is subjective and I assumed that environmental ethics is instrumental and non-foundational so the question becomes which shared values we want to implement to achieve a specified goal, including sustainability. In other words, whatever the source of environmental ethics, the focus on sustainability means looking for environmental ethics that lead to sustainable behaviour.

2.2 Remarks on Population

Over the centuries, population growth has been studied for its consequences on political institutions, economic development, and social systems to a greater extent than for its consequences on environmental degradation. Think of the extreme positions of the pessimistic Malthusians (i.e., the need to reduce population to avoid catastrophic prospects for humanity) and the optimistic Cornucopians (i.e., the faith in the powerful reproductive forces of nature).

In 1960s and 1970s, concerns about the impacts of world population growth on global environmental balance began to appear in the academic literature. Think of the book Silent Spring by Carson (1962) and the report Limits to Growth by Meadows et al. (1972). Next, the demographic growth was addressed in the political literature. Think of the report Our Common Future by Brundtland (1987).

From the 2000s onward, a progressive indifference of the public opinion and the academic literature has been observed about the relationships between population growth and environmental sustainability (Bergaglio, 2017). Next, both the past Millennium Development Goals and the new Sustainable Development Goals do not contain a specific and explicit target committed to the growth of the world population. Similarly, no mention to the relationship between population growth and the increase in the global average temperature is found in Paris Agreement.

However, the impact of population on the environment depends on its total size, structure and spatial distribution, as well as on the per capita consumption of resources (e.g., stock of renewable resources such as forest and fish; non-renewable resources as minerals and coals) and the per capita production of pollutants (i.e., stock of pollution to land as solid waste, pollution to air as GHG, pollution to water as plastic waste to the sea or chemical waste to groundwater).

From a theoretical perspective, Tables 2.1 and 2.2 in this chapter highlighted that the decisions of the current generation could affect the existence of future generations.

From an empirical perspective, Tables 2.1 and 2.2 in this chapter identified two main approaches to environmental sustainability. An approach based on achievement of average welfare (e.g., weak sustainability); an approach based on assurance of per capita use of resources and production of pollutants (e.g., strong sustainability). Note that population growth is a constraint for both weak sustainability (i.e., future generations must achieve at least the same level of welfare) and for strong sustainability (i.e., each individual in future generation should have access to the same amount of resources). These constraints aim at avoiding that the sustainability burden is on future generations.

In order to estimate if population growth affects the constraints in the average approach to a greater or smaller extent than in the per capita approach, let us assume that welfare achievement is linear in the use of resources and in the production of pollutants (i.e., U(E) = E): indeed, a concave relationship could be supported by an optimistic expectation of productivity growth, but a convex relationship could be based on a pessimistic expectation of consumption growth. Moreover, let us assume that the per capita approach distributes the Earth resources uniformly (i.e., E/popt for each individual with popt the total world population), whereas the average approach distributes the Earth resources according to a weighted average ((pops*((7.88*ress)/popt) + (1 − pops)*((7.88*(1 − ress))/popt))), where pops and ress are the share of world population living in and Earth resources used by developed countries, respectively (i.e., in terms of world population, pops = 0.18; in terms of Ecological Footprint EF, ress is 0.73 = 5.74/7.88 ha). Finally, let us standardise the per capita Ecological Footprint at the rounded 2020 level of world population POP (i.e., total available EF = 1.7 ha per capita times 7 Billion people).

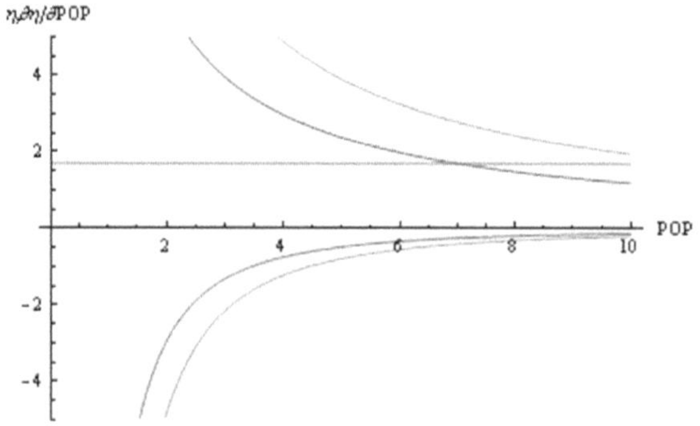

Fig. 2.1 Impacts of POP on levels and changes of sustainable EF at the current DC and LDC use of resources. The decreasing (blue) curve below = sustainable per capita EF, the decreasing (grey) curve above = average EF, the increasing (purple) curve above = derivative of sustainable per capita EF with respect to POP, the increasing (green) curve below = derivative of average EF with respect to POP. Abbreviations: POP = world population (billion people), η = the sustainable use of resources in terms of EF within the per capita or the average approach; EF = Ecological Footprint, DC = OECD countries, LDC = non-OECD countries. *Notes* Blue = level per capita EF; Grey = level of average EF; Purple = change per capita EF; Green = change of average EF

Figure 2.1 represents the impacts of the world population levels and changes on per capita and average Ecological Footprint at current share of world population and Earth resources between people in developed and developing countries. Note that the reduction of the use of resources to achieve sustainability within the average approach is −39%. Moreover, the population size affects the per capita approach to a greater extent than the average approach in terms of levels, but the opposite in terms of changes. Finally, the reduction of the use of resources to achieve sustainability within the per capita approach is −71 and −21% in developed and developing countries, respectively.

Figure 2.2 represents the impacts of the world population on the *levels* of per capita versus on average Ecological Footprint at *any* share of population and Earth resources between developed and developing countries for a world population at 10 billion people. Note that the population size could affect the per capita approach to a smaller extent than the average approach in terms of levels if the share of Earth resources in developed countries would be smaller than the current 0.73.

Figure 2.3 represents the impacts of the world population on the *changes* of per capita versus on average Ecological Footprint at *any* share of Earth resources between people in developed and developing countries. Note that the population size could affect the per capita approach to a larger extent than the average approach in terms of changes if the share of Earth resources in developed countries would be smaller than the current 0.73.

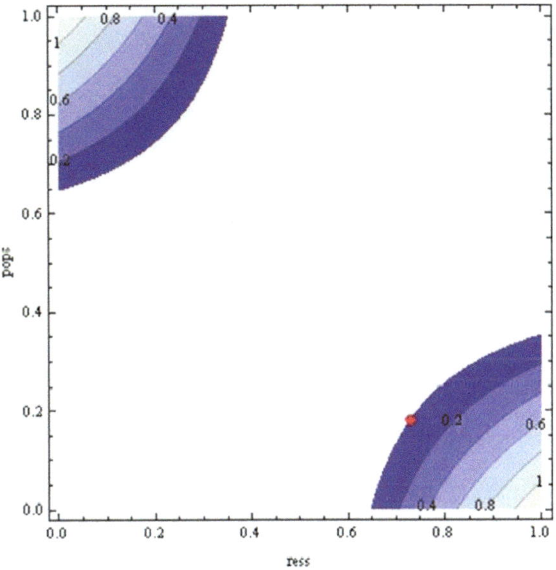

Fig. 2.2 Impacts of POP on levels of EF at any DC and LDC distribution of population and resources. Differences in levels of sustainable EF within per capita versus average approaches at a world population of 10 Billion and with all possible shares of population and Earth resources between developed and developing countries. Abbreviations: POP = world population, EF = Ecological Footprint, DC = OECD countries, LDC = non-OECD countries, pops = proportion of world population living in DC; ress = proportion of Earth resources used by DC. *Notes* White area = impact of POP on per capita EF levels is smaller than on average EF levels; Red point = the current distribution of world population and Earth resources

As a summary of this remark, population growth is currently disregarded as a sensitive ethical topic. However, it is relevant for any approach to sustainability, which is *also* an ethical topic. Note that Figs. 2.1, 2.2 and 2.3 statically and deterministically depicts the dynamic and stochastic interrelationships between the world population size, the share of resources used by developed versus developing countries and the share of individuals living in developed versus developing countries. However, all possible shares of population and resources among developed versus developing countries are depicted at the realistic world population in 2100.

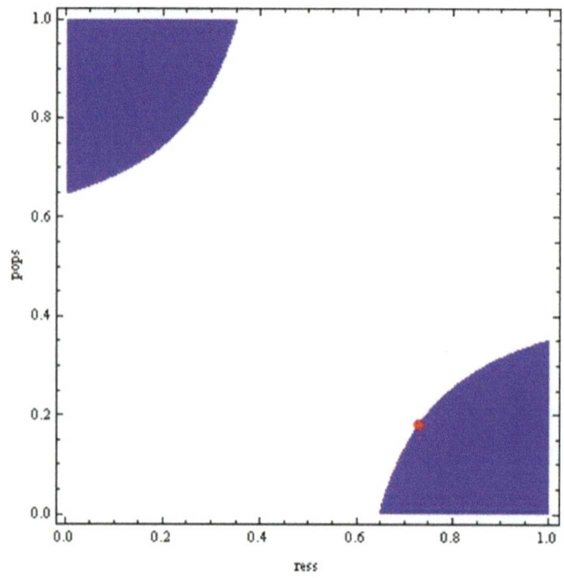

Fig. 2.3 Impacts of world POP on changes of EF at any DC and LDC distribution of population and resources. Differences in changes of sustainable EF with respect to POP within per capita versus average approaches at a world population of 10 Billion and with all possible shares of population and Earth resources between developed and developing countries. Abbreviations: POP = world population, EF = Ecological Footprint, DC = OECD countries, LDC = non-OECD countries, pops = proportion of world population living in DC; ress = proportion of Earth resources used by DC. *Notes* White area = impact of POP on per capita EF changes is larger than on average EF changes; Red point = the current distribution of world population and Earth resources

2.3 Exercises

Exercises below share the following conceptual Table 2.3. Note that climate change impacts on current generations too. Moreover, environmental issues involving stocks (i.e., biodiversity, wildness, climate, waste, forestation, nuclear) prevail on environmental issues involving flows (i.e., water, energy, migration, animal). Finally, population impacts on both POL and RES.

1. If biodiversity has a positive intrinsic (non-instrumental) value, since ecosystems rely on it to be resilient in the long-run, which approach (i.e., teleological vs. deontological), duty (i.e., to nature vs. to future generations) and perspective (i.e., anthropocentric, scientism, zoo-centric, bio-centric, eco-centric, speciesism) can support biodiversity conservation?
2. If wildness (i.e., non-human origin) has extrinsic (non-anthropocentric) value, since its protection or restoration favors the preservation of ecosystems, which approach (i.e., teleological vs. deontological), duty (i.e., to nature vs. to future generations) and perspective (i.e., anthropocentric, scientism, zoo-centric, bio-centric, eco-centric, speciesism) can support wildness restoration?

Table 2.3 Environmental behaviours in terms of ethical approaches, duties and perspectives

	Duty	Anthropocentric	Scientism	Zoo-centric	Bio-centric	Eco-centric	Speciesism
Teleological	NH		Animal	*Migration*			
	FG	Biodiversity *Wildness* **Climate** Population *Nuclear*				*Waste* *Forestation*	
Deontological	NH		*Animal*	Migration		**Biodiversity** Wildness Climate	Species *Climate*
	FG	**Biodiversity** Wildness **Climate** Forestation *Water* *Energy* Nuclear		*Migration*		Waste Forestation	*Species*

Table 2.3 summarises the main ethical approaches (i.e., teleological versus deontological), duties (i.e., to nature vs. to future generations) and perspectives (i.e., anthropocentric, scientism, zoo-centric, bio-centric, eco-centric, speciesism) which support the main ethical behaviours about the main environmental issues. *Abbreviations NH* Non-Humans, *FG* Future Generations, *Bold* strongly supported, *Italic* weakly supported; *Energy* Energy conservation, *Waste* Waste management, *Animal* Animal welfare, *Water* Water conservation are individual behaviours with local impacts; *Biodiversity* Biodiversity conservation, *Climate* Climate change mitigation, *Population* Population reduction, *Nuclear* Nuclear waste management, *Wildness* Wildness restoration, *Species* Species preservation, *Migration* assisted Migration, *Forestation* re-Forestation are global issues to be addressed by collective actions (e.g., protected areas, international agreements)

3. If climate change has a negative intrinsic (non-instrumental) value, since it affects resilience of ecosystems in the long-run, which approach (i.e., teleological vs. deontological), duty (i.e., to nature vs. to future generations) and perspective (i.e., anthropocentric, scientism, zoo-centric, bio-centric, eco-centric, speciesism) can support geo-engineering (i.e., intentionally manipulating the climate in response to climate change)?
4. If a species has a larger or a smaller value, according to its contribution to biodiversity, which approach (i.e., teleological vs. deontological), duty (i.e., to nature vs. to future generations) and perspective (i.e., anthropocentric, scientism, zoo-centric, bio-centric, eco-centric, speciesism) can support species preservation? Note that nature with intrinsic value does not imply nature with moral standing. Thus, humans should not be concerned with species extinctions (i.e., it is the natural law), unless extinctions are due to humans (e.g., urbanization).
5. Which approach (i.e., teleological vs. deontological), duty (i.e., to nature vs. to future generations) and perspective (i.e., anthropocentric, scientism, zoo-centric, bio-centric, eco-centric, speciesism) can support assisted migration? Note that nature with intrinsic value does not imply nature with moral standing. Thus, humans should not be concerned with deaths during migration (i.e., it is the natural law), unless pain or difficulty are due to humans (e.g., preservation or restoration of ecological corridors).
6. Which approach (i.e., teleological vs. deontological), duty (i.e., to nature vs. to future generations) and perspective (i.e., anthropocentric, scientism, zoo-centric, bio-centric, eco-centric, speciesism) can support the concern for the wild and domestic animals' welfare? Note that nature with intrinsic value does not imply nature with moral standing. Thus, humans should not be concerned with lions dying for starvation (i.e., it is the natural law), unless pain or difficulty are due to humans (e.g., preservation or restoration of wild spaces).
7. Which approach (i.e., teleological vs. deontological), duty (i.e., to nature vs. to future generations) and perspective (i.e., anthropocentric, scientism, zoo-centric, bio-centric, eco-centric, speciesism) can support re-forestation?
8. By leaving aside Catholic traditional thoughts (e.g., St. Augustine, St. Thomas, St. Francis, Pope Leo XIII, John Paul II), while focusing on happy sobriety in the recent Laudato Si' by Pope Francis, Christian Catholicism retains concepts from sacred texts on intergenerational solidarity and the value of a single organism; it assumes population growth and personal fulfilment as constraints; it relies on the intrinsic value of nature, based on respect for creation and on strong law enforcement, but based on universal communion; it supports frugality by referring to the sanctity of nature (like panentheism and Eastern Orthodoxy), and the same dignity of humans and non-humans (like pantheism or panentheism in Hinduism and Buddhism), but with an anthropocentric perspective (i.e., humans are above all other creatures), without an immanent God (i.e., no things are intrinsically bad). Is Christian Catholicism concerned about Energy conservation, Nuclear waste management, Population reduction, Waste management, Animal welfare, Water conservation, Biodiversity conservation, Climate change

mitigation, Wildness restoration, Species preservation, assisted Migration, re-Forestation? [**Energy conservation, Waste management**, Animal welfare, **Water conservation**, Biodiversity conservation, Climate change mitigation, *Wildness restoration*, Species preservation, *re-Forestation*, where Missing = no support, Bold = strongly support, Italics = weakly support]

9. In the Christian Orthodox religion, one should live in faith and love by following the contemplative life and withdrawing from the created world; God is omnipotent and independent, whereas the world is limited, dependent, and incomplete without God, although the Holy Spirit lets us combine the divine unity, divine transcendence, and God into creation. Is Christian Orthodox concerned about Energy conservation, Nuclear waste management, Population reduction, Waste management, Animal welfare, Water conservation, Biodiversity conservation, Climate change mitigation, Wildness restoration, Species preservation, assisted Migration, re-Forestation? [*Energy conservation, Nuclear waste management, Population reduction, Waste management, Animal welfare*, Water conservation, **Biodiversity conservation**, Climate change mitigation, Wildness restoration, **Species preservation**, assisted Migration, re-Forestation, where Missing = no support, Bold = strongly support, Italics = weakly support]
10. Christian Protestantism focuses on nature as a manifestation of God, on anthropocentrism, on an active life rather than a contemplative life, and on nature's beauty, with a traditional and recent stress on social and ecological justice, respectively. Is Christian Protestantism concerned about Energy conservation, Nuclear waste management, Population reduction, Waste management, Animal welfare, Water conservation, Biodiversity conservation, Climate change mitigation, Wildness restoration, Species preservation, assisted Migration, re-Forestation? [Nuclear waste management, Waste management, *Animal welfare*, Water conservation, Biodiversity conservation, **Climate change mitigation**, *Wildness restoration*, Species preservation, *assisted Migration*, **re-Forestation**, where Normal = no support, Bold = strongly support, Italics = weakly support]

References

Abbate, C. (2014). Virtues and animals: A minimally decent ethics for practical living in a non-ideal world. *Journal of Agricultural and Environmental Ethics, 27*, 909–929.

Aristotle. (2014). *Nicomachean ethics* (R. Crisp, Trans. and Ed.). *Texts in the History of Philosophy*. CUP.

Arneson, R. (1989). Equality and equal opportunity for welfare. *Philosophical Studies, 56*, 77–93.

Attfield, R. (1987). *The theory of value and obligation*. Croom Helm.

Azar, C., & Holmberg, J. (1995). Defining the generational environmental debt. *Ecological Economics, 14*, 7–19

Barry, B. (1977). Justice between generations. In P. M. S. Hacker & J. Raz (Eds.), *Law morality and society: Essays in honour of H.L.A. Hart*. Clarendon.

Becker, G. S. (1974). A theory of social interactions. *Journal of Political Economy, 82*, 1063–1093.

Bergaglio, M. (2017). The contemporary illusion: Population growth and sustainability. *Environment, Development and Sustainability, 19*, 2023–2038.
Callicott, J. B. (1985). Intrinsic value, quantum theory, and environmental ethics. *Environmental Ethics, 7*, 257–275.
Caney, S. (2018). Justice and future generations. *Annual Review of Political Science, 21*, 475–493
Carson, R. (1962). *Silent spring*. Houghton Mifflin.
Cochrane, A. (2012). *Animal rights without liberation*. Columbia University Press.
Cohen, C. (1986). The case for the use of animals in biomedical research. *The New England Journal of Medicine, 315*, 865–869.
Coyne, L. (2017). Phenomenology and teleology: Hans Jonas's philosophy of life. *Environmental Values, 26*, 297–315.
Dawkins, M. S. (1988). Behavioural deprivation: A central problem in animal welfare. *Applied Animal Behavioural Science, 20*, 209–225.
De George, R. (1981). The environment, rights, and future generations. In E. Partridge (Ed.), *Responsibilities to future generations: Environmental ethics* (pp. 157–166). Prometheus Books.
Dicks, H. (2017). Environmental ethics and biomimetic ethics: Nature as object of ethics and nature as source of ethics. *Journal of Agricultural and Environmental Ethics, 30*, 255–274.
Diehm, C. (2014). Darwin and deep ecology. *Ethics and the Environment, 19*, 73–93.
DRD. (1986). www.un.org
Dworkin, R. (1981). What is equality? Part 2: Equality of resources. *Philosophy and Public Affairs, 10*, 283–345.
Elliot, R. (1989). The rights of future persons. *Journal of Applied Philosophy, 6*, 159–169.
Frankfurt, H. (1987). Equality as a moral idea. *Ethics, 98*, 21–43.
Gansmo Jakobsen, T. (2017). Environmental ethics: Anthropocentrism and non-anthropocentrism revised in the light of critical realism. *Journal of Critical Realism, 16*, 184–199.
Griffith, A. M. (2017). The rights of future persons and the ontology of time. *Journal of Social Philosophy, 48*, 58–70.
Hardin, G. (1974). Living on a lifeboat. *Bioscience, 24*, 561–568.
Harsanyi, J. C. (1977). Morality and the theory of rational behaviour. *Social Research, 44*, 623–656.
Hayward, T. (2004). *Constitutional environmental rights*. OUP.
Hill, T. (1983). Ideals of human excellence and preserving natural environment. *Environmental Ethics, 5*, 211–224
Jaffe, A. (2015). Towards a Kantian moral psychology or the practical effects of self-predicating judgements of sublimity. *Critical Horizons 16,*, 88–106.
Johnson, L. (1991). *A morally deep world*. CUP.
Jonas, H. (1979). *Das prinzip verantwortung*. Insel.
Kant, I. (1996). *The metaphysics of morals* (M. J. Gregor, Trans. and Ed.) *Practical Philosophy*. Cambridge: CUP.
Kasperbauer, T. J. (2015). Naturalizing sentimentalism for environmental ethics. *Environmental Ethics, 37*, 221–237.
Keitsch, M. (2018). Structuring ethical interpretations of the sustainable development goals—Concepts, implications and progress. *Sustainability, 10*(829).
Leopold, A. (1949). *A sand county almanac*. OUP.
Lomasky, L. (1987). *Persons, rights, and the moral community*. OUP.
Lowe, B. J. (2019). Ethics in the Anthropocene: Moral responses to the climate crisis. *Journal of Agricultural and Environmental Ethics, 32*, 479–485.
Maklin, R. (1981). Can future generations correctly be said to have rights? In E. Partridge (Ed.), *Responsibilities to future generations: Environmental ethics* (pp. 151–156). Prometheus Books.
Meadows, D. H. et al. (1972). *The limits to growth*. Universe Book.
Minteer, B. A. (2012). *Refounding environmental ethics: Pragmatism, principle and practice*. Temple University Press.
Naess, A. (1973). The shallow and the deep, long-range ecology movements. *Inquiry, 16*, 95–100
Norton, B. (1991). *Toward unity among environmentalists*. OUP.

Nozick, R. (1974). *Anarchy, state and utopia.* Basic Books.

O'Connor, S., & Kenter, J. (2019). Making intrinsic values work: Integrating intrinsic values and more-than human nature through the Life Framework of Values. *Sustainability Science, 14*, 1247–1265.

Palmer, C. (2010). *Animal ethics in context.* Columbia University Press.

Palmer, C., et al. (2014). Environmental ethics. *Annual Review of Environmental, Resources, 39*, 419–442.

Parfit, D. (1984). *Reasons and persons.* Clarendon Press.

Perret, R. (2003). Future generations and the metaphysic of the self: Western and Indian philosophical perspectives. *Asian Philosophy, 13*, 29–37.

Rawls, J. (1971). *A theory of justice.* OUP.

Regan, T. (1983). *The case for animal rights.* University of California Press.

Reuther, R. R. (1975). *New woman, new Earth: Sexist ideologies and human liberation.* Seabury.

Rolston, H. (1979). Can and ought we follow nature? *Environmental Ethics, 1*, 7–30.

Sandler, R. (2007). *Character and environment: A virtue-oriented approach to environmental ethics.* Columbia University Press.

Schmidtz, D. (1998). Are all species equal? *Journal of Applied Philosophy, 15*, 57–67.

Schweitzer, A. (1987). *The philosophy of civilisation.* Prometheus.

Sen, A. (1999). *Development as freedom.* Knopf.

Sher, G. (1979). Compensation and transworld personal identity. *Monist, 62*, 378–391.

Singer, P. (1975). *Animal liberation: A new ethics of our treatment of animals.* New York Review.

Sterba, J. (1980). Abortion, distant people, future generations. *Journal of Philosophy, 77*, 424–440.

Taylor, P. (1986). *Respect for nature.* Princeton University Press.

Thompson, J. (2009). *Intergenerational justice. Rights and responsibilities in an intergenerational polity.* Routledge.

Thompson, J. (2017). The ethics of intergenerational relationships. *Canadian Journal of Philosophy, 47*, 313–326.

UDHR. (1948). www.un.org

Varner, G. (1998). *In nature's interests? Interests, animal rights, and environmental ethics.* OUP.

Webster, A. J. F. (2003). International standards for animal welfare: Science and values. *Veterinary Journal, 198*, 2–3.

Wilson, E. O. (2016). *Half-Earth: Our planet's fight for life.* Norton & Company.

Zagonari, F. (2019). Responsibility, inequality, efficiency, and equity in four sustainability paradigms: Insights for the global environment from a cross-development analytical model. *Environment, Development and Sustainability, 21*, 2733–2772.

Zagonari, F. (2020). Comparing religious environmental ethics to support efforts to achieve local and global sustainability: Empirical insights based on a theoretical framework. *Sustainability, 12* (2590).

Chapter 3
Environmental Sustainability

Chapter 2 highlighted two main secular environmental ethics: the maximisation of total welfare (i.e., efficiency) within the teleological duties to future generations, from an average person's perspective; and the minimisation of resource inequality (i.e., equity) within the deontological duties to future generations, from a per-capita perspective.

This chapter discusses the main sustainability paradigms by referring to these two main ethics and perspectives. In particular, it provides two *Figures* that depict a dynamic model based only on economic features (i.e., the Ramsey model) and a dynamic model based only on ecological features (i.e., the Holling model). I will combine these models in a comprehensive model that includes all relevant features about efficiency and equity. I will *then* simplify the model into two equations to discuss policies designed to achieve efficiency in Chap. 4. Moreover, I will characterise several sustainability paradigms that range from a purely economic framework to a purely ecological framework (e.g., weak sustainability, a-growth, de-growth, strong sustainability) in a *Table*. Finally, I will provide some remarks on sustainability metrics.

Sustainability can be defined as the maximization of social-ecological continuity (long-run) (e.g., human perpetuity if focused on social features) *and/or* the minimisation of social-ecological impacts (short-run) (e.g., pain alleviation if focused on social features) (Salas-Zapata et al., 2017). In other words, it is a matter to find a compromise between social and ecological continuity or impacts, where economic features are instrumental to social features (Khan, 2015).

Table 3.1 summarizes the main sustainability paradigms.

Note that both the RCK model (i.e., Ramsey (1928)–Cass (1965)–Koopmans (1965)) and the ADM model (i.e., ADM = Arrow and Debreu (1954)–McKenzie (1959)) in the first cell (i.e., maximization of social continuity with an increasing economy) would require an *unlikely* absolute decoupling to reduce impacts on ecology, but these models are not concerned about these impacts (Haberl et al., 2020; Luukkanen et al., 2019; Savona & Ciarli, 2019; Wiedenhofer et al., 2020).

F. Zagonari, *Environmental Ethics, Sustainability and Decisions*,
https://doi.org/10.1007/978-3-031-21182-9_3

Table 3.1 The main sustainability paradigms in terms of social/ecological continuity/impacts

	Social focus		Ecological focus	
	Increasing economy	Decreasing economy	Increasing economy	Decreasing economy
Max continuity	RCK model ADM model (forward markets) WS (capital investments to human needs)	AG (technological innovation)	ESS	ER
Min impacts	Bio economy Circular economy Green economy Three circles model Triple bottom-line model Three pillars model	DG (cultural change)	DE	SS (environment preservation to ecological equilibrium) Egg model Concentric rings model

This table characterise the main sustainability paradigms by referring to the definition of sustainability by Salas-Zapata et al. (2017). Abbreviations: RCK = Ramsey (1928)–Cass (1965)–Koopmans (1965); ADM = Arrow and Debreu (1954)–McKenzie (1959); ER = Ecological resilience by Holling (1973); WS = weak sustainability; SS = strong sustainability, AG = a-growth, DG = de-growth, ESS = Eco-System Services, DE = Doughnut economy

Moreover, weak sustainability is focused on human welfare, whereas strong sustainability is focused on ecological impacts. Finally, all models (i.e., Bio, Circular, Green economy; Three circles, Triple bottom-line, Three pillars model) in the second cell (i.e., minimization of social impacts with an increasing economy) rely on an unlikely absolute decoupling to reduce impacts on ecology, since the *likely* relative decoupling is insufficient and unfair (i.e., it rests on the assumption that poor people remain poor) (Juknys et al., 2014).

In particular, *green* economy (i.e., a sustainable development economy with a local and long-run social perspective focused on tourism and education sectors, where ecological innovation based on natural processes looks for nature conservation) (Ruggerio, 2021), *circular* economy (i.e., a sustainable development economy with a local and short-run economic perspective focused on urban and industrial sectors, where technological innovation looks for resource efficiency, productivity and decoupling) (Corona et al., 2019; Fan et al., 2019), *bio* economy (i.e., a sustainable development economy with a global and short-run economic perspective focused on health and rural sectors, where technical innovation based on transformation processes looks for making the basic building blocks for materials, chemical and energy being derived from renewable biological resources, by making use of residual biomass streams with the help of new technologies) (Bennich & Belyazid, 2017), all rely on economic growth concepts, although they complement each other (D'Amato et al., 2017). Specifically, Table 3.1 is focused on environmental sustainability (i.e., an end equilibrium state based on dynamic relationships between nature and humans) to a smaller extent than on sustainable development (i.e., a process or a set of means, including participation, to approach or achieve an end equilibrium state that *might* include environmental sustainability).

Moreover, the three pillars model (i.e., achievement in the economic, social and environmental dimensions), the triple bottom-line model (i.e., minimum achievements in the economic, social and environmental dimensions), the three circles model (i.e., maximum achievements in the economic, social and environmental dimensions, with partial overlapping between each dimension), the egg model (i.e., maximum achievements in the environmental dimension, with a social constraint, where people are the egg yellow included in ecosystems as the egg white), the concentric rings model (i.e., maximum achievements in the environmental dimension, with social and economic constraints, where the economic dimension is included in the social dimension which in turn is included in the environmental dimension), all lack the time dimension, although they highlight interrelationships (Ramcilovic-Suominen & Pulzl, 2018).

Finally, *doughnut* economy (i.e., a sustainable development economy that meets the social needs of people (see weak sustainability) in terms of healthy life years without overshooting Earth's ecological ceiling (see strong sustainability) in terms of global hectares consumption: Luukkanen et al., 2021) is intermediate with respect to weak sustainability and strong sustainability. Specifically, weak sustainability and strong sustainability will be discussed below together with eco-system services, a-growth and de-growth.

Four main Ecological Resilience ER definitions are suggested to represent ecological continuity:

i. The distance from other stable equilibria (Yi & Jackson, 2021)
ii. The amount of change or disruption that is required to transform a system (Holling, 1973)
iii. The rate at which a system returns to an equilibrium following disturbance (Pimm, 1984)
iv. The set of states which leads to an equilibrium (Li et al., 2020)

Note that all four definitions of resilience are non-linear in the environmental status (i.e., feasibility of decisions as potentials to achieve environmental sustainability should be considered before reliability of decisions as capacities to control environmental interventions), although definition (ii) better depicts the non-linear (and uncertain) consequences of climate change for pollution production or biodiversity loss for resource use.

Figure 3.1 presents a numerical example of alternative ecological resilience.

The ER framework in discrete time t can be formulated in terms of basins of attraction (i.e., the ii Ecological Resilience) as follows (Peterson et al., 1998):

$$\forall \varphi > 0 \exists \chi > 0 \text{ such that } \left|E(t_0) - E^*(t_0)\right| < \chi \rightarrow \forall t \geq t_1, \left|E(t) - E^*(t)\right| < \varphi$$

where E^* is a solution of the ecological system (i.e., $\partial E^*(t)/\partial t = 0$) and

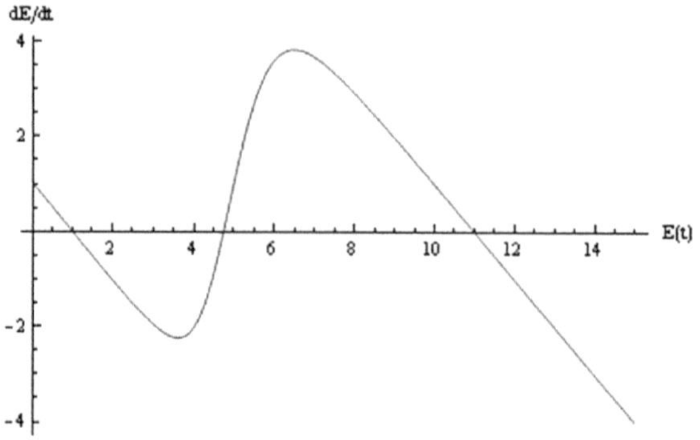

Fig. 3.1 Four definitions of ecological resilience based on the Holling (1973) model. This figure measures the four definitions of resilience for the dynamics of an ecological system dE/dt = $\varphi - E + \chi\,(E(t)^{\psi}/(E(t)^{\psi} + \omega^{\psi}))$ in terms of its status E(t) if $\varphi = 1$, $\chi = \psi = 10$, $\omega = 5$: (i) E3-E1: here 9.99; (ii) E3-E2: here 6.24; (iii) dE/dt at E2: here 3.93; (iv) $4.75 - 0 = 4.75$ for E1 and $15 - 4.75 = 10.25$ for E3. *Notes* E1 = 1 (stable), E2 = 4.75 (unstable) and E3 = 10.99 (stable) are the first, second and third equilibrium points (dE/dt = 0), respectively

$$\frac{\partial E(i)(\mathrm{t})}{\partial \mathrm{t}} = E(i)(\mathrm{t})\left[\psi_i - \sum_{j}^{n} \omega_{ij} E(\mathrm{j})(\mathrm{t})\right]$$

where t_0 and t_1 represent the time (t) at the start of the study period and at the return of the systems' equilibrium, respectively; χ represents the system's amplitude (i.e., the basin of attraction); φ depicts the system's resistance to small changes, and it is assumed that a circular (as opposed to other shapes) attractor basin and a deterministic (as opposed to a stochastic) model both exist (see Peterson et al., 2012 for an alternative basin shape and specification of stochastic models); ψ_i depicts the intrinsic growth rate of species i; and ω_{ij} represents the impact of species i on species j.

In particular, if $E(t) = (E_1(t), \ldots E_i(t), \ldots, E_I(t))$ and $E^* = (E_1{}^*, \ldots E_i{}^*, \ldots, E_I{}^*)$ are the vectors for existing species sizes at time t and in equilibrium (*), respectively, there are three implications: the *resistance* (i.e., the system's capacity of small changes in response to external pressures) is measured (i.e., the size of χ), no substitution between species is allowed (i.e., species i $\neq$ species j), and changes are considered to be detrimental (i.e., E* > E for any E). Alternatively, if $E(t) = (E_1(t), \ldots E_i(t), \ldots, E_I(t))$ and $E^* = (E^*, \ldots E_i{}^*, \ldots, E_I{}^*)$ are the vectors of *potential* species at time t and in equilibrium (E*) to preserve some given relationships between species, respectively, there are three implications. the *resilience* (i.e., the system's ability to retain its functional and structural organizations after perturbations) is measured (i.e., the size of χ), substitution between species is allowed (i.e., species i $\approx$ species j), and changes are considered to be neither detrimental nor beneficial (i.e., E* = E for some E). In other words, the above formula can depict both ecological resilience and ecological resistance.

Note that resistance is a narrower concept than resilience. For example, an invasive species could replace more than one current species, by preserving the same functional and structural roles within the ecosystem, where the replacement of one species by another implies that both the equilibrium E^* and the ω_{ij} parameters will change (i.e., nature applies an average perspective in the long-run). Moreover, the elasticity or recovery is the speed with which the system returns to equilibrium (i.e., the period t_1–t_0); and the inertia or persistence is the time period in which the system is within φ. For example, if species i could play a role in a desert ecosystem, but it is not present at time t_1, $E_i(t_1) = 0$, although this species could replace another species j in this role at time t_2 or subsequently. Finally, strong sustainability assumes $\psi = \omega = 0$ (i.e., $\partial E(i)(t)/\partial t = 0$ for any species i) and $E^* = E(0)$ (i.e., the goal is the ecological resistance for each species i).

The main features (from Chap. 2) of the ER framework can be summarized as follows:

- Unit = resilience for each ecosystem
- Efficiency = no
- Equity = human inter-generational, if E(t) = E* for each t, with each species or each role of species having the same importance (i.e., the focus on resistence or resilience)
- Substitution = no between nature and other human-made capitals (i.e., nature has intrinsic value), although some species can replace other species, if the pursued equilibrium is the ecological resilience

The main critiques (on feasibility) to the ER framework can be summarized as follows:

1. It is a metaphorical framework (Biesbroek et al., 2017; Carpenter et al., 2001)
2. The stability conditions depend on the characteristics of the ecosystems (Hoekstra et al., 2018; Scheffer et al., 2001)
3. Many possible solutions might be locally stable only (Cumming & Peterson, 2017; Rietkerk et al., 2004)

Some minor weaknesses (on transparency) of the ER framework can be summarized as follows:

1. Human impacts are represented by external shocks rather than by internal dynamics (De Luca Peña et al., 2022) (i.e., human impacts can be evaluated in terms of energy such as emergy or exergy, but social dynamics are disregarded)
2. ER depends on species richness (Grumbach & Hamant, 2020; Sterk et al., 2017)
3. Maintaining the current environmental status might not be a sufficient condition for ER (Bjorn et al., 2020)

Note that, in Table 3.1, the Ecological Resilience must be distinguished from the *Eco-System Services* approach (i.e., Millennium Ecosystem Assessment MEA, 2005), in which four main ecosystem service functions are identified (i.e., Provisioning, Regulating, Cultural, Supporting) and the ecological resilience of these four ecosystems is pursued, by disregarding other ecosystems: these choices have been widely criticized for mixing processes (means) and benefits (ends) (e.g., Boyd & Banzhaf, 2007). Alternatively, it would be possible to refer to the Eco-System Services definition proposed by The Economics of Ecosystem and Biodiversity project (TEEB, 2009): core ecosystem service processes (production, decomposition, nutrient and water cycling, hydrological and evolutionary processes, ecological interactions), beneficial ecosystem service processes (e.g., Regulating such as waste assimilation, water cycling and purification, climate regulation, erosion and flood control, ...; Supporting such as primary and secondary production, food web dynamics, species and genetic diversification, biogeochemical cycling, ...; Cultural such as pleasant scenery), beneficial Eco-System Services (e.g., Provisioning such as food, raw materials, energy, physical well-being, ...; Cultural such as psychological and social well-being, knowledge). Moreover, circular economy and bio economy refer to provisioning Eco-System Services, whereas green economy also to regulating and cultural Eco-System Services. Finally, in Table 3.1, the ecological resilience is linked to a decreasing economy, since all ecological indicators suggest to reduce the economic scale. For example, the average Ecological Footprint per capita for the world population should be 1.7 ha, while it is 5.74 ha in developed (i.e., OECD) countries (18% of the world population) and 2.14 ha in developing (i.e., non-OECD) countries (82% of the world population). Thus, in order to have a reduction in the average Ecological Footprint per capita, *either* one assumes the *likely* relative decoupling with an unchanged use of the Earth and distribution of the world population (i.e., an unfair scenario), *or* one assumes the *unlikely* absolute decoupling with a reduced use of the Earth resources and an increased and richer population in non-OECD countries (i.e., a risky scenario).

Two main models in economics can represent social continuity in equilibrium in a pure economic perspective: the Ramsey (1928)–Cass (1965)–Koopmans (1965) (RCK) model; the Arrow and Debreu (1954)–McKenzie (1959) (ADM) model.

These models refer to the following main definitions:

- Consequentialism (i.e., actions are chosen because of consequences).
- Welfarism (i.e., consequences are assessed in terms of welfare).
- Individualism (i.e., welfare refers to the individual taking actions).
- Rationality (i.e., each individual takes systematic and consistent actions aimed at maximising his/her welfare).
- Everyone's welfare is the satisfaction of rational, constant, well-informed and self-interested preferences (i.e., the best guide to what is beneficial to each individual). Note that this utilitarian perspective is different from the eudemonic and the virtue approaches.
- Equity amounts to equally weighting everyone's welfare to maximize social welfare as the number of individuals multiplied by the welfare level of the *representative* or average individual. Note that this concept is different from equal access to resources, actions or outcomes.
- Justice amounts to selecting social rules that maximize social welfare. Note that this concept is different from safeguarding rights in terms of actions (i.e., protecting or promoting a way of acting or of being treated) or in terms of

interests (i.e., protecting people's properties) and from safeguarding liberties (i.e., eliminating obstacles preventing or discouraging individuals from doing something).

- Efficiency amounts to Pareto efficiency (i.e., it is not possible to have winners without losers) or Kaldor-Hicks efficiency (i.e., the possible winners can more than compensate losers). Note that these concepts are different from technical efficiency (i.e., minimizing the inputs required to produce a given output) and from cost-efficiency or cost-effectiveness (i.e., minimizing the cost of producing a given output). For example, if scenario A (70 to rich, 20 to poor) and scenario B (80 to rich, 10 to poor) are possible, then A is Pareto and Kaldor-Hicks efficient if A is the status quo, B is Pareto and Kaldor-Hicks efficient if B is the status quo. Alternatively, if scenario A (70 to rich, 20 to poor), scenario B (80 to rich, 10 to poor) and scenario C (90 to rich, 10 to poor) are possible, then the status quo A is Pareto but it is not Kaldor-Hicks efficient, whereas the status quo B is neither Pareto nor Kaldor-Hicks efficient.

The Ramsey (1928)–Cass (1965)–Koopmans (1965) (RCK) model maximizes the social welfare for an infinitely-lived and representative individual:

$$Max \int_{t=0}^{\infty} u[q(t)]e^{-\sigma t}dt$$

Subject to

$$q(t) = f[k(t)] - (\nu + \delta)k(t) - \dot{k}(t)$$

where q(t) is the single good produced and consumed at time t, k is the per capita stock of capital and k dot is the change of k at time t (i.e., k dot $= \partial k/\partial t$), ν is the population growth rate, δ is the deprecation rate, f[k(t)] is the production function transforming a stock k into a flow q, u[q(t)] is the utility function, and σ is the social discount rate.

Note that f[k(t)] can be expressed in per capita terms (i.e., capital per worker) since it is based on a production function that satisfies the Inada conditions (i.e., f(0) = 0, f concave, lim $f' = \infty$ if $k \to 0$, lim $f' = 0$ if $k \to \infty$) in general and in particular is homogeneous of degree 1 such as $Q = F[L, K] = L^l K^m$ with $l > 0$, $m > 0$ and $l + m = 1$. Moreover, population growth is assumed to equal labor growth. Finally, q(t) is expressed in per capita terms (i.e., consumption per person).

The main assumptions of the RCK model are as follows:

- The amount of investment equals the amount of saving
- The stock of capital at time 0 is positive
- The utility function u[q(t)] is constant in time t and strictly increasing and concave in consumption q

The problem of the social planner's model of maximizing the inter temporal social welfare function can be solved by setting the following Hamiltonian function:

$$HAM = u(q) + \mu[f(k) - (\nu + \delta)k - q]$$

where μ is the co-state variable to be interpreted as the shadow price of q.

$$\frac{\partial HAM}{\partial q} = u'(q) - \mu = 0 \leftrightarrow u'(q) = \mu$$

$$\dot{\mu} = \sigma\mu - \frac{\partial HAM}{\partial k} = \sigma\mu - \mu\left[f'(k) - \nu - \delta\right] \geq 0 \leftrightarrow \mu\left[\sigma - f'(k) + \nu + \delta\right] \geq 0$$

$$\dot{k} = f(k) - (\nu + \delta)k - q \geq 0$$

where μ dot is the change of μ at time t (i.e., μ dot = ∂μ/∂t).

Let us assume the following transversality condition holds:

$$lim_{t\to\infty}\mu(t)k(t) = 0$$

where its meaning is closed to the boundary conditions (i.e., non negative μ(t) and k(t)), whereas it satisfaction is based on the decreasing marginal utility of consumption and on the decreasing marginal production of capital (i.e., u(q) concave and f(k) concave). In other words, consumption is very large (i.e., marginal utility of consumption tends to 0) only if production is very large (i.e., capital tends to infinity); similarly, consumption is very small (i.e., marginal utility of consumption tends to infinity) only if production is very small (i.e., capital tends to 0).

Let us assume $u(q) = q^{\alpha}$ so that $u' = \alpha\, q^{\alpha-1}$ with $\alpha < 1$. Similarly, $f(k) = k^{m}$ so that $f' = m\, k^{m-1}$ with $m < 1$ to depict the Inada conditions (i.e., f(k) is concave).

By using the first equality and the third inequality, one gets the equilibrium condition for k:

$$\mu = \alpha\left[k^{m} - (\nu + \delta)k\right]^{\alpha - 1}$$

By using the first equality and the second inequality, one gets the equilibrium condition for μ:

$$\sigma - mk^{m-1} + \nu + \delta = 0 \leftrightarrow k^{*} = [(1/m)(\sigma + \nu + \delta)]^{1/(m-1)}$$

Since $u'(q) = \mu > 0$ and terms in brackets sum up to 0.

Figure 3.2 depicts the phase diagram for the RCK model.

Let us focus on stability conditions for μ. If k is larger than in equilibrium (i.e., on the right of the vertical line) then q increases and $u' = \mu$ decreases. Let us focus on stability conditions for k. If μ is larger than in equilibrium (i.e., above the decreasing and increasing curve) or, in other words, q is smaller than the equilibrium q* (i.e.,

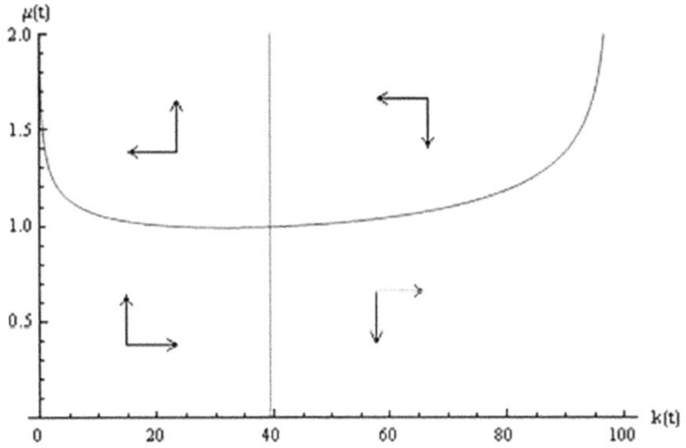

Fig. 3.2 The phase diagram for the Ramsey (1928) model. This figure shows the dynamic relationship between the stock of capital at time t k(t) and its shadow price at time t μ(t), where the k dot = 0 is the decreasing and increasing (blue) curve, while the μ dot = 0 is the vertical (grey) line, if ν = 0.2, δ = 0.2, σ = 0.2, m = 0.8, α = 0.8. *Notes* the solution of the purely economic model (i.e., k* = 39.24 and μ* = 0.99) is locally unstable (i.e., it is a saddle)

saving is larger than its equilibrium), then saving = capital k decreases. However, the stability condition for μ (i.e., the decreasing and increasing curve) tends to infinity for k = 0 and for

$$k^m - (\nu + \delta)k = 0 \leftrightarrow \overline{k} = (\nu + \delta)^{1/(m-1)}$$

Consequently, the phase diagram represents the solution provided that:

$$\overline{k} > k^* \leftrightarrow (\nu + \delta)^{\frac{1}{m-1}} > [(1/m)(\sigma + \nu + \delta)]^{\frac{1}{m-1}}$$

Or, equivalently,

$$m(\nu + \delta) < \sigma + \nu + \delta \leftrightarrow \sigma > (m - 1)(\nu + \delta)$$

In other words, a larger technological efficiency (i.e., m = 1) (Egli & Steger, 2007) requires a smaller inter-generational equity (i.e., σ > 0) (Figueres et al., 2010).

If the solution exists (i.e., a decreasing large k and a decreasing large μ; alternatively, an increasing small k and an increasing small μ), it is unique, it is (almost) globally stable (i.e., it is a saddle) and it is Pareto efficient. Note that the solution for an increasing small k and an increasing small μ will be neglected as unrealistic, since an increasing small μ requires a decreasing consumption q (i.e., the North East solution will be showed to depict weak sustainability). Moreover, a decreasing large k implies a realistically decreasing natural capital, since the per-capita capital k includes also environmental capital, and physical capital is realistically increasing. Finally, a social welfare function is needed if more than one representative individual is considered (i.e., to move from an ordinal utility to a cardinal utility).

The main features (from Chap. 2) of the RCK model can be summarized as follows:

- Unit = welfare or utility of humans
- Efficiency = inter-generational Pareto
- Equity = irrelevant for intra-generational (average individual), no inter-generational if $\sigma > 0$
- Substitution = total, since there is a single good q and a single capital k

The main critiques (on feasibility) to the RCK model can be summarized as follows:

1. The single solution might be locally stable only (i.e., k(t) cannot be too large and $\mu(t)$ must be far away from k dot = 0, where k(t) decreases and q(t) increases) (DiMaria, 2014)
2. There is no uncertainty (i.e., it is a deterministic model) (Hallegatte et al., 2007)
3. The single solution might not exist (i.e., m = 1 implies $\sigma > 0$) (DiMaria, 2019)

Some minor weaknesses (on transparency) of the RCK model can be summarized as follows:

1. There is no technological progress (i.e., there is no a dynamic m) (Chang et al., 2015)
2. There is a single good (i.e., resource use and pollution production are not distinguished from consumed goods and services) and a single capital (i.e., natural, produced, human and social capital are combined) (Richters & Siemoneit, 2017)
3. Natural capital included in k can be increased by decreasing the consumption level (i.e., saving is complementary to environment) (Pender, 1998)

Note that the price of the single good is standardized to 1.

The Arrow and Debreu (1954)–McKenzie (1959) (ADM) model searches for a set of n prices for each discrete time t from time 0 to time T such that individuals maximise welfare subject to current endowments and firms maximise profits subject to available technologies (i.e., it identifies choices that are mutually beneficial and consistent so that demand equals supply in all competitive markets; productive resources are Pareto efficiently allocated among firms and consumption goods among consumers). Note that it includes environmental resource X and pollution Y like other goods and services in terms of supply, demand, and prices in competitive markets. This model rests on the following assumptions:

- Complete information (i.e., no asymmetric information where asymmetric information is defined as follows: some actions by an individual are not perfectly observed by other bargaining individuals (moral hazard) or some characteristics of individuals or goods are known to some bargaining individuals only (adverse selection)).
- Perfect information (i.e., no uncertainty but risk is allowed where uncertainty is defined as circumstances where individuals cannot predict which events can occur in the future *or* they cannot attach probabilities to these events, whereas risk is

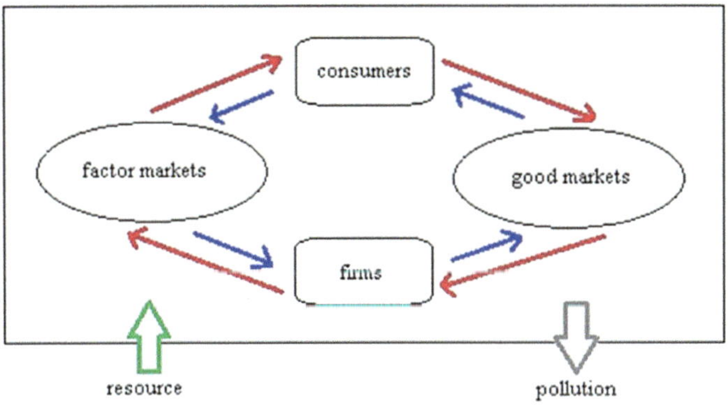

Fig. 3.3 The economic general equilibrium framework. This figure depicts a simplified economy with two roles (i.e., firms and consumers) and two markets (i.e., goods and factors) linked to nature through the use of resources and the production of pollution. *Notes* Blue arrows = real exchanges; Red arrows = monetary exchanges; box = the closed economy; green arrow = use of resources; grey arrow = production of pollution

defined as circumstances where individuals can predict which events can occur in the future *and* they can attach probabilities to these events).

- No externalities (i.e., production or consumption activities (or goods) by a firm or a consumer show positive (negative) impacts on production or consumption activities (or goods) by other firms or consumers *and* these impacts are not embedded in prices).
- Commodities are distinguished by where (i.e., it includes trade), when (i.e., there are no markets in the future, but forward markets for all goods at all states) and under which conditions (i.e., it is risk free by relying on the complete market assumption) will be delivered.
- Competitive bargaining (i.e., exchanges are not allowed until all individuals agree upon the n x t equilibrium prices).
- Competitive markets such that each firm assumes not to be in a position to alter the equilibrium price by modifying the quantity supplied (i.e., it is price-taker); the offered good is perceived as homogeneous by consumers (i.e., due to the complete and perfect information, they do not purchase from firms fixing prices higher than the equilibrium price); each supplier can entry and exit the market without costs in the long-run (i.e., it gets no profit in the long-run).

Note that (rational) expectation is not needed in the ADM model, since demand and supply of goods are decided at time 0 under complete and perfect information, based on a common definition of relevant events.

Figure 3.3 depicts a simplified Economic General Equilibrium economy, by focusing on two economic roles only (i.e., consumers and firms), while disregarding other roles (e.g., governmental agencies, foreign economies).

Note that the grey arrow depicts the II principle of thermodynamics (i.e., any transformation process produces some undesired outcomes such as pollution). Moreover, if each individual is depicted as a fixed set of preferences which can be represented by utility (i.e., non-use values are excluded, while altruism is included), then the equilibrium price observed in the single institution under consideration (i.e., markets) is a suitable compromise value for each good (i.e., the environment is included) (Vatn, 2009). Finally, the green arrow depicts the III principle of thermodynamics (i.e., no transformation process can be reverted to obtain the total amount of resources originally used so new resources are need for new transformation processes).

The main features (from Chap. 2) of the ADM model can be summarized as follows:

- Unit = welfare or utility of humans
- Efficiency = intra and inter-generational Pareto
- Equity = the same value is attached to each individual's welfare in current and future generations but a representative individual so neither intra nor inter-generational
- Substitution = perfect between different capitals (i.e., capitals are exchangeable, although at rates characterising the prevailing production functions), possible between different goods (i.e., goods are exchangeable, although at rates characterising the prevailing utility functions)

The main critiques (on feasibility) to the ADM model can be summarized as follows:

1. It is a conceptual framework (i.e., a useful lens) rather than a positive model (Salas-Zapata & Salas-Zapata, 2017)
2. The solution (i.e., a vector of equilibrium prices) might not be unique and might not be stable (Breschger, 2017)
3. The solution is based on stringent restrictions (Herrero Jauregui et al., 2018)

Some minor weaknesses (on transparency) of the ADM model can be summarized as follows:

1. Ecological features can be considered, but at market prices (i.e., it is consistent with total depletion of natural capital if its opportunity cost is sufficiently high) (Hediger, 2000) (i.e., ecological impacts can be evaluated in monetary terms, but ecological dynamics are disregarded)
2. Social relationships other than market exchanges are disregarded (Vouvaki & Xepapadeas, 2008)
3. It is a static model, although it refers to a sequence of discrete times (Hediger, 2009)

Note that environmental sustainability in economic models (i.e., both RCK and ADM models) is an opportunity cost (i.e., what is expected to renounce to obtain something else), since it is depicted as a constraint within a maximization problem (Burnett et al., 2014) (e.g., reduced consumption and increased saving to produce physical capital vs. natural capital in the RCK model; reduced production to reduce externalities from pollution vs. increased consumption in the ADM model). Moreover, attaching the same value to individuals' welfare in the current generation (e.g.,

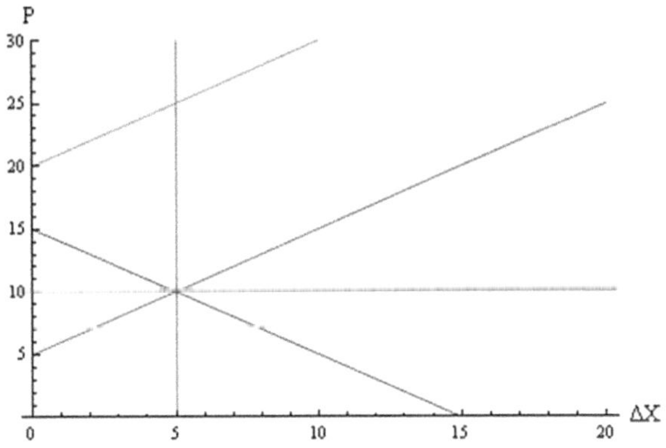

Fig. 3.4 A partial equilibrium within the economic general equilibrium framework. This figure depicts the partial equilibrium for the conservation of a resource X in terms of the positive change of its stock ΔX and the social value attached to its conservation P, where the (blue) decreasing line = the willingness to pay WTP for conservation ΔX, the (purple) increasing line below = the low Marginal Cost MC of conservation, the (grey) increasing line above = the high Marginal Cost MC of conservation. *Notes* the partial equilibrium is at $\Delta X = 5$ with low MC, it is at $\Delta X = 0$ with high MC

consumers, polluting firms, polluted people) leads to a single solution to the maximisation problem. Finally, a larger value attached to individuals' welfare in the current generation than to individuals' welfare in future generations could be depicted by equilibrium prices in future contingent markets (i.e., no inter-generational equity).

Figure 3.4 depicts the partial competitive equilibrium (i.e., in a single market) for a resource conservation X (e.g., white rinos).

Note that the equilibrium price represents the value that society attaches to ΔX: indeed, it is a compromise between the value that consumers attach to it in terms of welfare (i.e., if consumers are willingness to pay P then they expect from the use or enjoyment of ΔX at least a welfare level of P. In other words, P represents the translation of preferences into monetary terms by consumers themselves) and the value that firms attach to it in terms of costs (i.e., firms are willing to offer ΔX at price P if they reimburse all factors at their opportunity costs). Moreover, the competitive equilibrium is consistent with no conservation of ΔX, whenever the willingness to pay by consumers is smaller than the opportunity costs of firms at each level of conservation X. In particular, the social economic value of the first unit of ΔX in Fig. 3.4 is the willingness to pay for it (to be read on the demand curve) rather than the marginal cost to produce it (to be read on the supply curve) (i.e., 15 rather than 20), where the marginal cost does not include the fixed costs which are relevant in the long-run. Finally, the equilibrium quantity maximizes social welfare. Indeed, a larger ΔX would require opportunity costs which are larger than the welfare achieved by consumers, while a smaller ΔX would reduce the welfare achieved by consumers more than the saving in opportunity costs.

If the economic system is explicitly linked to the ecological system, the RCK and ADM models can be combined and reformulated in a sustainability model with discrete time t as follows:

$$Max \sum_{t=0}^{\infty} \frac{1}{(1+\sigma)^t} U_t\left[Q_t\left(Z_{phyt}, Z_{soct}, Z_{envt}\right), Z_{phyt}, Z_{soct}, Z_{envt}\right] \tag{3.1}$$

Subject to

$$Z_{phyt+1} + Z_{soct+1} + Z_{envt+1} \geq Z_{phyt} + Z_{soct} + Z_{envt}$$

where

$$\frac{\partial \partial Q_t}{\partial X_t \partial t} < 0 \text{ and } \frac{\partial \partial Y_t}{\partial Q_t \partial t} \geq 0$$

where, $Z_{\mathrm{phy}t}$, $Z_{\mathrm{soc}t}$, and $Z_{\mathrm{env}t}$ are the use of artificial/physical/manufactured, social, and environmental capitals (both stocks and flows and included) at time *t*, respectively, where Z_{envt} can be split into resources (X_t) and pollution (Y_t) at time *t*, σ is the social discount rate, and the constraints represent the II and III thermodynamic laws (i.e., the increase in entropy and the absence of total recycling, respectively) as a marginal increase in resource use and pollution production for a given level of goods and services. Note that the specification of U_t is uncertain, since future generations could attach a greater value to the environment (i.e., $\partial\partial U_t/\partial Z_{\mathrm{env}t}\partial t \geq 0$) (Krysiak & Krysiak, 2006). Moreover, inter-generational equity may compete with intra-generational equity unless U_t includes all current generations (Cairns & Van Long, 2006). Finally, the specification of U_t is uncertain, since future generations could attach a smaller value to consumption (i.e., $\partial\partial U_t/\partial Q_t\partial t \leq 0$) and rely on more efficient technologies (i.e., $\partial\partial Q_t/\partial Z_{\mathrm{env}t}\partial t \leq 0$) (Zagonari, 2015).

In other words, the sustainability model (1) is a RCK model with discrete time (i.e., it distinguishes the different forms of capitals, but it sums them up with the same non-decreasing constraint characterizing the RCK model), by including the II and III principle of thermodynamics, since it refers to the discounted social utility achieved from consumption of marketed and non-marketed goods, including environmental services and the discounted social utility of traded and non-traded capital stocks, including environmental stocks.

Let us assume that the dynamic problem (3.1) with an infinite time horizon can be split into an infinite number of two-period problems, in which *t* refers to the current (C) period and *t* + 1 to the future (F) period. This is consistent with Tables 2.1 and 2.2 in Chap. 2.

The sustainability model (3.1) becomes:

$$\begin{aligned} &Max\ U_C\left[Q_C\left(Z_{Cphy}, Z_{Csoc}, Z_{Cenv}\right), Z_{Cphy}, Z_{Csoc}, Z_{Cenv}\right] \\ &\quad + 1/(1+\sigma)U_F\left[Q_F\left(Z_{Fphy}, Z_{Fsoc}, Z_{Fenv}\right), Z_{Fphy}, Z_{Fsoc}, Z_{Fenv}\right] \end{aligned} \tag{3.2}$$

Subject to

$$Z_{Fphy} + Z_{Fsoc} + Z_{Fenv} \geq Z_{Cphy} + Z_{Csoc} + Z_{Cenv}$$

Note that the social discount rate for the median rather than the average agent could be used to adjust the wealth effect in the Ramsey rule (Emmerling et al., 2019). Moreover, the social discount rate amounts to the interest rate in competitive capital markets if there is no uncertainty (i.e., short-run *or* supply side are considered only). Finally, a smaller substitutability between welfare (i.e., satisfaction obtained from the different environmental services linked to the same natural capital such as sunbath with and without sunscreen lotion due to the ozone layer) should be coupled with a larger value of σ (Traeger, 2011).

Let us assume that the social discount factor is 0 and the concern is about environmental capital only, although production of goods and services uses all forms of capitals.

If both intra and intergenerational equity are considered, the welfare of the Northern and Southern current generation can be represented as follows:

$$Max\, U_{\mathrm{W}} = \left[(p_{\mathrm{N}} U_{\mathrm{N}})^{1-\varepsilon} + (p_{\mathrm{S}} U_{\mathrm{S}})^{1-\varepsilon}\right]^{\frac{1}{1-\varepsilon}} \tag{3.3}$$

Subject to

$$Z_{Fphy} + Z_{Fsoc} + Z_{Fenv} \geq Z_{Nphy} + Z_{Nsoc} + Z_{Nenv} + Z_{Sphy} + Z_{Ssoc} + Z_{Senv}$$

where

$$U_{\mathrm{N}} = \left\{\left[Q_{\mathrm{N}}{}^{\alpha\mathrm{N}} Z_{\mathrm{Nenv}}{}^{-\beta\mathrm{N}} U_{\mathrm{F}}{}^{\gamma\mathrm{N}} U_{\mathrm{S}}{}^{\upsilon\mathrm{N}}\right]^{1/(1-\zeta)} + U_{\mathrm{F}}{}^{1/(1-\zeta)}\right\}^{1-\zeta}$$

$$U_{\mathrm{S}} = \left\{\left[Q_{\mathrm{S}}{}^{\alpha\mathrm{S}} Z_{\mathrm{Senv}}{}^{-\beta\mathrm{S}} U_{\mathrm{F}}{}^{\gamma\mathrm{S}}\right]^{1/(1-\zeta)} + U_{\mathrm{F}}{}^{1/(1-\zeta)}\right\}^{1-\zeta}$$

$$U_{\mathrm{F}} = Q_{\mathrm{F}}{}^{\alpha\mathrm{F}}$$

$$Q_{\mathrm{N}} = \theta_{\mathrm{N}}\left(Z_{Nphy}, Z_{Nsoc}, Z_{Nenv}\right)$$

$$Q_{\mathrm{S}} = \theta_{\mathrm{S}}\left(Z_{Sphy}, Z_{Ssoc}, Z_{Senv}\right)$$

$$Q_{\mathrm{F}} = \theta_{\mathrm{F}}\left(Z_{Fphy}, Z_{Fsoc}, Z_{Fenv}\right)$$

Let us assume that $\varepsilon = \upsilon_{\mathrm{N}} = 0$ (i.e., the world is unconcerned with intragenerational equity and OECD countries do not care about non-OECD welfare) and production depends on environmental capital only (i.e., Z_{Fenv} becomes E_{F}, Z_{Cenv} becomes E_{C}). Thus, we can focus on the overall welfare for the current generation in OECD and non-OECD countries:

$$Max\, U_C = \left\{\left[Q_C{}^{\alpha} E_C{}^{-\beta} U_F{}^{\gamma}\right]^{1/(1-\zeta)} + U_F{}^{1/(1-\zeta)}\right\}^{1-\zeta} \tag{3.4}$$

Subject to

$$E_F \geq E_C$$

where

$$Q_C = \theta_C E_C \text{ and } Q_F = \theta_F E_F$$

and

$$U_F = Q_F{}^{\alpha F}$$

where, for simplicity, α_C, β_C and γ_C are replaced by α, β, γ, respectively.

Note that future technologies and consumption preferences are assumed to be certain (i.e., rational expectation).

Between pure economic and pure ecological frameworks, there are four main sustainability paradigms: weak sustainability WS, a-growth AG, de-growth DG, and strong sustainability SS.

The main features (from Chap. 2) of weak sustainability (i.e., a development that meets the needs of (representative individuals of) the present generation without compromising the ability of (representative individuals of) future generations to meet their own needs) (Dietz & Neumayer, 2007) can be summarized as follows:

- units = average human needs in at least three incommensurable dimensions (i.e., economic, social, and environmental)
- efficiency = intra-generational Pareto efficiency based on equilibrium prices and inter-generational Pareto efficiency based on constraints for current and future needs
- equity = each individual's needs in current and future generations have the same importance, but because the analysis is based on representative individuals, the comparison is only inter-generational (not intra-generational)
- substitution = perfect substitution is possible between natural, manufactured, human, and social capitals

The main critiques (on feasibility) of weak sustainability can be summarized as follows:

1. It is theoretically difficult to estimate, since it is difficult to predict future consumption preferences and production technologies (Zagonari, 2020)
2. It disregards transaction costs of policies or projects that might affect welfare (Munda, 2014)
3. It is empirically difficult to estimate, since the adjusted genuine saving does not estimate the well-being of current and future generations (Qasim & Grimes, 2021)

Some minor weaknesses (on transparency) of weak sustainability can be summarized as follows:

1. It is based on fixed rights to the environment in terms of pollution production and resource use (Munda, 1997).
2. It relies on unlikely absolute decoupling, since there are no constraints on population growth (Biely et al., 2018).
3. It aims at maximising welfare, but there is no intra-generational equity (i.e., it refers to total welfare, regardless of its distribution) and it is based on instrumental rationality (i.e., on average) (Hanley et al., 2015).

Note that weak sustainability cannot be criticised based on the observation that it does not preserve the environment, since this is not one of its goals (Hediger, 2006). Moreover, weak sustainability is close to the linear economy assumption that underlies neo-classical economics (Martins, 2016). Finally, introducing ecological constraints under the weak sustainability paradigm (e.g., tipping points, uncertainties, resilience) because the results are disliked in terms of environmental status is an ad hoc modification of its logical framework that hides the ethical approach behind the weak sustainability paradigm (Irwin et al., 2016).

Thus, WS can be depicted as follows:

$$Max\ U_{\mathrm{C}} = \left\{\left[(\theta E)^{\alpha} E^{-\beta} U_{\mathrm{F}}{}^{\gamma}\right]^{1/(1-\varsigma)} + U_{\mathrm{F}}{}^{1/(1-\varsigma)}\right\}^{1-\varsigma} \tag{3.4a}$$

where

$$U_{\mathrm{F}} = \left(\theta E^{*}\right)^{\alpha \mathrm{F}} \geq U_{\mathrm{C}}$$

The main features (from Chap. 2) of a-growth (i.e., an ecological and economic strategy that focuses on indifference to or neutrality about the economic (GDP) growth as a non-robust and unreliable indicator of social welfare and progress due to the many neglected non-market transactions (e.g., informal activities and relationships) and the many unpriced environmental effects (e.g., long-term impacts of nuclear power or plastic production)) (Van den Bergh, 2010) can be summarized as follows:

- unit = average human welfare
- efficiency = intra-generational Pareto efficiency based on equilibrium prices that include environmental externalities and inter-generational Pareto efficiency based on constraints for current and future welfare
- equity = each individual's welfare in current and future generations has the same importance, but because the analysis is based on representative individuals, the comparison is only inter-generational (not intra-generational)
- substitution = there is generally perfect substitution between natural, produced, human, and social capital

The main critiques (on feasibility) of a-growth can be summarized as follows (Shao, 2020):

1. The same environmental externalities may be evaluated differently by the current generation in different contexts, can produce different impacts on ecosystems and different impacts on the welfare of future generations.
2. Both rich and poor people will oppose policies that threaten their real incomes via an increase in prices.
3. Green GDP might not account for all environmental externalities, including overexploitation of resources or overproduction of pollution.

Some minor weaknesses (on transparency) of a-growth can be summarized as follows (Kallis, 2011):

1. Different increases in prices, which result from different levels of internalisation of environmental impacts, will produce different changes in production sectors and in rich and poor members of society.
2. Tradable environmental permits cannot be used for all national and trans-national environmental issues such as climate change, acid rain, and biodiversity losses due to a lack of consistent policies for some environmental issues.
3. Information campaigns to educate the adult population or environmental education to inform the young population can produce only long-run behavioural and structural changes, together with technological and scale changes, based on relative and absolute decoupling, respectively.

Thus, AG can be depicted as follows:

$$Max\ U_C = \left\{\left[\left(\theta E^*\right)^{\alpha} E^{*-\beta} U_F{}^{\gamma}\right]^{1/(1-\zeta)} + U_F{}^{1/(1-\zeta)}\right\}^{1-\zeta} \tag{3.4b}$$

where

$$U_F = \left(\theta_{AG} E^*\right)^{\alpha F} \geq U_C$$

With

$$\theta_{AG} = (\theta E)/E^* > 1$$

And $\zeta \geq 0$. In words, the current generation bears the costs of the transition to more efficient technologies (i.e., $\theta_{AG} > \theta$) by paying larger prices for the same consumption level (i.e., $\theta_{AG} E^* = \theta E$).

The main features (from Chap. 2) of de-growth (i.e., an ecological and economic perspective based on a socially sustainable and equitable reduction (and eventually stabilization) of the quantities of materials and energy that a society extracts, processes, transports, distributes, consumes, and returns back to the environment as waste) (Kallis et al., 2010) can be summarized as follows:

- unit = individual human welfare
- efficiency = unimportant since the focus is on the possible unequitable costs of the transition towards smaller quantities
- equity = each individual achieves the same minimum welfare level so that both intra- and inter-generational equity are achieved
- substitution = only some substitution between natural, produced, human, and social capital is possible

The main critiques (on feasibility) of de-growth can be summarized as follows (Hanacek et al., 2020):

1. There will be strong social opposition to this strategy by individuals or groups with vested interests in the status quo due to the current distribution of power.
2. The reduced investment in cleaner technologies in the short-run due to smaller production and profits will lead to a larger production of pollution in the long-run at the reduced economic scale.
3. The selection of produced capital to be reduced cannot be based on market forces or on voluntary choices by consumers or producers, so many private goods must be replaced by public goods.

Some minor weaknesses (on transparency) of de-growth can be summarized as follows (Van den Bergh, 2011):

1. It requires institutional changes towards eco-villages, co-housing, consumer–producer cooperatives, and non-monetary exchange systems (Cosme et al., 2017).
2. It requires environmental standards to limit (e.g., CO_2) or to stop (e.g., plastic) production of pollution as well as to limit (e.g., oil) or to stop (e.g., metals) resource extraction (Kallis et al., 2018).
3. It requires a decrease in working hours due to the smaller scale (excluding the health and education sectors), where the larger inequality will be constrained from above by a salary cap and from below by a basic income, whereas the smaller consumption due to dematerialisation will be compensated by a cultural change (Buch-Hansen & Koch, 2019).

Thus, DG can be depicted as follows:

$$Max\, U_C = \left\{\left[\left(\theta E^*\right)^{\alpha DG} E^{*-\beta} U_F{}^{\gamma}\right]^{1/(1-\zeta)} + U_F{}^{1/(1-\zeta)}\right\}^{1-\zeta} \tag{3.4c}$$

where

$$U_F = \left(\theta E^*\right)^{\alpha DG} \geq U_C$$

With

$$\alpha_{DG} = \alpha\left(\log[\theta E]/\log[\theta E^*]\right)$$

And $\zeta \geq 0$. In words, the current generation attaches a greater value to the current consumption level (i.e., $\alpha_{DG} > \alpha$) to achieve the same welfare at a sustainable consumption level (i.e., $(\theta E)^{\alpha} = (\theta E^{*})^{\alpha DG}$).

The main features (from Chap. 2) of strong sustainability (i.e., a development that allows (each individual in) future generations to access to the same amount of natural resources and the same status of the environment as (each individual in) the current generation, where natural and physical or social capitals are complementary but not interchangeable) (Jain & Jain, 2013) can be summarized as follows:

- unit = capitals in at least three main incommensurable dimensions (i.e., economic, social, and environmental)
- efficiency = disregarded because the environmental goals are considered more important than all other goals
- equity = each individual has access to the same amount of natural and other types of capital, so both intra- and inter-generational equity are achieved
- substitution = no substitution is allowed between natural capital and the produced, human, or social capitals due to the intrinsic value of nature

The main critiques (on feasibility) of strong sustainability can be summarized as follows:

1. The reference point (e.g., the current environmental status) for minimising impacts is arbitrary and could be inconsistent with a long-run ecological equilibrium (Oliveira Neto et al., 2019).
2. It is difficult to distinguish the critical capital that must be preserved from non-critical natural capital (Hickel, 2020).
3. It must be applied operationally by using imprecise norms such as the precautionary principle and rough indicators such as material-flow accounting (i.e., it is based on ecological indicators for both stocks and flows) (Boyd & Banzhaf, 2007).

Some minor weaknesses (on transparency) of strong sustainability can be summarized as follows (Haskell et al., 2021):

1. Command and control policies are independent of transaction costs because they are governed by a procedural rationality (i.e., it focuses on each individual).
2. Technological progress is disregarded, although this could imply a smaller sustainability burden.
3. It must be combined with measures to reduce inequality, since redistribution of rights to use resources will be required.

Note that strong sustainability tends to have a local focus, since most studies referred to single ecosystems. Moreover, strong sustainability cannot be criticised for requiring unnecessary costs, since cost reduction is not a criterion within its goals (Randall, 2020). Finally, ecology creates an opportunity cost under strong sustainability, provided at least social and economic features are considered.

Thus, SS can be depicted as follows:

$$Max\,Min\left[E, E^*\right] \tag{3.4d}$$

where

$$E \geq E^*$$

Thus

$$E = E^*$$

where E *also* includes social capital.

In summary, social continuity amounts to weak sustainability with efficiency (i.e., it maximises inter-generational total welfare) *if* the social discount rate is 0; in contrast, ecological continuity amounts to strong sustainability with equity (i.e., it minimises inter-generational resource inequality) *if* the reference environmental status E* meets the conditions for resilience. Note that population growth affects both the ecological equilibrium (i.e., total ecosystems E(t) could move from a stable to an unstable level in the short-run and towards an insufficient level in the long-run) and economic equilibrium (i.e., per capita capitals k(t) in the long-run could be too large in terms of their temporal discount rate and capital decay rate).

In particular, the ADM model amounts to weak sustainability without constraints or with non-binding constraints (i.e., sustainability within the ADM model is easier to achieve than weak sustainability due to the possible substitution between current and future welfare). Moreover, a discount factor σ at 0 amounts to an inter-generational inequality aversion of 1, although the former should be correctly applied to welfare, whereas the latter could be properly applied to resources. Finally, strong sustainability is equivalent to an ecological resilience framework for each time t and for each species (i.e., sustainability within an ecological resilience framework is easier to achieve than strong sustainability due to its references to resilience or resistance goals and, consequently, its larger flexibility of equilibria).

Note that weak sustainability (i.e., Max U s.t. $U_F \geq U_C$) resembles Judaism; a-growth (i.e., Min E s.t. $U_F \geq U_C$) is close to Islam; de-growth (i.e., Min X_i s.t. $U_i \geq u$, where u is the minimum individual welfare level) evokes the approach in Christianity (although Max U replaces Min X); strong sustainability (i.e., MaxMin E_i by including both natural and social capitals) resembles the precepts of Hinduism or Buddhism (although E_i refers to nature only).

Let us assume that $\gamma = \zeta = 0$ (i.e., concerns about future generations are introduced as constraints).

Let us proxy objectives (3.4a)–(3.4d) in logarithms.

If production depends on all capital forms, *weak sustainability* can be formulated as follows:

$$Max\,U_C = \alpha \text{Log}\left[\theta\left(W_{Cphy}Z_{Cphy} + W_{Csoc}Z_{Csoc} + W_{Cenv}Z_{Cenv}\right)\right] - \beta \text{Log}(Z_{Cenv}) \tag{3.5a}$$

Subject to

$$U_{\mathrm{F}} = \alpha \mathrm{Log}\left[\theta\left(W_{Fphy}Z_{Fphy} + W_{Fsoc}Z_{Fsoc} + W_{Fenv}Z_{Fenv}\right)\right] - \beta \mathrm{Log}(Z_{Fenv}) \geq U_{\mathrm{C}}$$

where W_C and W_F are the relative weights attached to economic, social and environmental features by current and future generations. Note that the use of many forms of capital combined with the assumption of perfect substitution between types of capital increases the risk for future generations (Figge, 2005).

Let us assume $\beta = 0$ and specify $U_F \geq U_C$ in terms of changes required for relative weights and for different capitals:

$$\begin{aligned} &Z_{phy}\Delta W_{phy} + W_{phy}\Delta Z_{phy} + Z_{soc}\Delta W_{soc} + W_{soc}\Delta Z_{soc} \\ &\quad + Z_{env}\Delta W_{env} + W_{env}\Delta Z_{env} \geq 0 \end{aligned}$$

where $\Delta Z_{env} = Z_{Fenv} - Z_{Cenv} = Z^* - Z_{Cenv} < 0$.

No changes are required for relative weights within weak sustainability:

$$W_{phy}\Delta Z_{phy} + W_{soc}\Delta Z_{soc} + W_{env}\Delta Z_{env} \geq 0$$

Thus, environmental conservation is *unlikely* pursued in weak sustainability, unless the *unlikely* absolute decoupling is achieved, since both physical and social capitals must increase.

A-growth can be represented as follows:

$$Max\ U_{\mathrm{C}} = \alpha \mathrm{Log}\left[\theta\left(W_{Cphy}Z_{Cphy} + W_{Csoc}Z_{Csoc} + W_{Cenv}Z_{Cenv}\right)\right] - \beta \mathrm{Log}(Z_{Cenv}) \tag{3.5b}$$

Subject to:

$$U_{\mathrm{F}} = \alpha \mathrm{Log}\left[\theta_{\mathrm{AG}}\left(W_{Fphy}Z_{Fphy} + W_{Fsoc}Z_{Fsoc} + W_{Fenv}Z_{Fenv}\right)\right] - \beta \mathrm{Log}\left(Z^*\right) \geq U_{\mathrm{C}}$$

With

$$\theta_{\mathrm{AG}} = \theta \frac{\left(W_{Cphy}Z_{Cphy} + W_{Csoc}Z_{Csoc} + W_{Cenv}Z_{Cenv}\right)}{\left(W_{Cphy}Z_{Cphy} + W_{Csoc}Z_{Csoc} + W_{Cenv}Z^*\right)} > 1$$

where Z^* identifies the long-run environmental equilibrium. Note that both linear constraints refer to flows (e.g., *total* welfare) by allowing for substitution between forms of capital.

Let us assume $\beta = 0$ and $\theta_{AG} = \theta$, by specifying $U_F \geq U_C$ in terms of changes required for relative weights and for different capitals:

$$\begin{aligned} &Z_{phy}\Delta W_{phy} + W_{phy}\Delta Z_{phy} + Z_{soc}\Delta W_{soc} + W_{soc}\Delta Z_{soc} \\ &\quad + Z_{env}\Delta W_{env} + W_{env}\Delta Z_{env} \geq 0 \end{aligned}$$

where $\Delta Z_{env} = Z_{Fenv} - Z_{Cenv} = Z^* - Z_{Cenv} < 0$.

A change in environmental technology is required for relative weights within a-growth:

$$Z_{phy} \Delta W_{phy} + W_{phy} \Delta Z_{phy} + W_{soc} \Delta Z_{soc} + W_{env} \Delta Z_{env} \geq 0$$

Thus, environmental conservation is *possibly* pursued in a-growth, due to *partial* decoupling (i.e., ΔW_{phy} together with $\Delta\theta > 1$ assumed to be 0 for simplicity), *although* both physical and social capitals must increase.

De-growth can be represented as follows:

$$Max\ U_C = \alpha_{DG} \mathrm{Log}\left[\theta\left(W_{Cphy} Z_{Cphy} + W_{Csoc} Z_{Csoc} + W_{Cenv} Z_{Cenv}\right)\right] - \beta \mathrm{Log}\left(Z^*\right) \quad (3.5c)$$

Subject to:

$$U_F = \alpha_{DG} \mathrm{Log}\left[\theta\left(W_{Fphy} Z_{Fphy} + W_{Fsoc} Z_{Fsoc} + W_{Fenv} Z_{Fenv}\right)\right] - \beta \mathrm{Log}\left(Z^*\right) \geq U_C$$

with

$$\alpha_{DG} = \alpha \frac{\mathrm{Log}[W_{Cphy} Z_{Cphy} + W_{Csoc} Z_{Csoc} + W_{Cenv} Z_{Cenv}]}{\mathrm{Log}[W_{Cphy} v_{Cphy} + W_{Csoc} Z_{Csoc} + W_{Cenv} Z^*]} > 1$$

where Z_{Ceco} and Z_{Feco} can depict both de-growth of production or GDP and decreased consumption or radical de-growth. Note that both linear constraints refer to flows (e.g., *individual* welfare) by allowing for substitution between forms of capital.

Let us assume $\beta = 0$, by specifying $U_F \geq U_C$ in terms of changes required for relative weights and for different capitals:

$$Z_{phy} \Delta W_{phy} + W_{phy} \Delta Z_{phy} + Z_{soc} \Delta W_{soc} + W_{soc} \Delta Z_{soc} + Z_{env} \Delta W_{env} + W_{env} \Delta Z_{env} \geq 0$$

where $\Delta Z_{env} = Z_{Fenv} - Z_{Cenv} = Z^* - Z_{Cenv} < 0$.

A change in cultural values is required for relative weights within de-growth:

$$W_{phy} \Delta Z_{phy} + Z_{soc} \Delta W_{soc} + W_{soc} \Delta Z_{soc} + Z_{env} \Delta W_{env} + W_{env} \Delta Z_{env} \geq 0$$

Thus, environmental conservation is *possibly* pursued in de-growth, due to *cultural* changes towards social and environmental issues, *and* some physical capitals could decrease (i.e., decreased consumption or radical de-growth).

Strong sustainability can be formulated as follows:

$$Max\ Min\left[Z_{Cphy} - Z_{Fphy}, Z_{Csoc} - Z_{Fsoc}, Z_{Cenv} - Z^*\right] \quad (3.5d)$$

Subject to:

$$Z_{Fphy} \geq Z_{Cphy}$$

$$Z_{Fsoc} \geq Z_{Csoc}$$

$$Z_{Fenv} = Z^* = Z_{Cenv}$$

where physical and social capitals are also included. Note that all constraints refer to stocks (e.g., *per capita* resources) by *not* allowing for substitution between forms of capital.

Thus, environmental conservation is *certainly* pursued in strong sustainability, since $Z_{Cenv} = Z^*$.

In other words, from the ADM model to weak sustainability, the constraint *total* $U_F \geq U_C$ is introduced, by relying on absolute decoupling; from weak sustainability to a-growth, some $Z_{Fphy} < Z_{Cphy}$ are allowed, by relying on partial decoupling ΔW_{phy}, while maintaining the constraint *total* $U_F \geq U_C$; from a-growth to de-growth, the constraint *individual* $U_F \geq U_C$ is introduced and many $Z_{Fphy} < Z_{Cphy}$ are allowed, by relying on cultural changes ΔW_{soc} and ΔW_{env}; from de-growth to strong sustainability, the constraint *individual* $Z_{Fenv} = Z_{Cenv} = Z^*$ is introduced, together with $Z_{Fphy} \geq Z_{Cphy}$ and $Z_{Fsoc} \geq Z_{Csoc}$; from strong sustainability to ecological resilience, Z^* is required to be resilient.

However, *individual* $U_F \geq U_C$ from weak sustainability, relative decoupling (i.e., ΔW_{phy}) from a-growth, cultural changes (i.e., ΔW_{soc} and ΔW_{env}) from de-growth, and individual constraints $Z_{Fenv} = Z_{Cenv} = Z^*$, $Z_{Fphy} \geq Z_{Cphy}$ and $Z_{Fsoc} \geq Z_{Csoc}$ with a *resilient* Z^* from strong sustainability, are all required conditions to achieve social and ecological continuity (Zagonari, 2022).

Note that all sustainability paradigms refer to some degree of inter-generational equity (i.e., $U_F \geq U_C$ in weak sustainability, a-growth, and de-growth; $Z_{Fphy} \geq Z_{Cphy}$ and $Z_{Fsoc} \geq Z_{Csoc}$ in strong sustainability) to avoid that the sustainability burden is moved to future generations by the current generation. However, an improvement in environmental technology (i.e., $\Delta W_{phy} > 0$) financed by current debt to be paid back by future generations (i.e., $\Delta Z_{soc} < 0$) unfavour inter-generational equity.

The first derivatives of U_C with respect to Z_{env} for weak sustainability is given by:

$$MPNB - MEC = \frac{\alpha W_{Cenv}}{W_{Cphy}Z_{Cphy} + W_{Csoc}Z_{Csoc} + W_{Cenv}Z_{Cenv}} - \frac{\beta}{Z_{Cenv}}$$

Let us proxy linearly the previous function by applying a Taylor expansion to weak sustainability for the current generation:

$$MPNB = \alpha \mathrm{Log}\left[\theta\left(\sum_k W_k Z_k\right)\right] + \frac{\alpha W_{env}}{\sum_k W_k Z_k}(Z - Z_{env})$$

$$MEC = \beta \mathrm{Log}(Z_{Cenv}) + \frac{\beta}{Z_{Cenv}}(Z - Z_{env})$$

With

$$\sum_k W_k Z_k = W_{phy} Z_{phy} + W_{soc} Z_{soc} + W_{env} Z_{env}$$

Note that MPNB depicts beneficial effects on welfare from consumption and production, whereas MEC represents detrimental effects on welfare from the use of the environment for all individuals. Moreover, weak sustainability aims at efficiency, but it can measure the produced equity. Finally, strong sustainability aims at equity, but it can measure the achieved efficiency.

3.1 Main Insights of Chap. 3

Weak and strong sustainability represent two crucial paradigms that should be properly linked to the achievement of efficiency and equity, respectively. In particular, weak sustainability is close to a purely economic model (i.e., the north-eastern part of the phase diagram in the Ramsey (1928) model) from the perspective of a representative individual, as it is based on the maximisation of welfare, with perfect substitution among different types of capital. In contrast, strong sustainability is close to a purely ecological model (i.e., the Holling (1973) model applied to both ecological and social resilience equilibria) from a per-capita perspective, and is based on minimising inequality in terms of environmental capital (i.e., resource use and pollution production) and of social capital, with no substitution among the different types of capital.

Note that within the literature on stochastic optimal control, the results are similar to deterministic approaches (e.g., maximisation of welfare in weak sustainability) if the expected value is maximised, whereas the results suggest variance minimisation (e.g., minimisation of inequality in strong sustainability) if the risk of a loss is minimized.

Moreover, although weak sustainability is based on maximising the total welfare, it achieves social continuity only if the social discount rate σ is at 0 (i.e., the utility of future generations $\geq$ the utility of the current generation, which ensures social continuity). Similarly, a-growth allows for different reductions in sectoral production due to the increased prices that arise to account for the detrimental impacts on nature by disregarding the impacts on equality that could jeopardize social continuity. In contrast, although strong sustainability is based on minimizing inter-generational inequality, it achieves ecological continuity only if the reference per-capita ecological status (use of Earth's resources, E^*) is resilient; that is, minimising impacts with respect to the current status ensures ecological continuity. Similarly, de-growth is concerned with the social inequity that arises from the reduction of sectoral production in the short-run by disregarding the smaller investment in cleaner technologies, which could jeopardize ecological continuity.

Finally, the purpose of the inter-generational constraint that characterises both weak sustainability (i.e., future generations must achieve the same level of welfare as the current generation, on average) and strong sustainability (i.e., each individual in future generations must have access to the same environmental status as each individual in the current generation) is to avoid having future generations pay the cost of moving the current generation towards an environmentally sustainable path. However, the large debt of the current generation to be invested in green technologies and to be paid by future generations sidesteps this inter-generational constraint.

3.2 Remarks on Metrics

A plethora of sustainability assessment systems have been suggested in the scientific literature, both theoretical and empirical literature (e.g., ecological footprint, carbon footprint, water footprint) and by governmental and non-governmental organisations (green net national product, sustainable human development index, pollution-sensitive human development index) (Liu et al., 2017). Moreover, many classifications could be used to distinguish sustainability indicators and indexes (i.e., aggregations of a suitable number of indicators): general vs. sectoral, human vs. ecosystem oriented, local vs. national, … (Liu et al., 2016). Finally, some indicators combine efficiency with economic, social and environmental resources in monetary terms (i.e., efficiency and equity) (Ang et al., 2011).

Sustainability indicators and indexes should be suggested to *consistently* evaluate sustainability paradigms presented in this chapter. Unfortunately, the selection of indicators to be developed and applied is often based on ex-ante value-judgements of what is to be sustained and for whom. In contrast, once highlighted the crucial importance of values to sustainability measurement and decision-making, these values should be made *explicit*, together with their underlying worldviews and the decision-making processes (Reid & Rout, 2020).

Note that most indicators are biased towards a short-term perspective (i.e., they are more suitable to adopt the impact approach than the continuity approach) (Liu et al., 2017). Moreover, many uses of sustainability indicators and indexes can be identified: instrumental (i.e., they directly lead to decisions), conceptual (i.e., they catalyse learning and understanding), tactical (i.e., they substitute for action and deflect criticism), and political (they support pre-determined decisions) (Morse, 2015). However, within the instrumental use applied here, the crucial importance of ethics in selecting sustainability indicators is independent on the technocratic paradigm (i.e., the systematic application of technology to all levels of human activity which enables the control of life by means of management techniques) or the mechanistic worldwide (i.e., the perspective that nature is like a machine that humans can control perfectly once understood its mechanics): even if it is assumed that humans perfectly measure and control nature (i.e., if not, objectives could be missed), alternative approaches to intervene over nature still exist. Finally, most indicators are adequate for decisions at national, regional or local (governmental and non-governmental) actors (i.e., they must be aggregated to operationalise the planetary boundaries concept) (Hayha et al., 2016).

As a summary of this remark, this chapter highlighted that environmental sustainability implies a compromise between economy and ecology. Thus, regardless of time scale, uses, and space scale, indicators should include both biophysical and socioeconomic dimensions within an ethical framework. In other words, indicators based only on human assessment (e.g., genuine investment) will never help find such a compromise.

3.3 Exercises

Exercises below share the following analytical model. Let QN = EN/txn the production function in DCs, with txn the technology transforming the environmental status into goods and services (i.e., txn = 5.74/36.727 = 0.156 ha EF/$ thousands). Similarly for LDC: QS = ES/txs (i.e., txs = 2.14/8.216 = 0.260 ha EF/$ thousands). Moreover, let UN = (QN^an) (EN^(−bn)) (UF^gn) (US^dn) the utility function of DCs, where an is the preference for consumption of goods and services, bn the concern for the environmental status, gn the concern for future generations, dn the concern for current generation in LDCs. Let an = 0.6, bn = 0.2, gn = 0.2, dn = 0. Similarly for LDCs: US = (QS^as) (ES^(−bs)) (UF^gn). Let as = 0.8, bs = 0.1, gs = 0.1. Finally, let QF = EF/txf and UF = QF^af the production function and the utility function for future generations, with txf = PN txn + PS txs, EF = 1.7 and af = PN an + PS as, with PN = 0.18 and PS = 0.82 the proportion of the world population in DCs and LDCs, respectively.

Numerical results depend on the following assumptions: 1. preference changes required by de-growth are in proportions of initial level of preferences in developed and developing countries; 2. technology improvements required by a-growth are in proportions of initial level of technologies in developed and developing countries; 3. the average perspective within weak sustainability and a-growth is depicted by referring to both developed and developing countries properly weighted in terms of relative populations, whereas the individual perspective within de-growth and strong sustainability is depicted by singularly referring to representative individuals in developed and developing countries; 4. both a sustainable consumption preference and sustainable production technology are implemented by the current generation; 5. there is no intra-generational equity for weak sustainability and a-growth, while inter-generational equity is in terms of welfare in weak sustainability, a-growth and de-growth (e.g., Aristotle or Harshani) and in terms of environment in strong sustainability (e.g., Kant or Rawls); 6. nature has instrumental value for weak sustainability, a-growth, and de-growth, but intrinsic value for strong sustainability; 7. weak sustainability and a-growth rely on technological improvements, while de-growth and strong sustainability require production reductions; 8. a-growth, de-growth and strong sustainability refer to sustainability in terms of resource use X and pollution production Y, while weak sustainability and de-growth in terms of utility of the current and future generations.

In summary, strong sustainability refers to sustainability in terms of individual Earth use (i.e., $EN \leq EF$, $ES \leq EF$); de-growth in terms of both individual welfare and Earth use (i.e., $EN \leq EF$, $ES \leq EF$, $UN \leq UF$, $US \leq UF$); a-growth in terms of both average welfare and Earth use (i.e., $PN\ EN + PS\ ES = EC \leq EF$, $PN\ UN + PS\ US = UC \leq UF$); weak sustainability in terms of average welfare (i.e., $UC \leq UF$). In other words, weak sustainability and strong sustainability differ to the greatest extent.

Note that exercises on climate change use tyn = 13.375/36.727 = 0.357 ton GHG/$ thousands and tys = 24.375/8.216 = 2.966 ton GHG/$ thousands. Moreover, exercises on biodiversity loss equate use of resources X and environmental status E. Finally, the mitigation approach refers to resource use X and pollution production Y, while the adaptation approach refers to goods and services obtained from X and Y.

1. Within weak sustainability, $PN\ UN + PS\ US = UF$, with PN and PS the proportion of population in DCs and LDCs, respectively. Is the current use of environment in DCS and LDCs (i.e., $EN = 5.74$, $ES = 2.14$) sustainable? [No, since $UF = 4.444 < 6.251 = UC$] What is ΔUC required for DCs and LDCs become sustainable? [$\Delta UC = -29\%$]
2. Within a-growth, improvements in technologies in DCs and LDCs should be achieved in order to meet $PN\ UN + PS\ US = UF$ and $PN\ UN + PS\ US = UF$. Is the current use of environment in DCS and LDCs (i.e., $EN = 5.74$, $ES = 2.14$) sustainable? [No, since $EF = 1.7 < 2.788 = EC$ and $UF = 4.444 < 6.251 = UC$] What is the improvement in tx required for both DCs and LDCs become sustainable? [$\Delta tx = -39\%$ to meet $EC = EF$, $\Delta tx = +86.85\%$ to meet $UC = UF$]
3. Within de-growth, changes in values in both DCs and LDCs should be achieved in order to meet $UN = US = UF$, where $EN = ES = EF$ are included. Is the current use of environment in DCS and LDCs (i.e., $EN = 5.74$, $ES = 2.14$) sustainable? [No, since $UN = 8.265 > 4.444 = UF$, $US = 5.808 > 4.444 = UF$, $EN = 5.74 > 1.7$ and $ES = 2.14 > 1.7 = EF$] What are the decreases in an and as required for both DCs and LDCs become sustainable? [$\Delta an = -64\%$, $\Delta as = -78\%$]
4. Within strong sustainability, $EN = ES = EF$. Is the current use of environment in DCS and LDCs (i.e., $EN = 5.74$, $ES = 2.14$) sustainable? [No, since $EN = 5.74 > 1.7$ and $ES = 2.14 > 1.7 = EF$] What are the reductions in EN and ES required for DCs and LDCs become sustainable? [$\Delta EN = -71\%$, $\Delta ES = -21\%$]
5. Within weak sustainability and a-growth, what is the reduction in population in DCs and LDCs for them to become sustainable? [$PN = 0.823$ from 1.35, $PS = 3.750$ from 6.15 based on $\Delta PN\ EN + \Delta PS\ ES = EF$ with $\Delta PN = \Delta PS = 0.609$]

6. Within strong sustainability and de-growth, what are the reductions in population in DCs and LDCs to become sustainable? [PN at 0.399 from EN = EF, PS at 4.885 from ES = EF]
7. Within a-growth, what is the improvement in tx required for both DCs and LDCs become sustainable if the concern for the environment (bn and bs) or for future generations (gn and gs) are increased by 50%? [Δtx = +27.43% < +86.85% with larger bn and bs; Δtx = +29.77% < +86.85% with larger gn and gs]
8. Within de-growth, what are the decreases in an and as required for both DCs and LDCs become sustainable if the concern for the environment (bn and bs) or for future generations (gn and gs) are increased by 50%? [Δan = −46% < −66% and Δas = −67% < −78% with larger bn and bs; Δan = −69% > −64% and Δas = −82% > −78% with larger gn and gs]
9. If climate change mitigation requires a reduction of YN and YS by 33%, what are the required decreases in QN and YN in DCs and in QS and YS in LDCs for them to become sustainable within an average perspective and an individual perspective? [ΔYN = −4.413 ton, ΔYS = −8.043 ton, ΔQN = −12.119 $ thousands and ΔQS = −2.711 $ thousands]
10. If biodiversity loss requires a reduction of XN and XS by 33%, what are the required decreases in QN and XN in DCs and in QS and XS in LDCs for them to become sustainable within an average perspective and an individual perspective? [ΔXN = −1.894, ΔXS = −0.706, ΔQN = −12.119 $ thousands and ΔQS = −2.711 $ thousands]

References

Ang, F., et al. (2011). An aggregate resource efficiency perspective on sustainability: A sustainable value application to the EU-15 countries. *Ecological Economics, 71*, 99–110.

Arrow, K., & Debreu, G. (1954). Existence of an equilibrium for a competitive economy. *Econometrica, 22*, 265–290.

Bennich, T., & Belyazid, S. (2017). The route to sustainability Prospects and challenges of the bio-based economy. *Sustainability, 9*, Art. No. 887.

Biely, K., et al. (2018). The idea of week sustainability is illegitimate. *Environment, Development and Sustainability, 20*, 223–232.

Biesbroek, R., et al. (2017). Explaining through causal mechanisms: Resilience and governance of social–ecological systems. *Current Opinion in Environment Sustainability, 28*, 64–70.

Bjorn, A., et al. (2020). Review of life-cycle based methods for absolute environmental sustainability assessment and their applications. *Environmental Research Letters, 15*, Art. No. 083001.

Boyd, J., & Banzhaf, S. (2007). What are ecosystem services? The need for standardised environmental accounting units. *Ecological Economics, 63*, 616–626.

Breschger, L. (2017). Is the environment compatible with growth? Adopting an integrated framework for sustainability. *Annual Review of Resource Economics, 9*, 185–207.

Buch-Hansen, H., & Koch, M. (2019). De-growth through income and wealth caps? *Ecological Economics, 160*, 264–271.

Burnett, K., et al. (2014). Islands of sustainability in time and space. *International Journal of Sustainable Society, 6*, 9–27.

Cairns, R. D., & Van Long, N. (2006). Maximin: A direct approach to sustainability. *Environment and Development Economics, 11*, 275–300.

Carpenter, S., et al. (2001). From metaphor to measurement: Resilience of what to what? *Ecosystems, 4*, 765–781.
Cass, D. (1965). Optimum growth in an aggregative model of accumulation. *Review of Economic Studies, 3*, 233–240.
Chang, Y., et al. (2015). Sustainability in Asian countries: Perspectives from economics. *Nature and Culture, 10*, 81–102.
Corona, B., et al. (2019). Towards sustainable development through the circular economy—A review and critical assessment on current circularity metrics. *Resources, Conservation and Recycling, 151,* Art. No. 104498.
Cosme, I., et al. (2017). Assessing de-growth discourse: A review and analysis of academic de-growth policy proposals. *Journal of Cleaner Production, 149*, 321–334.
Cumming, G. S., & Peterson, G. D. (2017). Unifying research on social-ecological resilience and collapse. *Trends in Ecology & Evolution, 32*, 695–713.
D'Amato, D., et al. (2017). Green, circular, bio economy: A comparative analysis of sustainability avenues. *Journal of Cleaner Production, 168*, 716–734.
De Luca Peña, L. V., et al. (2022). Towards a comprehensive sustainability methodology to assess anthropogenic impacts on ecosystems: Review of the integration of life cycle assessment, environmental risk assessment and ecosystem services assessment. *Science of the Total Environment, 808,* Art. No. 152125.
Dietz, S., & Neumayer, E. (2007). Weak and strong sustainability in the SEEA: Concepts and measurement. *Ecological Economics, 61*, 617–626.
DiMaria, C. H. (2014). Sustainability matters. *Quality & Quantity, 48*, 1257–1269.
DiMaria, C. H. (2019). Sustainability, welfare and efficiency of nations. *Quality & Quantity, 53*, 1141–1163.
Egli, H., & Steger, T. M. (2007). A dynamic model of the environmental Kuznets curve: Turning point and public policy. *Environmental & Resource Economics, 36*, 15–34.
Emmerling, J., et al. (2019). The role of discount rate for emission pathways and negative emissions. *Environmental Research Letters, 14,* Art. No. 104008.
Fan, Y. V., et al. (2019). Cross-disciplinary approaches towards smart, resilient and sustainable circular economy. *Journal of Cleaner Production, 232*, 1482–1491.
Figge, F. (2005). Capital substitutability and weak sustainability revisited: The conditions for capital substitution in the presence of risk. *Environmental Values, 14*, 185–201.
Figueres, C., et al. (2010). Sustainable development: Between moral injections and natural constraints. *Sustainability, 2*, 3608–3622.
Grumbach, S., & Hamant, O. (2020). How humans may co-exist with Earth? The case for suboptimal systems. *Anthropocene, 30,* Art. No. 100245.
Haberl, H., et al. (2020). A systematic review of the evidence on decoupling of GDP, resource use and GHG emissions, Chapter 3: Synthesizing the insights. *Environmental Research Letters, 15,* Art. No. 065003.
Hallegatte, S. et al. (2007). Why economic dynamics matter in assessing climate change damages: illustration on extreme events, Ecological Economics 62: 330–340.
Hanacek, K., et al. (2020). Ecological economics and de-growth: Proposing a future research agenda from the margins. *Ecological Economics, 169,* Art. No. 106495.
Hanley, N., et al. (2015). Genuine savings and sustainability. *Journal of Economic Surveys, 29*, 779–806.
Haskell, L., et al. (2021). Social innovation related to ecological crises: A systematic literature review and a research agenda for strong sustainability. *Journal of Cleaner Production, 325,* Art. No. 129316.
Hayha, T., et al. (2016). From planetary boundaries to national fair share of the global safe operating space—How can the scale be bridges? *Global Environmental Change, 40*, 60–72.
Hediger, W. (2000). Sustainable development and social welfare. *Ecological Economics, 32*, 481–492.
Hediger, W. (2006). Weak and strong sustainability, environmental conservation and economic growth. *Natural Resource Modeling, 19*, 359–395.

Hediger, W. (2009). Sustainable development with stock pollution. *Environment and Development Economics, 14*, 759–780.

Herrero Jauregui, C., et al. (2018). What we talk about when we talk about social-ecological systems? A literature review. *Sustainability, 10,* Art. No. 2950.

Hickel, J. (2020). The sustainable development index: measuring the ecological efficiency of human development in the Anthropocene. *Ecological Economics, 167,* Art. No. 106331.

Hoekstra, A. Y., et al. (2018). The control versus resilience rationale for managing systems under uncertainty. *Environmental Research Letters, 13,* Art. No. 103002.

Holling, C. S. (1973). Resilience and stability of ecological systems. *Annual Review of Ecological Systems, 4*, 1–23.

Irwin, E. G., et al. (2016). Welfare, wealth, and sustainability. *Annual Review of Resource Economics, 8*, 77–98.

Jain, P., & Jain, P. (2013). Sustainability assessment index: A strong sustainability approach to measure sustainable human development. *Journal of Sustainable Development and World Ecology, 20*, 116–122.

Juknys, R., et al. (2014). Sustainability of catch-up growth in the extended European Union. *Journal of Cleaner Production, 63*, 54–63.

Kallis, G. (2011). In defence of de-growth. *Ecological Economics, 70*, 873–880.

Kallis, G., et al. (2010). The economics of de-growth. *Ecological Economics, 84*, 172–180.

Kallis, G., et al. (2018). Research on de-growth. *Annual Review of Environment and Resources, 43*, 291–316.

Khan, A. M. (2015). Putting "good society" ahead of growth and/or "development": Overcoming neoliberalism's growth trap and its costly consequences. *Sustainable Development, 23*, 65–73.

Koopmans, T. C. (1965). On the concept of optimal economic growth. In J. Johansen (Ed.), *The econometric approach to development planning*. North-Holland.

Krysiak, F. C., & Krysiak, D. (2006). Sustainability with uncertain future preferences. *Environmental & Resource Economics, 33*, 511–531.

Li, T., et al. (2020). A review of social-ecological system resilience: Mechanism, assessment and management. *Science of the Total Environment, 723,* Art. No. 138113.

Liu, G., et al. (2017). Enhancing the sustainability narrative through a deeper understanding of sustainable development indicators. *Sustainability, 9,* Art. No. 19078.

Liu, X., et al. (2016). Comparing national environmental and economic performances thorough emergy sustainability indicators: Moving environmental ethics beyond anthropocentrism towards ecocentrism. *Renewable and Sustainable Energy Reviews, 58*, 1532–1542.

Luukkanen, J., et al. (2019). Green economic development in Lao PDR: A sustainability window analysis of green growth productivity and the efficiency gap. *Journal of Cleaner Production, 211*, 818–829.

Luukkanen, J., et al. (2021). Quantification of doughnut economy with sustainability window method: Analysis of development in Thailand. *Sustainability, 13,* Art. No. 847.

Martins, N. O. (2016). Ecosystems, strong sustainability and the classical circular economy. *Ecological Economics, 129*, 32–39.

McKenzie, L. W. (1959). On the existence of general equilibrium for a competitive market. *Econometrica, 27*, 54–71.

Millennium Ecosystem Assessment (MEA). (2005). www.millenniumassessment.org

Morse, S. (2015). Developing sustainability indicators and inches. *Sustainable Development, 23*, 84–95.

Munda, G. (1997). Environmental economics, ecological economics, and the concept of sustainable development. *Environmental Values, 6*, 213–233.

Munda, G. (2014). On the use of shadow prices for sustainable well-being measurement. *Social Indicators Research, 118*, 911–918.

Oliveira Neto, G. C., et al. (2019). A framework of actions for strong sustainability. *Journal of Cleaner Production, 196*, 1629–1643.

Pender, J. L. (1998). Population growth, agricultural intensification, induced innovation and natural resource sustainability: An application of neoclassical growth theory, Agricultural Economics 19: 99–112.

Peterson, G., et al. (1998). Ecological resilience, biodiversity, and scale. *Ecosystems, 1*, 6–18.

Peterson, T. J., et al. (2012). Analytical methods for ecosystem resilience: A hydrological investigation. *Water Resources Research, 48*, 1–16.

Pimm, S. L. (1984). The complexity and stability of ecosystems. *Nature, 307*, 321–326.

Qasim, M., & Grimes, A. (2021). Sustainability and wellbeing: The dynamic relationship between subjective wellbeing and sustainability indicators. *Environment and Development Economics, 27*, 1–19.

Ramcilovic-Suominen, S., & Pulzl, H. (2018). Sustainable development—A "selling point" of the emerging EU bioeconomy policy framework? *Journal of Cleaner Production, 172*, 4170–4180.

Ramsey, F. (1928). A mathematical theory of saving. *The Economic Journal, 38*, 543–559.

Randall, A. (2020). On intergenerational commitment, weak sustainability, and safety. *Sustainability, 12,* Art. No. 5381.

Reid, J., & Rout, M. (2020). Developing sustainability indicators—The need for radical transparency. *Ecological Indicators, 110,* Art. No. 105941.

Richters, O., & Siemoneit, A. (2017). Consistency and stability analysis of models of a monetary growth imperative. *Ecological Economics, 136*, 114–125.

Rietkerk, M., et al. (2004). Self-organised patchiness and catastrophic shifts in ecosystems. *Science, 305*, 1926–1929.

Ruggerio, C. A. (2021). Sustainability and sustainable development: A review of principles and definitions. *Science of the Total Environment, 786,* Art. No. 147481.

Salas-Zapata, W. A., et al. (2017). Social-ecological resilience and the quest for sustainability as object of science. *Environment, Development and Sustainability, 19*, 2237–2252.

Salas-Zapata, W. A., & Salas-Zapata, L. (2017). Contributions of sustainability science to the study of environmental health problems. *Environment, Development and Sustainability, 19*, 347–367.

Savona, M., & Ciarli, T. (2019). Structural changes and sustainability: A selected review of the empirical evidence. *Ecological Economics, 159*, 244–260.

Scheffer, M., et al. (2001). Catastrophic shifts in ecosystems. *Nature, 413*, 591–596.

Shao, Q. (2020). Paving ways for a sustainable future: A literature review. *Environmental Science and Pollution Research, 27*, 13032–13043.

Sterk, M., et al. (2017). How to conceptualize and operationalize resilience in socio-ecological systems? *Current Opinion in Environment Sustainability, 28*, 108–113.

The Economics of Ecosystem and Biodiversity (TEEB). (2009). www.teebweb.org

Traeger, C. P. (2011). Sustainability, limited substitutability, and non-constant social discount rates, Journal of Environmental Economics and Management 62 (2): 215–228.

Van den Bergh, J. C. J. M. (2010). Externality or sustainability economics? *Ecological Economics, 69*, 2047–2052.

Van den Bergh, J. C. J. M. (2011). Environment versus growth—A criticism of "de-growth" and a plea for "a-growth." *Ecological Economics, 70*, 881–890.

Vatn, A. (2009). An institutional analysis of methods for environmental appraisal, Ecological Economics 68: 2207–2215

Vouvaki, D., & Xepapadeas, A. (2008). Changes in social welfare and sustainability: Theoretical issues and empirical evidence. *Ecological Economics, 67*, 473–484.

Wiedenhofer, D., et al. (2020). A systematic review of the evidence on decoupling of GDP, resource use and GHG emissions, Chapter 2: Bibliometric and conceptual mapping. *Environmental Research Letters, 15,* Art. No. 063002.

Yi, C., & Jackson, N. (2021). A review of measuring ecosystem resilience to disturbance. *Environmental Research Letters, 16,* Art. No. 053008.

Zagonari, F. (2015). Technology improvements and value changes for sustainable happiness: A cross-development analytical model. *Sustainability Science, 10*, 687–698.

Zagonari, F. (2020). Environmental sustainability is not worth pursuing unless it is achieved for ethical reasons. *Nature—Palgrave Communications, 6,* Art. No. 108.

Zagonari, F. (2022). De-growth and a-growth together to achieve both strong and weak sustainability: Empirical results based on a theoretical model. *Futures* (under review).

Chapter 4
Environmental Decisions

Chapter 3 demonstrated that social continuity can be depicted by weak sustainability (which is close to a pure economic model) in terms of maximisation of the total welfare (i.e., efficiency), provided that the social discount rate σ is at 0. In contrast, ecological continuity can be depicted by strong sustainability (which is close to a pure ecological model) in terms of minimisation of individual resource and pollution inequality (i.e., equity), provided the reference status E^* is ecologically resilient. Note that environmental sustainability is an opportunity cost for the current generation under weak sustainability because nature, like other substitutable forms of capital, produces utility so that nature can be preserved at the cost of reducing other forms of capital and reducing welfare arising from these other forms of capital. For example, school tables could be made of plastic instead of wood to preserve forests, but this would require the use of plastic that could be used for other purposes. Similarly, sustainability is an opportunity cost for the current generation under strong sustainability because preserving nature might require a reduction of other non-substitutable forms of capital due to the negative impacts of these forms of capital on nature. For example, school tables made of wood are not produced to preserve forests and school tables made of plastic are not produced to avoid air and water pollution during plastic production.

This chapter discusses the main environmental decisions by referring to weak sustainability in terms of efficiency and to strong sustainability in terms of equity, under the assumption that some real features do not meet the sustainability criteria required by weak sustainability (if this sustainability paradigm is adopted) or strong sustainability (if this sustainability paradigm is adopted). In particular, it consists of two sections; the first discusses environmental policy measures and the second discusses environmental investment projects.

I will show that both policy measures and investment projects can be designed to achieve efficiency and equity. Policy measures to achieve efficiency can be focused on pollution and on renewable and non-renewable resources, in terms of both flows and stocks. I assume that investment projects to achieve efficiency are based on cost–benefit analysis. Policy measures to achieve equity can be focused on inequalities

F. Zagonari, *Environmental Ethics, Sustainability and Decisions*,
https://doi.org/10.1007/978-3-031-21182-9_4

between (groups of) individuals in terms of national laws and regulations or focused on inequalities between (groups of) countries in terms of bilateral or multilateral agreements. I assume that investment projects to achieve equity are based on multi-criteria analysis.

Note that I will discuss sustainability decisions by enterprises (e.g., technology choices) within the investment projects.

4.1 Environmental Policy Measures

The main purpose of this book is to present the simplest theoretical models to highlight some empirical issues in the literature on environmental sustainability (i.e., it is a textbook with a research outlook). To do so, I had two options: first, to provide a comprehensive model (e.g., the Economic General Equilibrium Model discussed in Chap. 3) that can be simplified for a given issue; second, to slightly complicate the basic model (e.g., the marginal private net benefits–marginal external costs model that I discuss in Chap. 3) to obtain the most suitable model for each issue. I chose the second option. In particular, policies to achieve efficiency are summarized in two *Tables* (for pollution and resources), whereas policies to achieve equity are summarized in one *Figure* and two *Tables* (for equity of national and international policies and agreements among individuals and for equity of international policies and agreements among countries).

Note that each demonstration or example rely on positive existential statements (i.e., in logical symbols, $\exists\ x \in X \mid P(x)$—there is a case x such that statement P is true—so many other demonstrations or examples could be provided). Consequently, students are not expected to read all demonstrations or examples once they understand the main logic behind this book.

4.1.1 Environmental Policy Measures to Achieve Efficiency

In this Section, I will refer to the Economic General Equilibrium Model, in which a set of *n* prices exists such that decisions by consumers to maximize their utility are consistent with decisions by producers to maximize their profit if there are no externalities, no asymmetric information, no uncertainty, and no imperfect competition and if individuals are consequentialist (i.e., they decide based on the consequences), welfarist (i.e., they decide based on improving welfare), individualist (i.e., they decide based on their own welfare), and rational (i.e., they take systematic and consistent decisions aimed at maximising their welfare). I can then discuss and systematize environmental policies to achieve efficiency. Indeed, even if the assumption of no environmental externalities is eliminated, all main policies are equivalent in terms of efficiency if all other assumptions behind the Economic General Equilibrium Model hold. Note that the presence of ecological issues could coexist with the absence of environmental externalities by implementing some kind of monetary compensation (e.g., direct payments, modified prices). In contrast, even if the assumptions of rationality, consequentialism, welfarism, and individualism are retained, the main policies might *not* be equivalent in terms of efficiency in the case of asymmetric information, uncertainty, or imperfect competition.

Note that the introduction of pollution production and resource use within the Economic General Equilibrium Model is required by the principles of thermodynamics (i.e., economy cannot be decoupled from nature). Moreover, imperfect competition for renewable resources will depict a preferential access for one agent rather than sharing of access among multiple stakeholders (i.e., a single access rather than an open access), whereas this preferential access is assumed for non-renewable resources to depict exploitation rights. Finally, the policies I describe to achieve efficiency in Sect. 4.1.1 cover all theoretically relevant policies. These policies include taxes, standards, subsidies, permits for pollution; taxes, standards, subsidies, exploitation rights, and protected areas for renewable resources; and taxes, standards, subsidies, exploitation rights, and technological support for non-renewable resources. Indeed, if a governmental agency wants to implement an environmental action, then:

- It can impose this action (e.g., by developing standards for pollutants; standards, protected areas and exploitation rights for renewable resources; standards and exploitation rights for non-renewable resources) and can fine agents that do not adhere to the policy, with a guarantee that these agents will be fined in the context of complete and perfect information.
- It can provide incentives to adhere to this policy (e.g., subsidies for pollutants and renewable resources; subsidies and technological support for non-renewable resources) so that agents who choose not to adhere to the policy renounce the incentive and bear an opportunity cost. Note that this is theoretically impossible if agents are rational.
- It can provide incentives for agents to avoid alternative actions (e.g., taxes for pollutants and resources) so that agents who refuse to adhere to the policy bear an actual cost. Note that this is theoretically impossible if agents are rational.
- It can implement a combination of all three measures.

Note that standards, exploitation rights, and protected areas for renewable resources are usually called command-and-control policies. Moreover, taxes, subsidies, and technological support for non renewable resources are usually called market-based policies. Finally, market-based permits are a type of standard in which the fine for violating a standard is replaced by the equilibrium price for a permit, which arises from the free interactions among agents in the competitive market for permits.

4.1.1.1 Pollution

This Section will identify optimal pollution policies (i.e., taxes TAX, standards STA, subsidies SUB and permits PER) in a simplified framework. Table 4.1 suggests the main articles on pollution policies to efficiency in more complicated contexts (i.e., asymmetric information ASY, uncertainty UNC, imperfect competition IMP). Let us assume that there is no time (i.e., what matters is the amount of pollution at each time t). Moreover, there is no space (i.e., what matters is the total amount of pollution regardless of its spatial distribution). Finally, there is no decisional interactions (i.e., decisions taken by a governmental agency do not impact on welfare in other countries and consequently on decisions by other governmental agencies).

Reducing Quantity: Taxes, Standards, Subsidies

Let us assume that P = MR is the marginal revenue and Q = MC is the marginal cost in a single product competitive market. In other words, MPNB = MR − MC introduced in Chap. 3 becomes:

$$\text{MPNB} = P - \text{Q}$$

Let us assume that β is the concern for the amount of pollution Y and d is the good production Q for unit of pollution Y (i.e., Q = d Y). In other words, MEC introduced in Chap. 3 becomes:

$$\text{MEC} = (\beta/d)Q$$

Note that MPNB represents benefits both from consumption (consumer surplus) and production (producer surplus) (i.e., the area below MPNB measures the total net benefits), whereas MEC represents costs for polluted people (i.e., the area below MEC measures the total external costs). Moreover, the number of competitive firms N is standardized at 1. Finally, MPNB = 0 (i.e., P = MC = Q) depicts the profit maximization for competitive firms. Thus, the optimal quantity Q* is given by:

$$Q^* = \frac{dP}{\text{d} + \beta}$$

Note that the optimal quantity Q* maximises the social welfare SW:

$$Max_Q SW = PQ - FC - (1/2)Q^2 - (\beta/d)(1/2)Q^2$$

where P Q depicts the total revenues from production, FC is the fixed costs in production, the third term depicts the variable costs in production, and the last term is the total external cost from production due to the related pollution.

Table 4.1 Main articles on pollution policies to efficiency

	Taxes	Standards	Subsidies	Permits
UNC	Aronsson (1999) ERE (DYN), Soretz (2007) ERE, Aronsson and Blomquist (2003) ERE	Chen et al. (2013) EnEc, Stavins (1996) JEEM (& TAX), Gren et al. (2012) EnEc	Lefebvre et al. (2020) ERE	Haiata-Falah (2016) JRE, Durand-Lasserve et al. (2010) EP, Xu et al. (2016) EnEc, Stranlund et al. (2019) JEEM
ASY	D'Amato and Dijkstra (2015) REE (& PER), Abrell et al. (2018) JEEM	Selden and Terrones (1993) JEEM, Rege (2000) ERE, Huang et al. (2018) JCP	Arguedas and Van Soest (2009) JEEM	MacKenzie and Ohndorf (2012) JEEM, Chesney and Taschini (2012) AMF, Fischer (2005) EP
IMP	Schoonbeek and De Vries (2009) JRE, Carlsson (2000) ERE, Bayinder-Upmann (2004) ERE, Ebert and Von Dem Hagen (1998) ERE, Brecard (2011) ERE, Hart (2004) JEEM, Heijnen and Kooreman (2006) JTEP, Vetter (2009) BEJTE, Orlov and Grethe (2012) EP	Ebert (1998) JE, Barcena-Ruiz and Campo (2017) ERE (& TAX), Maloney and Yandle (1984) JEEM (& PER)	Chu and Lai (2014) JEDC (& TAX) (DYN), Shao et al. (2019) JTEP (& TAX)	Fang and Ma (2020) EnEc, Bernard et al. (2008) JEDC, Rubin (1996) JEEM, Schennach (2000) JEEM, Boom and Dijkstra (2009) ERE, Lee (2011) MASGC, Sunnevag (2003) ERE, Berger et al. (1992) REE (& TAX)
UNC & ASY				Taschini (2010) APFM
UNC & IMP	Rentschler et al. (2018) IEEP, Sartzetakis and Tsigars (2009) EDE			Haurie and Vigueier (2003) EMA
ASY & IMP	Antelo & Louriero (2009) EE (& SUB)	Cavaliere (2000) ERE, Kirchhoff (2000) ERE	Nannerup (1998) ERE (& STA)	Meunier (2011) ERE

Abbreviations: UNC = uncertainty, ASY = asymmetric information, IMP = imperfect competition, TAX = taxes, STA = standards, SUB = subsidies, PER = permits: see the list of acronyms for journals. *Notes* The main references for TAX is Pigou (1920) The economics of welfare, for SUB is Baumol and Oates (1988) The theory of environmental policy, for PER is Montgomery (1972) Journal of Economic Theory

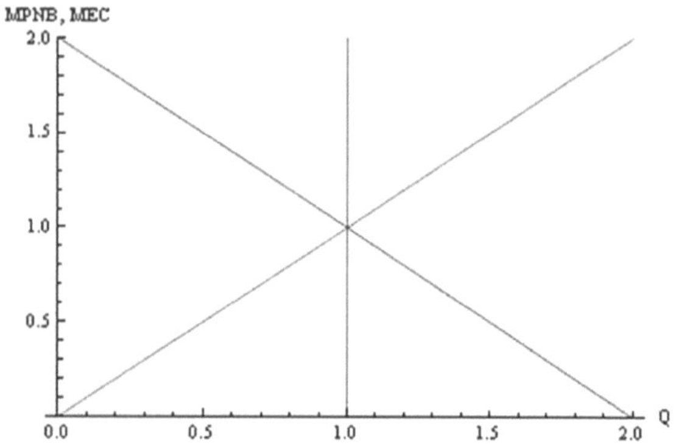

Fig. 4.1 Optimal polluting quantity. The decreasing (blue) line = Marginal Private Net Benefit MPNB, the increasing (purple) line = Marginal External Cost MEC. *Notes* The optimal polluting quantity is at Q* = 1 if $P = 2$, $\beta = d = 1$

Figure 4.1 depicts an optimal polluting quantity. Note that competitive firms produce Q such that P = MC or MPNB = 0 if there are no rights for polluted people. Moreover, identifying the Q level that makes MR = MC is a reliable marginal procedure to maximise the total welfare in a static analysis, since MR and MC are monotonically decreasing and increasing, respectively. Finally, the Coase (1960) theorem (i.e., the starting status, as the distribution of rights between polluting and polluted people, do not affect the final status, as the optimal amount of pollution, provided that few and identified people can barging over the amount of pollution without transaction costs in a context of complete and perfect information) can be depicted by MPNB = MEC in a dynamic analysis where the starting statuses are either Q such that MPNB = 0 (i.e., rights to polluting firms) or Q such that MEC = 0 (i.e., rights to polluted people).

Taxes TAX

Since competitive firms choose to produce Q such that P = MC, in the simplified context introduced above, in order for firms to choose the optimal quantity Q*, instead of Q = P, a tax could be introduced such that MR = P − tax. Thus, the optimal tax* such that MPNB = P − tax − Q* = MEC = (β/d) Q* is given by:

$$tax^* = \frac{\beta P}{d + \beta}$$

Indeed, searching for Q that maximises social welfare and Q that maximises firm profits leads to:

$$MPNB = MEC \leftrightarrow P - Q^* = (\beta/d)Q^* \leftrightarrow P = Q^* + (\beta/d)Q^*$$

$$MPNB = 0 \leftrightarrow P - tax = Q^* \leftrightarrow P = Q^* + tax$$

$$tax^* = \frac{\beta}{d}Q^* = \frac{\beta P}{d + \beta}$$

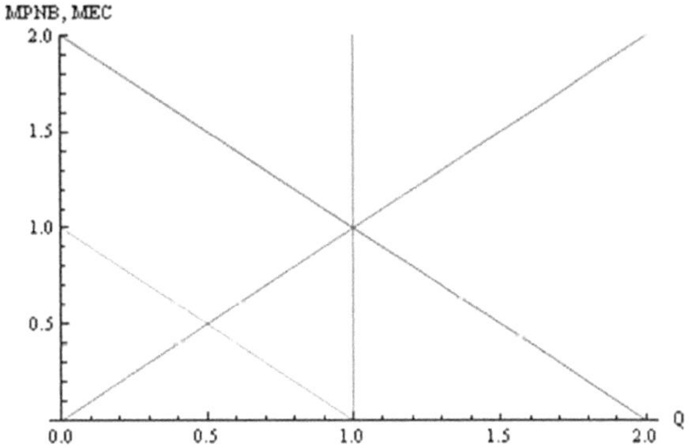

Fig. 4.2 Optimal tax on pollution production. The decreasing (blue) line above = Marginal Private Net Benefit MPNB, the increasing (purple) line = Marginal External Cost MEC, the decreasing (grey) line below = MPNB − tax*. *Notes* The optimal tax is at $tax^* = 1$ if $P = 2$, $\beta = d - 1$

Note that *tax** maximises SW. Moreover, tax revenue TR = P × tax* is irrelevant for SW, since it represents a transfer from two groups of people belonging to the same society (e.g., polluting firms and polluted people). Finally, *tax** = MEC at Q*. Figure 4.2 depicts an optimal tax on pollution.

Note that if Q = N q with q being the quantity produced by each competitive firm (i.e., q* = Q*/N), then tax* = P − q* = [P (N d + β N − d)]/[N (d + β)].

Standards STA

Let us assume that the optimal quantity Q* is implemented by introducing a standard sta such that a firm that produces more than Q* is fined.

The optimal fine *fin** to enforce the optimal standard *sta** at *Q** is given by:

$$fin^* = \frac{\beta P}{d + \beta}$$

Indeed, at any Q > Q*, MPNB = P − Q − fin < 0, since firms are fined for sure due to the assumption of complete and perfect information, whereas firms do not pay any fine if Q ≤ Q*.

Figure 4.3 depicts an optimal standard on pollution.

Note that fin* = tax*.

Taxes versus standards in the case of asymmetric information

Let us assume that the government is misinformed by firms about the true MPNB and consequently it fixes pollution policies such as taxes and standards by relying on biased information. In particular, let us realistically assume that firms hide profits to the government and consequently the government takes its decisions by referring to an under valuated MPNB: False MPNB = True MPNB − asy, where asy measures the amount of asymmetric information ASY. The tax and sta levels fixed by the government are calculated by applying to the False MPNB the same procedures introduced in the previous Section. Thus, the quantity levels produced by firms if these biased sta and tax are introduced by the misinformed government are given by:

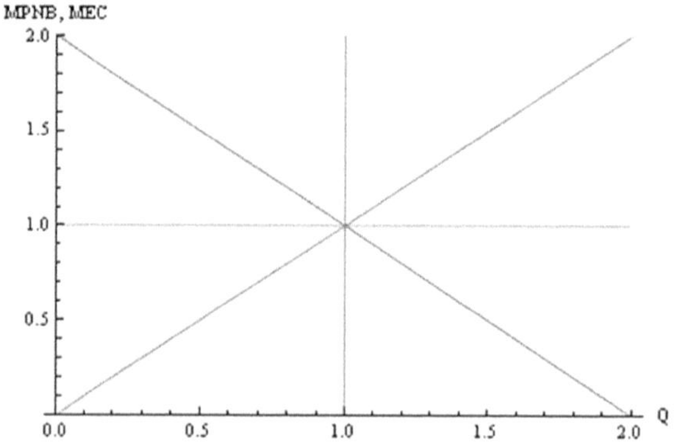

Fig. 4.3 Optimal standard and optimal fine on pollution production. The decreasing (blue) line = Marginal Private Net Benefit MPNB, the increasing (purple) line = Marginal External Cost MEC, the vertical (grey) line = the standard sta, the horizontal (grey) line = the fine fin applied with sta. *Notes* The optimal fine is at fin* = 1 for the optimal standard at sta* = 1 if $P = 2$, $\beta = d = 1$

$$P - asy - Q = \frac{\beta}{d}Q \leftrightarrow Q_{sta}^{asy} = \frac{d(P - asy)}{d + \beta}$$

$$P - \frac{\beta(P - asy)}{d + \beta} - Q = 0 \leftrightarrow Q_{tax}^{asy} = \frac{d\,P + as\,\beta}{d + \beta}$$

While the socially optimal quantity level is still given by:

$$Q^* = \frac{d\,P}{d + \beta}$$

Note that Q* depends on P.

Of course, both tax and sta based on the False MPNB do not maximise the social welfare SW. The question is which policy leads to a smaller social welfare loss. In particular, a sta calculated in terms of the False MPNB leads to underproduction with respect to the optimal quantity Q*, while a tax calculated in terms of the False MPNB leads to overproduction with respect to the optimal quantity Q*:

$$Q_{sta}^{asy} < Q^* \leftrightarrow \frac{d\,as}{d + \beta} > 0$$

$$Q_{tax}^{asy} > Q^* \leftrightarrow \frac{\beta\,as}{d + \beta} > 0$$

The related consequences on social welfare can be estimated by calculating the area size of the trapezoid below the True MPNB from Qsta and Q* for sta (i.e., the total social loss due to underproduction) and the area size of the trapezium below the True MPNB from Q* and Qtax for tax (i.e., the total social loss due to overproduction).

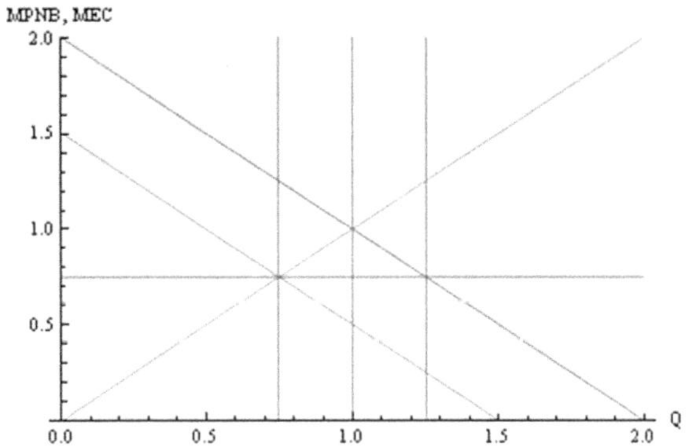

Fig. 4.4 Tax versus standard on pollution production in the case of asymmetric information. The decreasing (purple) line above = True Marginal Private Net Benefit MPNB, the decreasing (green) line below = False MPNB, the increasing (grey) line = Marginal External Cost MEC, the horizontal (blue) line = tax or fin fixed by referring to the False MPNB. *Notes* If $P = 2$, $\beta = d = 1$, asy = 0.5, the under production from sta is at $Q_{sta} = 0.75$, the optimal Q* is at $Q^* = 1$, the over production from tax is at $Q_{tax} = 1.25$, and tax and sta produce the same social welfare loss

$$SW_{tax} > SW_{sta} \leftrightarrow \frac{as^2(d - \beta)}{2d} > 0$$

Figure 4.4 compares a tax with a standard on pollution under asymmetric information.

Thus, tax should be preferred to sta whenever d > β (i.e., the marginal net loss for consumers and producers is larger than the marginal loss for polluted people).

Subsidies SUB

Let us assume that the government pays a unit subsidy sub to firms producing $q \le q^*$. Note that Q = N q and Q* = N q* with N the number of firms. Thus, the total cost function TC, the marginal cost function MC, and the average cost function AC for each identical firm are given by:

$$TC = FC + (1/2)q^2 - sub(q^* - q)$$

$$MC - q + sub$$

$$AC = FC/q + (1/2)q - sub(q^* - q)/q$$

Note that MC with sub is larger than without sub, since producing more than q* means renouncing to a potential subsidy (i.e., sub at q > q* is an opportunity cost). Moreover, AC with sub can be larger or smaller than AC without sub (i.e., it depends on FC and q*). Finally, N Q = q and Q* = q* if N is standardised to 1 as for tax and sta in the previous Sections.

In the *short-run*, the number of firms is fixed with $N_{SR} = 1$ and the decision rule for each competitive firm is represented by P = MC for any positive, null or negative profit level. Thus, each firm produces $q_{SR}* = P - sub_{SR}$ with a total production of $Q* = q_{SR}*$.

Consequently, the optimal subsidy $sub_{SR}*$ for a fixed number of firms $N_{SR} = 1$ is given by:

$$sub^*_{SR} = \frac{\beta P}{d + \beta}$$

Note that $sub_{SR}* = fin* = tax*$.

In the *long-run*, the number of firms changes, with N_{LR} increasing if profits are positive in the short-run and N_{LR} decreasing if profits are negative in the short-run. The market equilibrium is represented by P = min AC, where the decision rule for each competitive firm is still represented by P = MC. Thus, each firm produces:

$$q_{LR} = \sqrt{2}\sqrt{FC - sub_{LR} q^*}$$

With

$$q^* = Q^*/N_{LR} = [(\mathrm{d}\, P_{LR})/(\mathrm{d} + \beta)]/N_{LR}$$

Consequently,

$$P_{LR} = \min AC = sub_{LR} + \sqrt{2}\sqrt{\mathrm{FC} - q^* sub_{LR}}$$

Note that each firm produces a smaller amount with $sub_{LR} > 0$ (i.e., q_{LR} decreases with an increasing sub_{LR}). Moreover, $sub_{LR} < FC/q*$. In other words, sub_{LR} can be larger if FC is larger, P_{LR} is smaller, β is larger, d is smaller, N_{LR} is larger. Finally, P_{LR} increases with an increasing N_{LR}. However, $Q_{LR} > Q*$ if $P_{LR} < P*$ and $Q_{LR} < Q*$ if $P_{LR} > P*$ for any decreasing demand function, where $P* = (Q*/d)/(d + \beta)$.

Figures 4.5, 4.6 and 4.7 depict optimal short-run and long-run subsidies on pollution within alternative cost structures and pollution targets.

Consequently, in order to implement the long-run optimal quantity $Q* = (d\, P)/(d + \beta)$ also in the long-run, $sub_{LR}*$ must ensure that $P_{LR} = P$:

$$sub^*_{LR} = P - q^* + \sqrt{2FC - 2Pq^* + q^{*2}}$$

And

$$q^*_{LR} = \sqrt{2}\sqrt{FC - sub^*_{LR} q^*}$$

where:

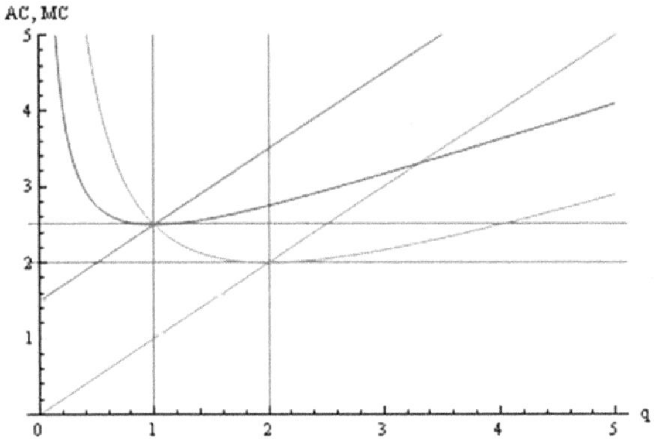

Fig. 4.5 Short and long-run impacts of a subsidy on pollution production if long larger than short-run prices. The increasing (green) line below = marginal cost MC with no subsidy sub, the increasing (purple) line above = MC with subsidy sub* = 1.5, the decreasing and increasing (grey) curve = average cost AC with no sub, the decreasing and increasing (blue) curve = AC with sub* = 1.5. *Notes* If $P = 2, \beta = d = 1, FC = 5$, the optimal quantity with no sub is $q^* = 2$, the optimal quantity with sub is $q^*_{sub} = 1$, the equilibrium price in the short-run is $P_{SR} = 2$, the equilibrium price in the long-run is $P_{LR} = 2.5$, the number of firms in the short-run is $N_{SR} = 1$, the number of firms in the long-run $N_{LR} = 1.25$

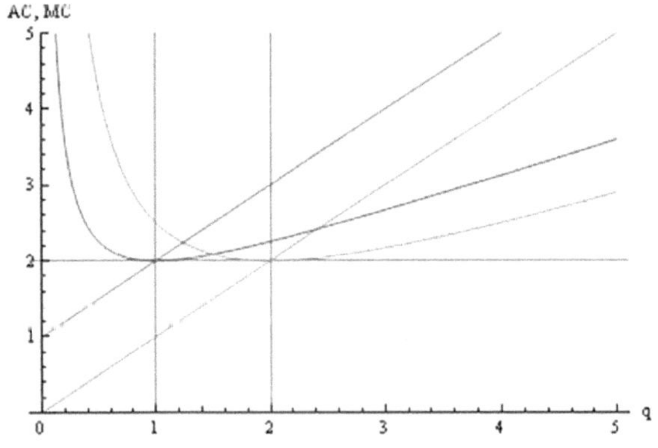

Fig. 4.6 Short and long-run impacts of a subsidy on pollution production if long equal short-run price. The increasing (green) line below = marginal cost MC with no sub, the increasing (purple) line above = MC with subsidy sub* = 1, the decreasing and increasing (grey) curve = average cost AC with no sub, the decreasing and increasing (blue) curve = AC with sub* = 1. *Notes* If $P = 2, \beta = d = 1, FC = 2$, the optimal quantity with no sub is $q^* = 2$, the optimal quantity with sub is $q^*_{sub} = 1$, the equilibrium price in the short-run is $P_{SR} = 2$, the equilibrium price in the long-run is $P_{LR} = 2$, the number of firms in the short-run is $N_{SR} = 1$, the number of firms in the long-run $N_{LR} = 1$

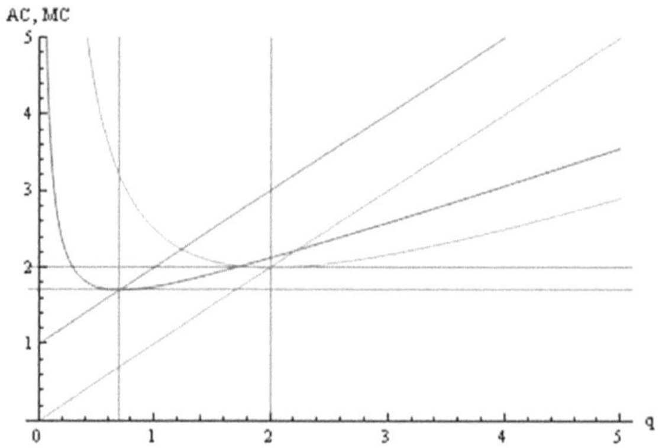

Fig. 4.7 Short and long-run impacts of a subsidy on pollution production if long smaller than short-run prices. The increasing (green) line below = marginal cost MC with no sub, the increasing (purple) line above = MC with subsidy sub* = 1, the decreasing and increasing (grey) curve = average cost AC with no sub, the decreasing and increasing (blue) curve = AC with sub* = 1. *Notes* If $P = 2$, $\beta = d = 1$, FC = 2, the optimal quantity with no sub is $q^* = 2$, the optimal quantity with sub is $q^*_{sub} = 0.7$, the equilibrium price in the short-run is $P_{SR} = 2$, the equilibrium price in the long-run is $P_{LR} = 1.75$, the number of firms in the short-run is $N_{SR} = 1$, the number of firms in the long-run $N_{LR} = 0.875$

$$N^*_{LR} = \frac{dP}{d+\beta}\frac{1}{q^*_{LR}}$$

Thus, sub_{LR} could be feasible in the long-run, but at a smaller price and at a larger total quantity prevailing in the short-run equilibrium.

Taxes versus subsidies in the case of imperfect competition

Let us assume that the demand function is represented by $Q = Q_k - P$ or, alternatively, $P = Q_k - Q$. Since MC = Q, the quantities produced by a monopolistic firm (i.e., MR = MC) and by competitive firms (i.e., P = MC) are given by:

$$Q^*_{com} = \frac{Q_k}{2} \text{ and } P^*_{com} = \frac{Q_k}{2};\; Q^*_{mon} = \frac{Q_k}{3} \text{ and } P^*_{mon} = \frac{2Q_k}{3}$$

Figures 4.8 and 4.9 compare a tax with a subsidy on pollution in case of imperfect competition in alternative concerns for the amount of pollution.

Since MEC = (β/d) Q, the socially optimal quantity is obtained by satisfying Q + (β/d) Q = MC + MEC = P = Q_k − Q. Consequently:

$$Q^*_{soc} = \frac{d\,Q_k}{2d+\beta} \text{ and } P^*_{soc} = \frac{(d+\beta)Q_k}{2d+\beta}$$

In order to make the monopolistic firm choose the socially optimal quantity, a tax could be introduce that satisfies MR = MC + tax:

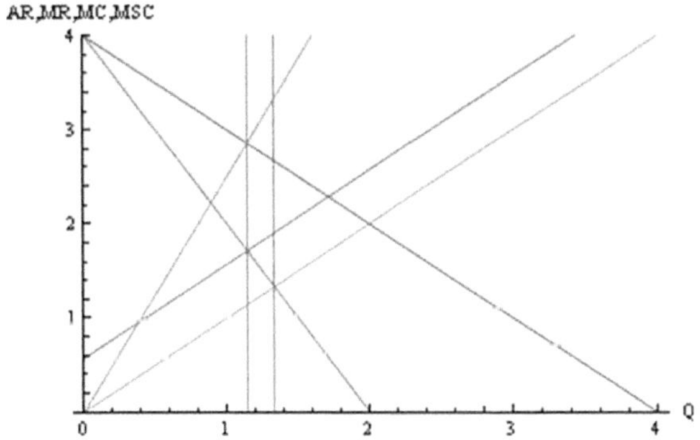

Fig. 4.8 Positive tax versus subsidy on pollution production in the case of imperfect competition. The decreasing (blue) curve above = Average Revenue AR, the decreasing (purple) line below = Marginal Revenue MR, the increasing (grey) line below = Marginal Cost MC, the increasing (green) steeper line = Marginal Social Cost MSC = MC + MEC, the increasing (blue) line above = MC + *tax*. *Notes* If AR is $P = 4 - Q$ and $d = 1$, $\beta = 1.5$, the optimal tax is $tax^* = 0.6$, the quantity produced by a monopolistic firm without tax is Q = 1.33, the socially optimal quantity with tax^* is $Q^* = 1.13$

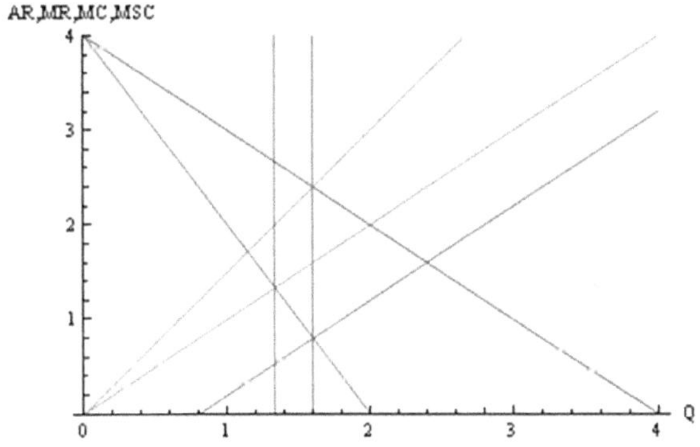

Fig. 4.9 Negative tax versus subsidy on pollution production in the case of imperfect competition. The decreasing (blue) curve above = Average Revenue AR, the decreasing (purple) line below = Marginal Revenue MR, the increasing (grey) line below = Marginal Cost MC, the increasing (green) steeper line = Marginal Social Cost MSC = MC + MEC, the increasing (blue) line above = MC + *tax*. *Notes* If AR is $P = 4 - Q$ and $d = 1$, $\beta = 0.5$, the optimal tax is $tax^* = -0.8$, the quantity produced by a monopolistic firm without tax is Q = 1.33, the socially optimal quantity with tax^* is $Q^* = 1.6$

$$\mathrm{MC} + tax = Q^*_{soc} + tax = \mathrm{MR} = P - 2\,Q^*_{soc} \leftrightarrow tax^*_{mon} = \frac{Q_k(\beta - d)}{2d + \beta} > 0 \text{ if } \beta > d$$

In order to make the monopolistic firm choose the socially optimal quantity, a sub could be introduce that satisfies MR = MC + sub:

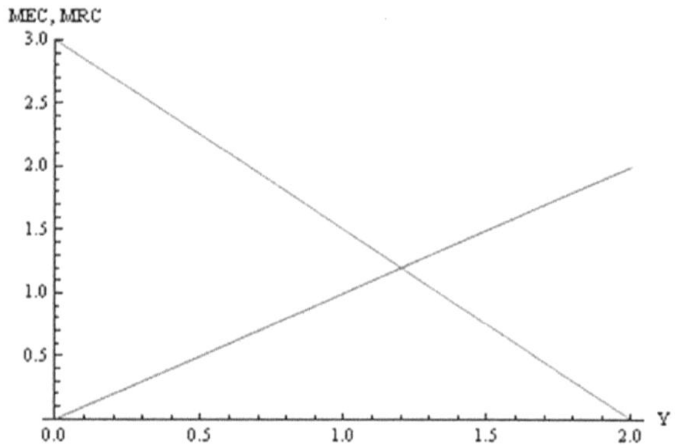

Fig. 4.10 Optimal pollution. The decreasing (purple) line = Marginal Reduction Costs MRC, the increasing (blue) line = Marginal External Cost MEC. *Notes* If $c_Y = 1.5$, $\beta = 1$, and $Y_0 = 2$, the socially optimal pollution is $Y^* = 1.25$

$$\mathrm{MC} + sub = Q^*_{soc} + sub = \mathrm{MR} = P - 2\,Q^*_{soc} \leftrightarrow sub^*_{mon} = \frac{Q_k(d-\beta)}{2d+\beta} > 0 \text{ if } \beta < d$$

Thus, sub_{mon} can be both positive or negative (i.e., a tax). Indeed, it might be impossible to solve two problems (i.e., externalities due to pollution and underproduction due to imperfect competition) with a single policy.

Improving Technology: Permits

Let us assume that it is possible to reduce pollution Y by investing in a greener technology instead of reducing production Q. In particular, the marginal reduction cost MRC is assumed to be linear in Y, where MRC is larger if the reduced pollution is larger with respect to the initial level Y_0 (i.e., $Y_0 - Y$) and MRC is smaller if a firm is technologically more efficient (i.e., c_Y is smaller):

$$\mathrm{MRC} = c_Y(Y_0 - Y)$$

Let us define MEC in terms of pollution rather than in terms of production:

$$\mathrm{MEC} = \beta Y$$

The socially optimal pollution level is given by satisfying MEC = MRC:

$$Y^* = \frac{c_Y Y_0}{c_Y + \beta}$$

Indeed, it is worth implementing a greener technology if its (marginal) cost in terms of investment is smaller than its perceived (marginal) benefit in terms of reduced pollution.

Figure 4.10 depicts an optimal amount of pollution.

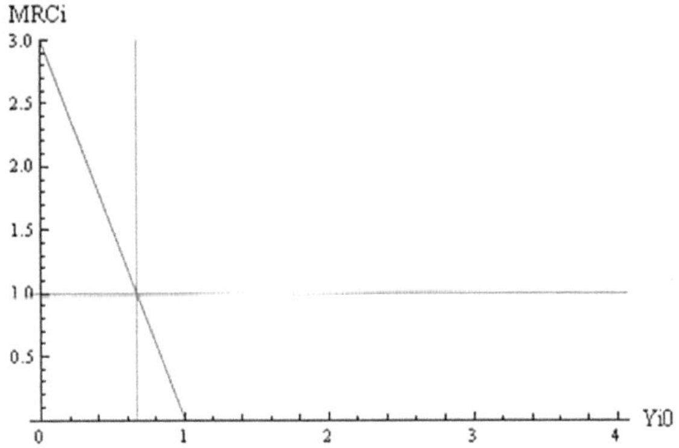

Fig. 4.11 Demand for permits by a firm at a given price of permits. The decreasing (blue) line = Marginal Reduction Costs for firm i MRC_i = Demand for permits by firm i DP_i. *Notes* If $c_Y{}^i = 3$, the initial pollution level by firm i is $Y_0{}^i = 1$ and the price of permits $P_p = 1$, the demand for permits by firm i $DP_i = 0.66$

Permits PER

Let us assume that there are two identical firms i and j, consistently with the assumption of complete and perfect information:

$$\mathrm{MRC}_i = c_Y^i\left(Y_0^i - Y^i\right)$$

$$\mathrm{MRC}_j = c_Y^j\left(Y_0^j - Y^j\right)$$

Let us assume that firms can either reduce pollution by investing in a greener technology or buy a permit that allows them to produce the specified amount of pollution. In particular, let assume that one permit allows the production of one unit of pollution. Since firms choose the cheaper option, the demand for permits DP by firm i and j are given by:

$$\mathrm{DP}_i = Y_0^i - \frac{P_P}{c_Y^i};\ \mathrm{DP}_j = Y_0^j - \frac{P_P}{c_Y^j}$$

where P_p is the price of permits. Figure 4.11 depicts a demand for permits by a single firm.

Consequently, the equilibrium price of permits P_p* for a given supply of permits SP is given by:

$$\mathrm{DP}_i + \mathrm{DP}_j = \frac{c_Y^i Y_0^i - P_P}{c_Y^i} + \frac{c_Y^j Y_0^j - P_P}{c_Y^j} = SP$$

$$P_P{}^* = \frac{c_Y^i c_Y^j (Y_0^i + Y_0^j - SP)}{c_Y^i + c_Y^j}$$

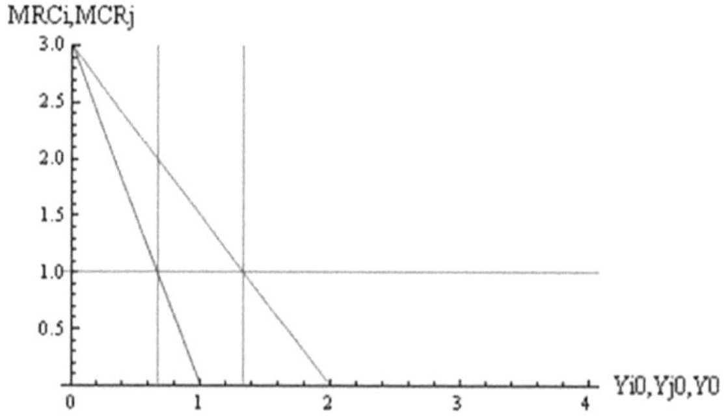

Fig. 4.12 Equilibrium price of permits. The decreasing (blue) line below = Marginal Reduction Costs for firm i MRC_i = Demand for permits by firm i DP_i, the decreasing (grey) line above = $MRC_i + MRC_j$ = Demand for permits by firm i DP_i and firm j DP_j at any price of permits. *Notes* If $c_Y{}^j = c_Y{}^i = 3$, $Y_0{}^i = Y_0{}^j = 1$ with a total initial pollution $Y_0 = 2$, and the supply of permits SP = 1.33 (vertical grey line), the equilibrium price of permits $P_p{}^* = 1$

where $DP_i + DP_j$ is the total demand for permits (i.e., by firm i and firm j). Figure 4.12 depicts an equilibrium price of permits in the case of demand for permits by two firms.

Note that the distribution of permits between firms (e.g., auction or grandfathering) does not affect the price of permits in equilibrium under the assumption of perfect competition of all markets (i.e. capital markets included), since what is relevant is the total opportunity cost, not the individual monetary cost.

Permits versus standards in the case of uncertainty

Let us assume that firms are different, due to uncertainty about the (stricter or looser) environmental policy adopted by the future government based on unpredictable technological progresses (i.e., this policy is also unknown to the current government, otherwise it would be a case of asymmetric information) so that one firm is more technologically advanced (i.e., it expects a smaller level of pollution allowed in the future and it paid costs to reduce its current pollution for each production unit) and one firm is producing a larger amount of pollution (i.e., it does not expect a smaller level of pollution allowed in the future and it saved costs to reduce its current pollution for each production unit). For concreteness, think of investment in electrical cars. Note that all four possible combination of technology advances and pollution levels can be relevant (i.e., one firm more efficient and less polluting than the other firm, one firm more efficient but more polluting than the other firm, one firm less efficient but less polluting than the other firm, one firm less efficient and more polluting than the other firm).

The total reduction cost to be borne by firms i and j if a sta = S is introduced can be calculated by measuring the size of the area below the MRC for each firm (i.e., base $Y_0 - S$ multiplied by height c_Y $(Y_0 - S)$ divided by 2):

$$TRC_i(sta) = \left(c_Y^i/2\right)(Y_0^i - S)(Y_0^i - S)$$

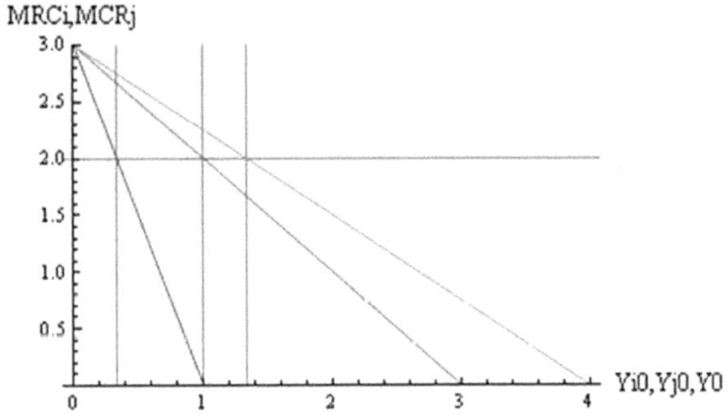

Fig. 4.13 Equilibrium price of permits under uncertainty. The decreasing (blue) line below = Marginal Reduction Costs for firm i MRC_i = Demand for permits by firm i DP_i, the decreasing (purple) middle line = Marginal Reduction Costs for firm j MRC_j = Demand for permits by firm j DP_j, the decreasing (grey) line above = $MRC_i + MRC_j$ = Demand for permits by firm i and firm j at any price of permits. *Notes* If $c_Y{}^j = c_Y{}^i = 3$, $Y_0{}^i = 1$, $Y_0{}^j = 3$ with a total initial pollution $Y_0 = 4$, and the supply of permits SP = 1.33 (vertical grey line), the equilibrium price of permits $P_p{}^* = 2$, $DP_i{}^* = 0.33$, $DP_j{}^* = 1$

$$TRC_j(sta) = \left(c_Y^j/2\right)(Y_0^j - S)(Y_0^j - S)$$

Similarly, the total reduction costs to be borne by firms if they can buy permits are given by:

$$TRC_i(per) = \left(Pp^*/2\right)(Y_0^i - S)$$

$$TRC_j(per) = \left(Pp^*/2\right)(Y_0^j - S)$$

Let us assume that the supply of permits is SP = 2 S. Consequently, the total cost of reduction is smaller with per than with sta. Indeed, TRCi (sta) + TRCj (sta) − TRCi (per) − TRCj (per) > 0 for any S. Figure 4.13 depicts an equilibrium price of permits under uncertainty in the case of demand for permits by two firms.

Thus, more polluting firms gain from permits, while less polluting firms loose from permits, although there is a social benefit from permits with respect to standards (i.e., gains of the gainer firms are larger than losses of the looser firms). However, unless the current government, in fixing the current supply of permits, perfectly predicts the environmental policy adopted by the future government, permits do not maximise the total welfare under uncertainty (i.e., permits reduce the social welfare loss with respect to standards, but they do not eliminate it).

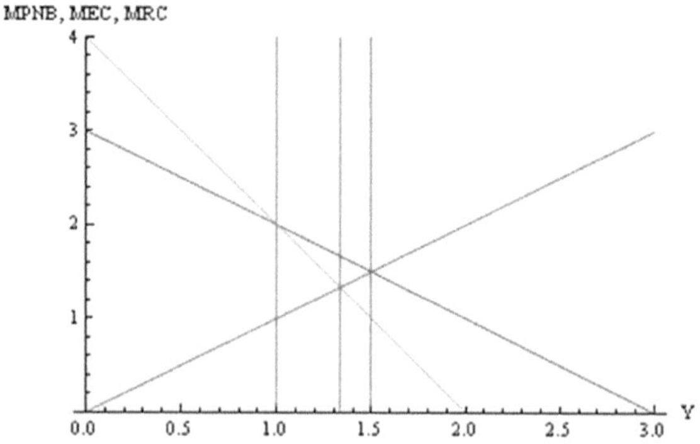

Fig. 4.14 Optimal polluting quantity versus optimal pollution. The decreasing (blue) flatter line = Marginal Private Net Benefit MPNB, the decreasing (grey) steeper line = Marginal Reduction Costs MRC, the increasing (purple) line = Marginal External Cost MEC. *Notes* If $P = 3$, $\alpha = \gamma = \delta = 1$, $\beta = 2$, $Y_0 = 2$, the socially optimal pollution $Y^* = 1.33$ is identified by MEC = MRC for MEC < MPNB

Reducing Quantity Versus Improving Technology

Let us assume that it is possible to choose whether reducing pollution by introducing a greener technology or by decreasing production. If MPNB is formalised in terms of pollution rather than in terms of production, the first option is represented by MRC = MEC, while the second option by MPNB = MEC:

$$\text{MPNB} = P - d\,Y$$

$$\text{MEC} = \beta\,Y$$

$$\text{MRC} = c_Y(Y_0 - Y)$$

Since firms choose the cheaper option, the social welfare is larger by reducing production if MEC > MPNB, while the social welfare is larger by introducing a greener technology if MEC < MPNB. Thus, the optimal pollution is given by:

$$\frac{c_Y Y_0}{c_Y + \beta} \text{ if } \frac{c_Y\,\beta\,Y_0}{c_Y + \beta} < \frac{\beta\,P}{d + \beta} \leftrightarrow Y_0 < \frac{(c_Y + \beta)P}{c_Y(d + \beta)}$$

$$\frac{P}{d + \beta} \text{ if } \frac{c_Y\,\beta\,Y_0}{c_Y + \beta} < \frac{\beta\,P}{d + \beta} \leftrightarrow Y_0 > \frac{(c_Y + \beta)P}{c_Y(d + \beta)}$$

Figure 4.14 compares an optimal amount of pollution if pollution can be reduced by introducing a greener technology or by decreasing production.

Thus, the range of pollution levels such that one option is cheaper than the other should be identified; the optimal amount of pollution by equating the cheaper option to reduce pollution with the perceived benefit of this reduction should *then* be calculated.

To summarise this Section, if one (or more) assumptions behind the Economic General Equilibrium Model EGEM identified in Chap. 3 are not met (i.e., asymmetric information ASY, uncertainty UNC, imperfect competition IMP), pollution policies might not achieve the socially optimal quantity (i.e., efficiency EFF is missed). Indeed, both taxes TAX and standards STA are inefficient in case of asymmetric information ASY; taxes TAX could be replaced by subsidies SUB in case of imperfect competition IMP; permits PER based on a supply of permits at a *non* optimal level due to uncertainty UNC are inefficient if MRC > MPNB, although the social loss from permits PER is smaller than the social loss from standards STA.

Remarks on pollution stocks with/without interactions

Let us assume the amount of pollution produced by one decision-maker (a country hereafter) affects the welfare of another decision-maker (another country hereafter). In other words, the marginal external cost MEC or the disutility D in one country depends on the total pollution S produced in one country Y and the other country y. For concreteness, let us refer to developed countries DC and less developed countries LDC (Zagonari, 1998). In particular, interactions between DC and LDC in a static context can be represented as follows:

$$S = Y + y$$

$$\max W = U(Q) - D(S)$$

$$\max w = u(q) - d(S)$$

where W and w are the welfare levels in DC and LDC, respectively, U and u are the utility levels from each consumption Q and q, in DC and LDC, respectively, D and d are the disutility levels from total pollution level S, in DC and LDC, respectively.

Interactions in a dynamic context can be represented as follows:

$$\dot{S} = Y + y - \xi S$$

$$\max \int_0^\infty W(t)e^{-\Sigma t}dt = \int_0^\infty \left[U(Q) - D(S)\right]e^{-\Sigma t}dt$$

$$\max \int_0^\infty w(t)e^{-\sigma t}dt = \int_0^\infty \left[u(q) - d(S)\right]e^{-\sigma t}dt$$

where Σ and σ are the social discount rates in DC and LDC, respectively, and ξ is the pollution decay rate.

Let $Q = Y$, $q = y$, $U(Q) = A\,Q - (1/2)\,Q^2$, $u(q) = \alpha\,q - (1/2)\,q^2$, $D(S) = (1/2)\,B\,S^2$ and $d(S) = (1/2)\,\beta\,S^2$ with A, α, B and β positive parameters.

Let us assume that DC and LDC choose the total amounts of pollution cooperatively in order to maximize the total welfare. Thus, the *cooperative* static context can be represented as follows:

$$\max_{Y,y} A\,Y - \left(\frac{1}{2}\right)Y^2 - \left(\frac{1}{2}\right)B\,S^2 + \alpha\,y - \left(\frac{1}{2}\right)y^2 - \left(\frac{1}{2}\right)\beta\,S^2$$

Subject to

$$S = Y + y$$

Therefore

$$S^*_{CS} = \frac{A + \alpha}{1 + 2(B + \beta)}$$

$$\lambda^*_{CS} = \frac{(A + \alpha)(B + \beta)}{1 + 2(B + \beta)}$$

Indeed, the Lagrange equation is given by:

$$\text{LAG} = A\,Y - \left(\frac{1}{2}\right)Y^2 - \left(\frac{1}{2}\right)B\,S^2 + \alpha\,y - \left(\frac{1}{2}\right)y^2 - \left(\frac{1}{2}\right)\beta\,S^2 - \lambda(Y + y - S)$$

So

$$Y = A - \lambda,\ y = \alpha - \lambda,\ S = Y + y$$

Note that the shadow price of the total amount of pollution S with interaction is different from the sum of the shadow prices of pollution in each country without interaction:

$$\lambda^*_{CS} = \frac{(A + \alpha)(B + \beta)}{1 + 2(B + \beta)} \neq \lambda^*_{IND,S} = \frac{AB}{1 + B} + \frac{\alpha\beta}{1 + \beta}$$

Indeed, if each country disregards decision interactions, while considering stock issues:

$$\max A\,Y - \left(\frac{1}{2}\right)Y^2 - \left(\frac{1}{2}\right)B\,S^2 - \lambda(Y - S)$$

Therefore, for DC:

$$\lambda^*_{DC,S} = \mathrm{A} - \mathrm{Y} = \mathrm{A} - \mathrm{S} = \mathrm{A} - \frac{A}{1+B}$$

Similarly for the other country. Therefore

$$\lambda^*_{IND,S} = \lambda^*_{DC,S} + \lambda^*_{LDC,S} = \mathrm{A} - \mathrm{Y} + \alpha - y = \frac{AB}{1+B} + \frac{\alpha\beta}{1+\beta}$$

Indeed, if each country disregards both decision interactions and stock issues:

$$\max_Y \mathrm{A\ Y} - \left(\frac{1}{2}\right)Y^2 - \left(\frac{1}{2}\right)\mathrm{B}\ Y^2$$

Therefore, for DC:

$$Y^*_S = \frac{A}{1+B}$$

Note that this amounts to the optimal quantity of pollution obtained in the previous Sections (i.e., Y = P/(1 + B/d)), since d = 1 (i.e., Q = Y) and A = P (i.e., the value of marginal utility U′(Q) = A − Q is A at Q = 0). Similarly, for LDC.

Next, the *cooperative* dynamic context can be represented as follows:

$$\max \int_0^\infty \left[\mathrm{A\,Y} - \left(\frac{1}{2}\right)Y^2 - \left(\frac{1}{2}\right)\mathrm{B}\,S^2 + \alpha\,\mathrm{y} - \left(\frac{1}{2}\right)y^2 - \left(\frac{1}{2}\right)\beta\,S^2\right]e^{-\varpi t}dt$$

Subject to

$$\dot{S} = Y + y - \xi\,S$$

where

$$\iota\Sigma + (1-\iota)\sigma = \varpi$$

Therefore

$$S^*_{CD} = \frac{(A+\alpha)[\varpi + \xi]}{2(B+\beta) + \xi[\varpi + \xi]}$$

$$\mu^*_{CD} = -\frac{(A+\alpha)(B+\beta)}{2(B+\beta) + \xi[\varpi + \xi]}$$

Indeed, the Hamilton function is given by:

$$\text{HAM} = \text{A}\,\text{Y} - \left(\frac{1}{2}\right)Y^2 - \left(\frac{1}{2}\right)\text{B}\,S^2 + \alpha\,\text{y} - \left(\frac{1}{2}\right)y^2 - \left(\frac{1}{2}\right)\beta\,S^2 + \mu(Y + y - \xi\,S)$$

Thus

$$\frac{\partial HAM}{\partial Y} = 0 \leftrightarrow Y = A + \mu$$

$$\frac{\partial HAM}{\partial y} = 0 \leftrightarrow y = \alpha + \mu$$

$$\dot{\mu} = \sigma\mu - \frac{\partial HAM}{\partial S} = \sigma\mu + (B + \beta)S + \mu\,\xi = 0 \leftrightarrow \mu = -\frac{(B + \beta)S}{\sigma + \xi}$$

$$\dot{S} = Y + y - \xi\,S = 0 \leftrightarrow Y + y = \xi\,S$$

where transversality conditions are assumed to hold.

Note that the shadow price of the total amount of pollution S with interaction is different from the sum of the shadow prices of pollution in each country without interaction:

$$\mu^*_{CD} = -\frac{(A + \alpha)(B + \beta)}{2(B + \beta) + \xi[\sigma + \xi]} \neq \mu^*_{IND,D} = -\frac{AB}{B + \xi[\Sigma + \xi]} - \frac{\alpha\beta}{\beta + \xi[\sigma + \xi]}$$

Indeed, if each country disregards decision interactions, while considering stock issues:

$$\max \int_0^\infty \left[\text{A}\,\text{Y} - \left(\frac{1}{2}\right)Y^2 - \left(\frac{1}{2}\right)\text{B}\,S^2\right] e^{-\Sigma t} dt$$

Subject to

$$\dot{S} = Y - \xi\,S$$

The Hamilton equation is given by:

$$\text{HAM} = \text{A}\,\text{Y} - \left(\frac{1}{2}\right)Y^2 - \left(\frac{1}{2}\right)\text{B}\,S^2 + \mu(Y - \xi\,S)$$

Thus

$$\dot{\mu} = \Sigma\mu - \frac{\partial HAM}{\partial S} = \Sigma\mu + BS + \mu\xi = 0 \leftrightarrow \mu = -\frac{BS}{\Sigma + \xi}$$

$$\dot{S} = Y - \xi\,S = 0 \leftrightarrow Y = \xi\,S$$

Thus

$$Y^*_{DC,D} = \frac{A}{1 + B/[\xi(\Sigma + \xi)]} = \frac{A\xi(\Sigma + \xi)}{B + \xi(\Sigma + \xi)}$$

$$\mu^*_{DC,D} = -A + \frac{A\,\xi(\Sigma + \xi)}{B + \xi(\Sigma + \xi)} = -\frac{AB}{B + \xi[\Sigma + \xi]}$$

Note that $Y^*_{DC,D}$ becomes $A/(1 + B)$ if $\Sigma = 0$ and $\xi = 1$.

Let us assume that the two countries choose their amount of pollution non-cooperatively in order to maximize their own welfare. In particular, since one country cannot choose the amount of pollution of the other country, the Nash equilibrium well depicts these interactions.

Thus, the non-cooperative static context can be represented as follows:

$$\max\ A\,Y - \left(\frac{1}{2}\right)Y^2 - \left(\frac{1}{2}\right)B\ (Y + y)^2$$

$$\max\ \alpha\,y - \left(\frac{1}{2}\right)y^2 - \left(\frac{1}{2}\right)\beta(Y + y)^2$$

Thus

$$A - Y - B(Y + y) = 0$$

$$\alpha - y - \beta(Y + y) = 0$$

Therefore

$$S^*_{NS} = \frac{A + \alpha}{1 + B + \beta}$$

$$\lambda^*_{NS} = \frac{(A + \alpha)(B + \beta)}{2(1 + B + \beta)}$$

Next, the non-cooperative dynamic context can be represented as follows:

$$\Sigma\,V = \max\left[A\,Y - \left(\frac{1}{2}\right)Y^2 - \left(\frac{1}{2}\right)B\,S^2 + V'(Y + y - \xi\,S)\right]$$

$$\sigma\,V = \max\left[\alpha\,y - \left(\frac{1}{2}\right)y^2 - \left(\frac{1}{2}\right)\beta\,S^2 + v'(Y + y - \xi\,S)\right]$$

where $V' = \partial V/\partial S$ and $v' = \partial v/\partial S$.

Let $Y^* = A + V'$ and $y^* = \alpha + v'$.

$$\Sigma\,\mathrm{V} = \left(\frac{1}{2}\right)A^2 - \left(\frac{1}{2}\right)\mathrm{B}\,S^2 - \left(\frac{1}{2}\right)V'^2 + V'\left(A + V' + \alpha + v' - \xi\,S\right)$$

$$\sigma\,\mathrm{v} = \left(\frac{1}{2}\right)\alpha^2 - \left(\frac{1}{2}\right)\beta\,S^2 - \left(\frac{1}{2}\right)v'^2 + v'\left(A + V' + \alpha + v' - \xi\,S\right)$$

Since utility functions are quadratic, let us assume that

$$\mathrm{V} = \Omega - \Psi\,\mathrm{S} - \frac{1}{2}\Phi\,S^2$$

$$\mathrm{v} = \omega - \psi\,\mathrm{S} - \frac{1}{2}\varphi\,S^2$$

Thus

$$\mathrm{Y} = \mathrm{A} - \Psi - \Phi\,\mathrm{S}$$

$$\mathrm{y} = \alpha - \psi - \varphi\,\mathrm{S}$$

Let $\beta = 0$ (i.e., LDC are not concerned about the environment). Thus, for DC,

$$\Sigma\,\Omega - \frac{1}{2}A^2 - \frac{1}{2}\Psi^2 + \mathrm{A}\,\Psi + \alpha\,\Psi - \psi\,\Psi = 0$$

$$-\Sigma\,\Psi - \Psi\,\Phi + \mathrm{A}\,\Phi - \varphi\,\Psi + \alpha\,\Phi - \psi\,\Phi - \xi\,\Psi = 0$$

$$-\frac{1}{2}\Sigma\,\Phi - \frac{1}{2}\Phi^2 - \xi\,\Phi + \frac{1}{2}\mathrm{B} - \varphi\,\Phi = 0$$

Similarly, for LDC,

$$\sigma\,\omega - \frac{1}{2}\alpha^2 - \frac{1}{2}\psi^2 + \alpha\,\psi + \mathrm{A}\,\psi - \psi\,\Psi = 0$$

$$-\sigma\,\psi - \psi\,\varphi + \alpha\,\varphi - \Phi\,\psi + \mathrm{A}\,\varphi - \Psi\,\varphi - \xi\,\psi = 0$$

$$-\frac{1}{2}\sigma\,\varphi - \frac{1}{2}\varphi^2 - \xi\,\varphi + \frac{1}{2}\beta - \varphi\,\Phi = 0$$

The solution values Ω^*, ω^*, Φ^*, φ^*, Ψ^*, ψ^*, can be obtained by substituting one equation in the previous one for both DC and LDC. Thus

$$\dot{S} = A + \alpha - \Psi^* - \psi^* - \left(\Phi^* + \varphi^* + \xi\right)S$$

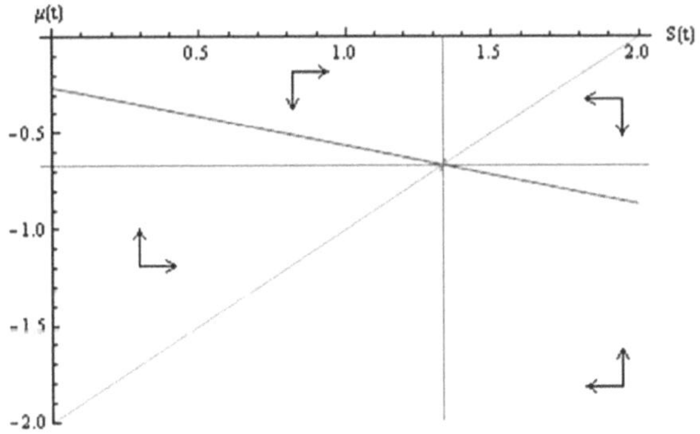

Fig. 4.15 The diagram phase of linear cooperative and non-cooperative solutions for pollution production. Figure 4.15 shows the dynamic relationship between the stock of pollution at time t S(t) and its shadow price at time t μ(t), where the S dot = 0 is the increasing (grey) line, while the μ dot = 0 is the decreasing (blue) line, if A = α = B = ξ = σ = 1 and β = 0. *Notes* The long-run globally stable equilibrium is at S* = 4/3 = 1.33 and μ* = −2/3 = −0.66 (i.e., it is a sink)

$$S(t) = \frac{A + \alpha - \Psi^* - \psi^*}{\Phi^* + \varphi^* + \xi}\left(1 - \exp\left[-(\Phi^* + \varphi^* + \xi)t\right]\right)$$

$$S^*_{ND} = \frac{A + \alpha - \Psi^* - \psi^*}{\Phi^* + \varphi^* + \xi}$$

The linear solution for the non-cooperative stock of pollution is given by:

$$S^*_{ND} = \frac{(A + \alpha)(\xi + \Sigma)}{B + \xi(\xi + \Sigma)}$$

Note that the shadow price is given by:

$$\mu^*_{ND} = -\frac{(A + \alpha)B}{B + \xi(\xi + \Sigma)}$$

Figure 4.15 presents the phase diagram for μ and S, where the equilibrium conditions for S and μ are given, respectively:

$$\mu = (A + \alpha) + \xi S$$

$$\mu = \frac{B[(A + \alpha)(\Sigma - \S) - 2S\#]}{(\Sigma + 2\xi + \S)\#}$$

With:

$$\S = \sqrt{4B + (\Sigma + 2\xi)^2}$$

Table 4.2 Optimal flows and stocks of pollution production with and without interactions

	Static (flows are relevant)		Dynamic (stocks are relevant)	
REL	$S^*_{NS} = \frac{A+\alpha}{1+B+\beta}$ $\lambda^*_{NS} = \frac{(A+\alpha)(B+\beta)}{2(1+B+\beta)}$	$S^*_{CS} = \frac{A+\alpha}{1+2(B+\beta)}$ $\lambda^*_{CS} = \frac{(A+\alpha)(B+\beta)}{1+2(B+\beta)}$	$S^*_{ND} = \frac{(A+\alpha)(\xi+\Sigma)}{B+\xi(\xi+\Sigma)}$ $\mu^*_{ND} = -\frac{(A+\alpha)B}{B+\xi(\xi+\Sigma)}$	$S^*_{CD} = \frac{(A+\alpha)[\sigma+\xi]}{2(B+\beta)+\xi[\sigma+\xi]}$ $\mu^*_{CD} =$ $-\frac{(A+\alpha)(B+\beta)}{2(B+\beta)+\xi[\sigma+\xi]}$
IND	$Y^*_S + y^*_S = \frac{A}{1+B} + \frac{\alpha}{1+\beta}$ $\lambda^*_S = \frac{AB}{1+B} + \frac{\alpha\beta}{1+\beta}$		$Y^*_D + y^*_D = \frac{A}{1+B/[\xi(\Sigma+\xi)]} + \frac{\alpha}{1+\beta/[\xi(\sigma+\xi)]}$ $\mu^*_D = -\frac{AB}{B+\xi[\Sigma+\xi]} - \frac{\alpha\beta}{\beta+\xi[\sigma+\xi]}$	

Abbreviations: REL = interaction; IND = no interaction; N = Nash equilibrium; C = cooperative equilibrium; S = static; D = dynamic. *Notes* $S_{NS} > S_{CS}$, $Y_S{*} + y_S{*} > S_{CS}$, $Y_S{*} + y_S{*} > S_{NS}$; β is assumed to be 0 for S_{ND} and μ_{ND}

$$\# = B + \xi(\Sigma + \xi)$$

Note that the optimal pollution production in a dynamic context equals that in a static context, whenever $\xi = 1$ and $\sigma = \Sigma = 0$ (i.e., crucial issues are interactions and pollution stocks, both in a static and a dynamic context). Moreover, $B = \beta = 0$ to depict a lack of concern for stock issues implies, both in static and dynamic contexts, that stocks in equilibrium amount to the sum of marginal utilities in developed and developing countries, whereas shadow prices in equilibrium are null. Finally, shadow prices in both the static and dynamic contexts refer to marginal utilities from consumption and to marginal disutilities from pollution.

Table 4.2 summarises the optimal flows and stocks of pollution with and without interactions.

As a summary of this remark, if interactions REL *or* pollution stocks STO are relevant, shadow prices (i.e., both static λ and dynamic μ) are different from equilibrium market prices (i.e., the marginal utility at the equilibrium quantity).

4.1.1.2 Resources

This Section will identify optimal renewable and non-renewable resource policies (taxes TAX, standards STA, subsidies SUB, protected areas GRO and exploitation rights RIG) in a simplified framework. Table 4.3 summarises the main articles on renewable and non-renewable resource policies to efficiency in more complicated contexts.

Note that comparing Tables 4.1 and 4.3 suggests that the classification of policies based on the Economic General Equilibrium Model EGEM assumptions better fits pollution policies than resource policies.

Table 4.3 Main articles on renewable and non-renewable resource policies to efficiency

REN	TAX	STA	SUB	RIG	GRO
		Caparros (2009) CC	Ibanez-Lopez and Moratilla-Soria (2017) ENY, Koch et al. (2017) REE	Bene and Doyen (2000) ERE, Abbott and Wilen (2011) JEEM	Elevitch and Johnson (2020) EM, Martinet and Blanchard (2009) EE, Singh and Weninger (2009) JEEM, Hedenus and Azar (2009) BD
UNC		Rogers et al. (1997) JEM		Singh et al. (2006) JEEM	Ferguson (2017) NZJFS
ASY					
IMP					BenDor et al. (2009) EE
NON	TAX	STA	SUB	RIG	TEC
		Cappellan-Perez et al. (2015) SS			Kopittke et al. (2019) EI
UNC	Tsur and Zemel (2008) ERE				
ASY	Briggs (2011) REE				
IMP	Wirl (2014) REE, Matsumura and Yamagishi (2017) EL, Shen et al. (2021) IJPR	Konishi (2011) REE (& TAX)	Reichenbach and Requate (2012) ERE	Hagem et al. (2006) ERE, Piga (2003) ERE	

Abbreviations: REN = renewable resources, NON = non-renewable resources, UNC = uncertainty, ASY = asymmetric information, IMP = imperfect competition, TAX = taxes, STA = standards, SUB = subsidies, RIG = exploitation rights, GRO = protected areas: see the list of acronyms for journals. *Notes* SUB includes payment for ecosystems, STA includes harvest protocols

Renewable Resources

Let us define a renewable resource as a resource characterized by a natural growth rate G[X(t)] at time t that depends on its stock X(t) at time t:

$$\frac{\partial[X(t)]}{\partial t} = G[X(t)]$$

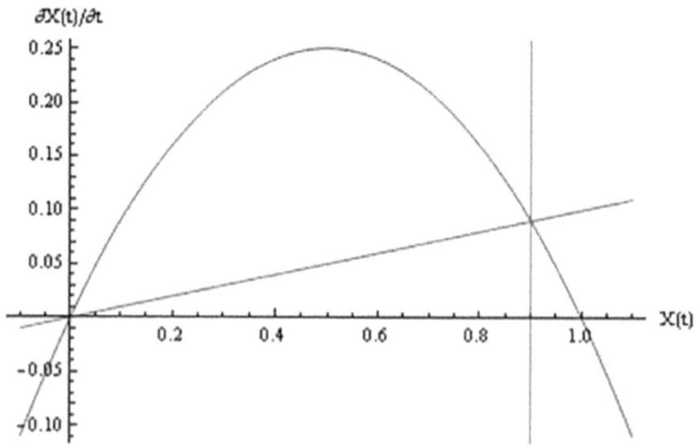

Fig. 4.16 A renewable resource dynamics and its steady state. The increasing and decreasing (blue) curve = the natural growth rate $G[X(t)] = a\,X(t) - b\,X(t)^2$; the increasing (purple) line = the harvest rate $H(t) = L\,X(t)$, where X(t) is the stock of the renewable resource at time t. *Notes* If a = b = 1 and L = 0.1, the globally stable steady state (i.e., $\partial X(t)/\partial t = 0$) is at $X^* = 0.9$

For concreteness, think of fish or forest. In particular, let us assume that the typical dynamics G[X(t)], suggested by ecologists, can be specified as $G(X) = a\,X - b\,X^2$ with a and b positive parameters, where the minimum X is at X = 0, the maximum X is at X = a/b and the Maximum Sustainable Yield (i.e., the maximum $\partial X(t)/\partial t$) is at X = (1/2)(a/b).

Let us define H as the amount of harvested resource for each period t. Thus:

$$\frac{\partial[X(t)]}{\partial t} = G[X(t)] - H$$

For example, let H = L X depict the harvest production function, with L the amount of labour.

Figure 4.16 depicts a steady state equilibrium for X (i.e., X(t) such that G(X) = H).

Note that different labour efforts will lead to different steady states.

Monopolistic firm: economic and ecological policies

Let us consider a monopolistic firm that has the exclusive right to harvest this renewable resource. Thus, it maximises profits from time 0 to infinity:

$$Max_{H(t)} \int_0^{\infty} \{P - C[X(t)]\}H(t)e^{-rt}dt$$

$$\frac{\partial[X(t)]}{\partial t} = G[X(t)] - H(t)$$

where r is the prevailing interest rate in the capital market (i.e., under the assumption of competitive markets, the active interest rate equals the passive interest rate), P is the fixed price (i.e., average price = marginal price) of the harvested flow (e.g., tons of fish per month; tons of wood per month), C[X(t)] is average cost of harvesting which depends on the available stock at time t (e.g., a larger amount of fish in the sea implies that it takes a shorter period to catch the same amount of fish; a larger amount of trees in a forest implies that it takes a shorter period to cut the same amount of wood), H(t) is the flow of resource harvested per month.

The Hamiltonian (HAM) is given by:

$$\text{HAM} = \{P - C[X(t)]\}H(t) + \mu(t)\{G[X(t)] - H(t)\}$$

$$\frac{\partial[HAM]}{\partial H(t)} = 0 \leftrightarrow \{P - C[X(t)]\} = \mu(t)$$

$$\frac{\partial[\mu(t)]}{\partial t} = \dot{\mu} = \mu(t)r - \frac{\partial[HAM]}{\partial X(t)} = \mu(t)r + H(t)\frac{\partial C[X(t)]}{\partial X(t)} - \mu(t)\frac{\partial G[X(t)]}{\partial X(t)}$$

$$\frac{\partial[X(t)]}{\partial t} = \dot{X} = G[X(t)] - H(t)$$

where the transversality condition is assumed to hold.

Optimal *rule* if $\dot{X} = 0$ and $\dot{\mu} = 0$:

$$r = \frac{\partial G[X(t)]}{\partial X(t)} - \frac{G[X(t)]\{\partial C[X(t)]/\partial X(t)\}}{P - C[X(t)]} = G' - \frac{GC'}{P - C}$$

The logics behind this optimal use of a renewable resource can be easily understood by assuming that C′ = C = 0. Indeed, in this simplified context, the monopolistic firm has two options: (i) harvesting the resource, sell it, put the revenue in a bank and obtain an increase of its money by r percent in one year; (ii) not harvesting the resource and rely on its natural growth rate by G′ percent. A marginal thinking suggests that these two options must produce the same benefit in order to maximise the total benefit (i.e., r > G′ would suggest to harvest to a greater extent, while G′ > r would suggest to harvest to a smaller extent).

Let us assume H = L X (i.e., a Cobb–Douglas harvest function with production coefficients set at 1), C = W L = W(H/X) (i.e., a variable cost function with W representing all variable costs) and G(X) = a X − b X^2.

The optimal *solution* if W ≠ 0 (i.e., the opportunity cost of labour is positive in a Developed Country DC) can be obtained as follows:

$$\frac{\partial[HAM]}{\partial H(t)} = 0 \leftrightarrow [P - W(H/X)] = \mu$$

$$\dot{\mu} = \mu r - H\frac{WH}{X^2} - \mu(a - 2bX)$$

$$\dot{X} = aX - bX^2 - H$$

Thus

$$\dot{\mu} = 0 \leftrightarrow \mu = \frac{W(a - bX)^2}{-a + r + 2bX}$$

$$\dot{X} = 0 \leftrightarrow \mu = P - aW + bWX$$

Thus

$$X_{mon} = \frac{W(a - r) - 2P + \sqrt{4P^2 + (a + r)^2 W^2)}}{2bW}$$

$$\mu_{mon} = \frac{1}{2}\left[-W(a + r) + \sqrt{4P^2 + (a + r)^2 W^2)}\right]$$

In particular,

$$\dot{\mu} < 0 \leftrightarrow \mu r + H C' - \mu G' < 0 \leftrightarrow \mu(r - G') < -H C' \text{ with } G' = a - 2bX, C' = -\frac{W H}{X^2}$$

Since $r < G'$ is consistent with an increasing X and a decreasing μ, then:

$$\dot{\mu} < 0 \leftrightarrow \mu > \frac{W(a - bX)^2}{-a + r + 2bX}$$

Similarly,

$$\dot{X} = 0 \leftrightarrow \mu = P - C = P - \frac{WH}{X}$$

Since $G > H$ is consistent with an increasing X and a decreasing μ, then

$$\dot{X} > 0 \leftrightarrow \mu > P - \frac{WH}{X} \leftrightarrow \mu > P - aW + bWX$$

Figure 4.17 presents the phase diagram for μ and X in case of a monopolistic use in a DC.

Note that the optimal solution if $W = 0$ (i.e., the opportunity cost of labour is 0 in a Less Developed Country LDC so $C = C' = 0$) can be obtained as follows:

$$\frac{\partial[HAM]}{\partial H(t)} = 0 \leftrightarrow P = \mu$$

$$\frac{\partial[\mu(t)]}{\partial t} = \dot{\mu} = \mu r - \mu(a - 2bX)$$

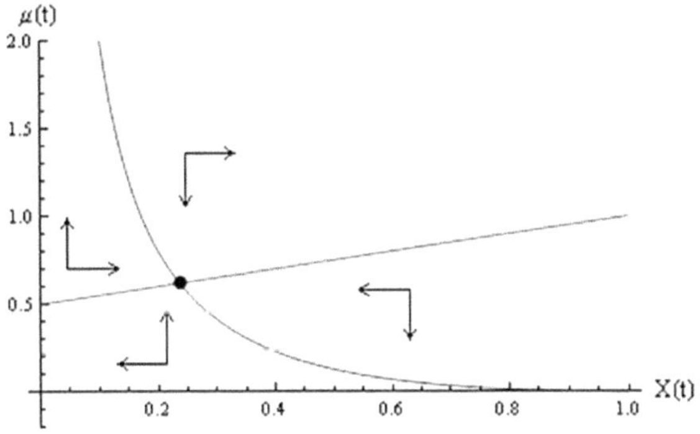

Fig. 4.17 The phase diagram for a monopolistic use of a renewable resource in a DC. Figure 4.17 shows the dynamic relationship between the stock of a renewable resource at time t X(t) and its shadow price at time t μ(t), where the X dot = 0 is the increasing (purple) line, while the μ dot = 0 is the decreasing (blue) curve. Abbreviations: DC = developed countries where W > 0. *Notes* If P = r = a = b = 1 and W = 0.5, the long-run equilibrium (i.e., X* = 0.23 and μ* = 0.61) is globally stable (i.e., it is a spiral sink)

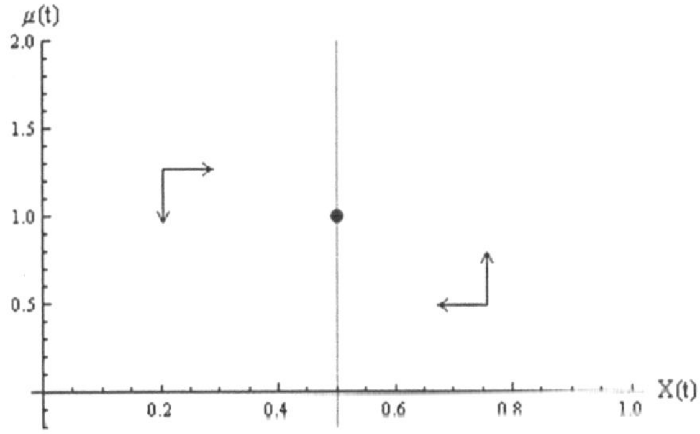

Fig. 4.18 The phase diagram for a monopolistic use of a renewable resource in a LDC. Figure 4.18 shows the dynamic relationship between the stock of a renewable resource at time t X(t) and its shadow price at time t μ(t), where the μ dot = 0 is the vertical (grey) line. Abbreviations: LDC = less developed countries where W = 0. *Notes* If P = r = a = b = 1 and W = r = 0, the long-run equilibrium (i.e., X* = 0.5 and μ* = 1) is globally unstable (i.e., it is a source)

$$\frac{\partial[X(t)]}{\partial t} = \dot{X} = aX - bX^2 - H$$

Thus

$$X_{mon} = \frac{1}{2}\frac{a-r}{b} \text{ and } \mu_{mon} = P$$

Figure 4.18 presents the phase diagram for μ and X in case of a monopolistic use in a LDC.

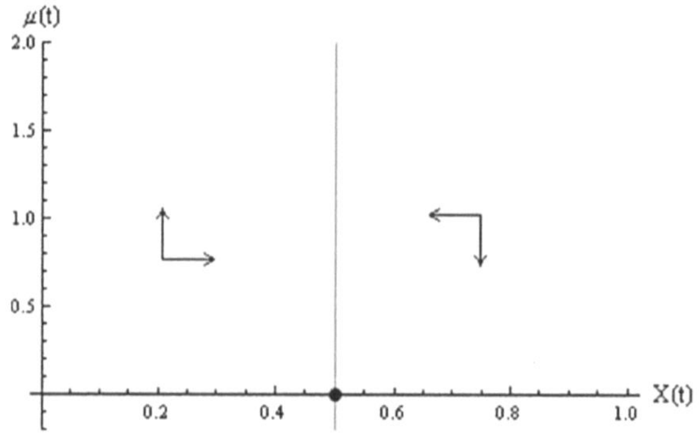

Fig. 4.19 The phase diagram for a competitive use of a renewable resource in a DC. Figure 4.19 shows the dynamic relationship between the stock of a renewable resource at time t X(t) and its shadow price at time t μ(t), where the X dot = 0 is the vertical (grey) line. Abbreviations: DC = developed countries where W > 0. *Notes* If W = r = a = b = 1 and P = 2, the long-run equilibrium (i.e., X* = 0.5 and μ* = 0) is globally stable (i.e., it is a spiral sink)

Note that the stock in the long-run X_{mon} is at the MSY = a/b if r = 0, whereas the shadow-price in the long-run μ_{mon} = P, although it is unstable.

Competitive firms: economic policies

Let us consider many competitive firms that all have the right to harvest this renewable resource. Thus, they will make no profits:

$$P = C \leftrightarrow \mu = 0$$

Optimal rule if $\dot{X} = 0$ and $\dot{\mu} = 0$:

$$G[X(t)]\frac{\partial C[X(t)]}{\partial X(t)} = GC^{'} = 0$$

Optimal solution

$$X_{com} = \frac{W}{P} \text{ and } \mu_{com} = 0$$

Figure 4.19 presents the phase diagram for μ and X in case of a competitive use in a DC.

Figure 4.20 presents the phase diagram for μ and X in case of a competitive use in a LDC.

Note that in order to obtain an X < 1, we need to refer to the stable conditions for X only.

$$\dot{X} = aX - bX^2 - H = 0$$

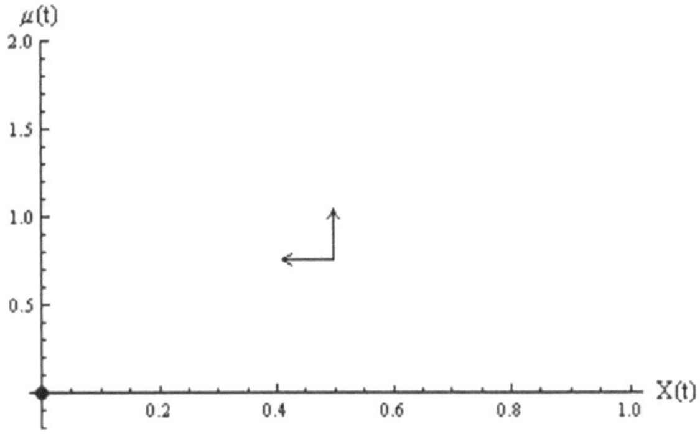

Fig. 4.20 The phase diagram for a competitive use of a renewable resource in a LDC. Figure 4.20 shows the dynamic relationship between the stock of a renewable resource at time t X(t) and its shadow price at time t μ(t), where the X dot = 0 is the μ (t) axis. Abbreviations: LDC = less developed countries where W = 0. *Notes* If W = r = a = b = 1 and W = 0, the long-run equilibrium (i.e., X* = 0 and μ* = 0) is globally unstable (i.e., it is a source)

$$H = LX$$

$$X = \frac{a}{b} - \frac{1}{b}L \text{ and } H = \frac{a}{b}L - \frac{1}{b}L^2$$

The profit function π is given by:

$$\pi = P\left[\frac{a}{b}L - \frac{1}{b}L^2\right] - WL$$

Optimal solution for single access is given by:

$$Max_L\pi \leftrightarrow L^*_{mon} = \frac{a}{2} - \frac{b}{2}\frac{W}{P} \leftrightarrow X^*_{mon} = \frac{1}{2}\left(\frac{a}{b} + \frac{W}{P}\right)$$

Optimal solution for open access is given by:

$$\pi = 0 \leftrightarrow L^*_{com} = a - b\frac{W}{P} \leftrightarrow X^*_{com} = \frac{W}{P}$$

Table 4.4 summarises stocks and shadow-prices of a renewable resource with and without interactions.

Let us assume that MEC in terms of the stock of renewable resources is represented by:

$$\text{MEC} = \beta(Xmax - X)$$

Table 4.4 Optimal stocks and shadow-prices of a renewable resource with and without interactions

	$W \neq 0$	$W = 0$
REL	$X_{com} = \frac{W}{P}$ $\mu_{com} = 0$	$X_{com} = 0$ $\mu_{com} = 0$
IND	$X_{mon} = \frac{1}{2}\frac{a-r}{b} - \frac{2P-\sqrt{4P^2+(a+r)^2W^2)}}{2bW}$ $\mu_{mon} = \frac{1}{2}\left[-W(a+r)+\sqrt{4P^2+(a+r)^2W^2)}\right]$	$X_{mon} = \frac{1}{2}\frac{a-r}{b}$ $\mu_{mon} = P$

Abbreviations: REL = interaction; IND = no interaction; mon = single access as monopolistic conditions, com = open access as competitive conditions, dyn = dynamic conditions. *Notes* In DC (i.e., $W \neq 0$), $X_{com} > X_{mon}$ if $b > [p\,w\,(a-r) - 2\,p^2 + p\,\surd(4\,p^2 + (a+r)^2\,w^2)]/(2\,w^2)$; in LDC (i.e., $W = 0$), $X_{mon} > X_{com} = 0$ if $a > r$

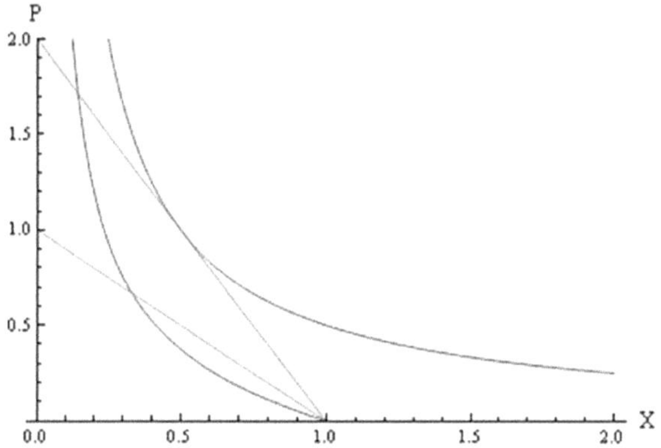

Fig. 4.21 The socially optimal stock of a renewable resource in equilibrium with and without interaction. The decreasing (purple) curve = MPNB with interaction, the decreasing (blue) curve = MPNB without interaction, decreasing (green) line above = Marginal External cost with $\beta = 2$, decreasing (grey) line below = MEC with $\beta = 1$. *Notes* If a = b = Xmax = r = 1 and W = 0.5, X = 1 is a stable and inefficient equilibrium with no interaction and $\beta = 2$ (i.e., MEC > MPNB); X = 0.33 is an unstable but efficient equilibrium with no interaction and $\beta = 1$; X = 0.5 is an unstable but efficient equilibrium with interaction and $\beta = 2$; X = 0 is a stable and inefficient equilibrium with interaction and $\beta = 1$ (i.e., MEC < MPNB)

where β represents the concern for resource depletion. Note that MEC is at its maximum MEC = β Xmax if the renewable resource is totally depleted, while MEC is at its minimum MEC = 0 if the renewable resource is totally preserved.

Figure 4.21 shows that a socially optimal stock of a renewable resource, if it exists, it might be positive or null, according to the value of β and the presence or absence of interactions between firms (i.e., $P = W[(b\,X - a)(r + b\,X)]/(a - r - 2bX)$ amounts to MPNB without interaction, whereas $P = W/X$ amounts to MPNB with interaction, where MPNB depicts the interests of both consumers and producers).

Note that MEC < MPNB for each P, both with and without interaction, implies an overexploitation of X (i.e., it is socially inefficient). Moreover, P is the observed price of a flow (e.g., price of fish, price of wood): although P affects the stock of a resource in equilibrium (i.e., its existence and its value), it is *not* the value of this resource. Finally, under the assumption that people concerned about preservation can increase the demand for X, an equilibrium identified by MPNB = MEC with MPNB > MEC at smaller X and MPNB < MEC at larger X is efficient but unstable (i.e., a tiny change in X or P will make the society moving away from the equilibrium), whereas an equilibrium identified by MPNB = MEC with MPNB < MEC at smaller X and MPNB > MEC at larger X is efficient and stable (i.e., a tiny change in X or P will make the society moving back to the equilibrium).

The optimal policies (i.e., ecological or economic interventions to achieve the optimal amount of renewable resource in the long-run equilibrium) can be obtained by solving the following conditions for a and b (to depict ecological policies such as harvest protocols or protected areas) or for P and W (to depict economic policies such as taxes on products obtained from harvesting or factors used to harvest), either in the absence of interaction between firms (i.e., P = W[(b X − a)(r + b X)]/(a − r − 2bX and MEC = β (Xmax − X)) or in the presence of interaction between firms (i.e., P = W/X and MEC = β (Xmax − X)). However, formulas are quite large to be detailed and quite uninformative. Indeed, the first derivatives of the stock of X in the long-run equilibria are univocally positive or negative (i.e., if a socially optimal stock of X exists, then it can be achieved by each policy). In particular, in the absence of interaction between firms:

$$\frac{\partial X_{mon}^{dyn}}{\partial a} = \frac{W(a+r) + \sqrt{\#}}{2b\sqrt{\#}} > 0$$

$$\frac{\partial X_{mon}^{dyn}}{\partial b} = \frac{2P - W(a+r) - \sqrt{\#}}{2b^2 W} < 0 \; if \; a > \frac{rP}{P + rW}$$

$$\frac{\partial X_{mon}^{dyn}}{\partial P} = \frac{2P - -\sqrt{\#}}{2b\sqrt{\#}} < 0$$

$$\frac{\partial X_{mon}^{dyn}}{\partial W} = \frac{2P^2 + P\sqrt{\#}}{2W^2\sqrt{\#}} > 0$$

where

$$\# = 4P^2 + a^2W^2 + 2a\,r\,W^2 + r^2W^2$$

Next, in the presence of interaction between firms:

$$\frac{\partial X_{com}^{dyn}}{\partial P} = \frac{1}{P} > 0$$

$$\frac{\partial X_{com}^{dyn}}{\partial W} = -\frac{W}{P^2} < 0$$

Note that ecological policies based on a and b show different impacts in the absence and presence of interactions, because an increase in the parameter a affects the steepness of the natural growth function, but not on its shape (i.e., if G(X) = a X − b X^2, then G′ = a − 2 b X and G″ = −2 b so ∂G′/∂a = 1, ∂G″/∂a = 0), whereas an increase in the parameter b affects both the steepness and the shape of the natural growth function (i.e., ∂G′/∂b = −2 X, ∂ G″/∂b = −2).

Non-renewable Resources

Let us define a non-renewable resource as a resource characterized by a natural growth rate fixed at 0 at each time t:

$$\frac{\partial[X(t)]}{\partial t} = G[X(t)] = 0$$

For concreteness, think of oil or diamonds. Let us consider a monopolistic firm that has the exclusive right to extract this resource so that it maximizes profits from time 0 to infinity:

$$Max_{H(t)} \int_0^{\infty} \{P[X(t)] - C[X(t)]\}H(t)e^{-rt}dt$$

$$\frac{\partial[X(t)]}{\partial t} = -H(t)$$

where r is the prevailing interest rate in the capital market (i.e., under the assumption of competitive markets, the active interest rate equals the passive interest rate), P is the price of the extracted flow (e.g., tons of oil per month, tons of diamonds per month) (i.e., products obtained from the resource) which depends on the stock of resource (e.g., P decreases if X increases due to discovery of additional reserves), C[X(t)] is the average cost of extraction which depends on the available stock at time t (e.g., a larger amount of oil implies a smaller cost to extract the same amount of oil), H(t) is the flow of resource extracted per month.

The Hamiltonian is:

$$HAM = \{P[X(t)] - C[X(t)]\}H(t) - H(t)\mu(t)$$

$$\frac{\partial[HAM]}{\partial H(t)} = 0 \leftrightarrow \{P[X(t)] - C[X(t)]\} = \mu(t)$$

$$\frac{\partial[\mu(t)]}{\partial t} = \dot{\mu} = \mu(t)r - \frac{\partial[HAM]}{\partial X(t)} = \mu(t)r + H(t)\left\{\frac{\partial C[X(t)]}{\partial X(t)} - \frac{\partial[P[X(t)]]}{\partial X(t)}\right\}$$

The optimal rule is:

$$r = \frac{H(t)\{\partial P[X(t)]/\partial X(t)\}}{P - C[X(t)]} - \frac{H(t)\{\partial C[X(t)]/\partial X(t)\}}{P - C[X(t)]} = \frac{H(P' - C')}{P - C}$$

Let us assume that $H = 1$ is the standardized extraction per period (i.e., the focus is on the final time T). Let us assume that $C(X) = C'(X) = 0$ so that P represents the marginal surplus and the resource royalty (i.e., $P = P - C$). Thus, the optimal rule becomes:

$$r = \frac{P'(t)}{P(t)} \leftrightarrow P(t) = P(0)e^{rt}$$

Note that the optimal rule is based on μ dot = 0, while X dot < 0 since H = 1.

Let us name X0 and P0 the initial stock and price, respectively; $Q^{sup}(t)$ the quantity supplied at time t, T the final time at which the resource is not extracted anymore, XT and PT the final stock and price, respectively. Let us specify the quantity demanded at time t as $Q^{dom} = Qk - P(t)$, with Qk the (maximum) quantity demanded if $P = 0$.

Thus, the optimal extracting solution satisfies the following two non-linear conditions which depend on 4 variables (i.e., T, P0, PT and XT) for 3 given parameters (i.e., Qk, r, and X0):

$$X(t) = X0 - \int_0^t Q^{sup}(t)dt$$

$$Q^{sup}(t) = Q^{dem}[P(t)] = Qk - P(t)$$

$$P(t) = P0\ e^{rt}$$

In particular, the optimal solution is given by:

$$X(T) = X0 - \int_0^T Qk - P(t)dt$$

$$P(T) = P0\ e^{rT}$$

Specifically, the optimal solution is given by:

$$XT = \frac{-\text{P0} + \text{e}^{rT}\text{P0} - \text{Qk}\ r\ T + r\ \text{X0}}{r}$$

$$\text{Pb} = \text{P0} * e^{rT}$$

Note that μ(t) = P(t), but it is globally unstable.

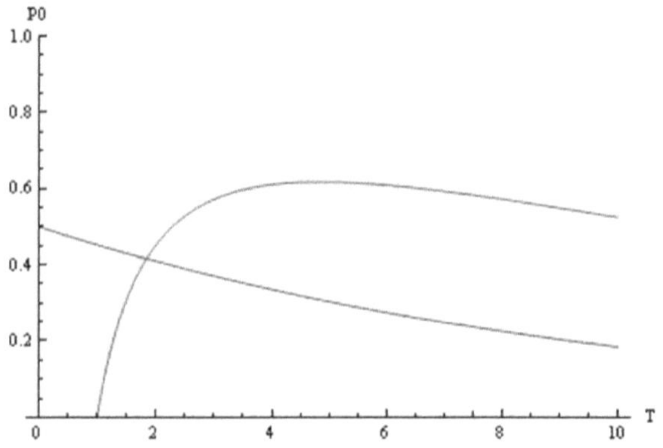

Fig. 4.22 Optimal final time and initial price for a non-renewable resource. The decreasing (blue) curve = P0 based on the optimal price, the increasing and decreasing (purple) curve = P0 based on the optimal final stock. Abbreviations: P0 = initial price, T = final time. *Notes* If Qk = 1, X0 = 1, PT = 0.5, r = 0.1, the solution is T* = 1.84 and P0* = 0.5

Max profits: no policies

Let us assume the goal is to maximise profits. Think of oil companies. Since there is no reason to have some resource not extracted at final time T, let us assume that XT = 0.

The solution for P0 and T in terms of X0, r, PT and Qk is given by:

$$P0 = \frac{r(X0 - Qk\ T)}{e^{rT} - 1}$$

$$\mathrm{P0} = \mathrm{PT} * e^{-rT}$$

The non-linear solution is depicted in Fig. 4.22.

If r tends to 0, these conditions become:

$$P0 = \frac{X0 - Qk\ T}{\mathrm{T}}$$

$$P0 = \mathrm{PT}$$

Thus, the solution if r = 0 is given by:

$$P0 = PT = \frac{Qk\ T - X0}{T}$$

$$T = \frac{X0}{Qk - PT}$$

Min pollution: economic policies

Let us assume that the goal is to reduce the pollution related to the resource use. Think of GHG emissions from fossil fuels. Since replacing the use of polluting resources

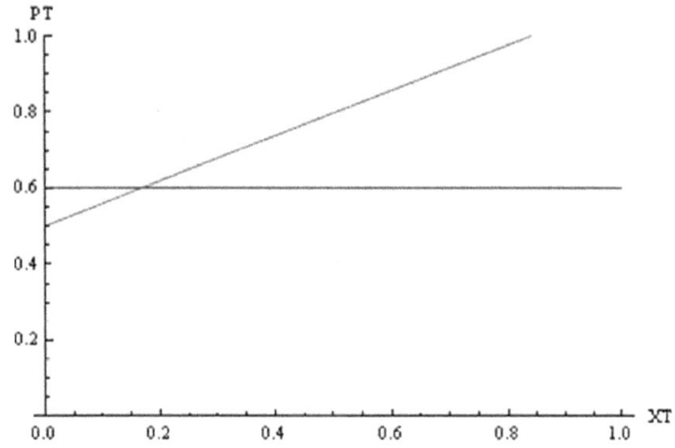

Fig. 4.23 Optimal final stock and price values for a non-renewable resource. The horizontal (blue) line = PT based on the optimal price, the increasing (purple) line = P based on the optimal final stock. Abbreviations: PT = final price, XT = final stock. *Notes* If Qk = 1, X0 = 1, T = 1.84, P0 = 0.5, r = 0.1, the solution is XT* = 0.17 and PT* = 0.60

with alternative sources producing the same goods and services can be represented by some polluting resource not extracted and the price of alternative technologies at final time T, let us assume that XT > 0 and look for PT.

The solution for PT and XT in terms of X0, r, P0 and Qk is given by:

$$PT = \frac{e^{rT} r(\mathrm{Qk}\, T - \mathrm{X0} + \mathrm{XT})}{e^{rT} - 1}$$

$$\mathrm{PT} = \mathrm{P0}\ e^{rT}$$

The non-linear solution is depicted in Fig. 4.23.

If r tends to 0, these conditions become:

$$PT = \frac{\mathrm{Qk}T - \mathrm{X0} + \mathrm{XT}}{T}$$

$$\mathrm{PT} = \mathrm{P0}$$

The solution if r = 0 is given by:

$$PT = P0$$

$$\mathrm{XT} = \mathrm{T\,P0} - \mathrm{Qk}\,T + \mathrm{X0}$$

Let us combine the two conditions for the general solutions by eliminating T:

$$\mathrm{T} = (1/\mathrm{r})\mathrm{Log}(\mathrm{PT}/\mathrm{P0})$$

Thus, MPNB representing the interests of both producers and consumers is given by:

$$XT = \frac{\mathrm{PT - P0 + r\,X0 - Qk\,Log(PT/P0)}}{\mathrm{r}}$$

Let us assume that MEC in terms of the stock of renewable resources is represented by:

$$\mathrm{MEC} = \beta(Xmax - XT)$$

Or

$$\mathrm{XT} = Xmax - \mathrm{MEC}/\beta$$

where β represents the concern for resource depletion. Note that MEC is at its maximum MEC = β Xmax if the non-renewable resource is totally extracted (i.e., XT = 0), while MEC is at its minimum MEC = 0 if the non-renewable resource is totally preserved (i.e., XT = Xmax).

Figure 4.24 shows that a socially optimal stock of a non-renewable resource at T, if it exists, might be positive or null, according to the value of β and the other parameters characterising the extraction sector such as X0, Qk and r. In particular, MEC < MPNB for each PT implies an underpreservation of X (i.e., it is socially inefficient). Moreover, under the assumption that people concerned about preservation can increase the demand for XT, an equilibrium identified by a *decreasing* MPNB = MEC with MPNB > MEC at smaller XT and MPNB < MEC at larger XT is efficient and unstable (i.e., a tiny change in XT or PT will make the society moving away from the equilibrium). Finally, an equilibrium identified by an *increasing* MPNB = MEC with MPNB < MEC at smaller XT and MPNB > MEC at larger XT is efficient and stable (i.e., a tiny change in XT or PT will make the society moving back to the equilibrium).

Note that P is the observed price of a flow (e.g., price of oil, price of diamonds): although P affects the stock of a resource in equilibrium (i.e., its existence and its value), it is *not* the value of this resource. Moreover, an increase in β (i.e., concern for the environment) leads to a larger XT (i.e., a larger stock not extracted at the final time T) and a smaller PT (i.e., a cheaper alternative technology at final time T). Finally, there is no a socially optimal equilibrium if r is large enough (i.e., MEC < MPNB).

The optimal policies (i.e., economic interventions to achieve the optimal amount of non-renewable resource at the final time T) can be obtained by solving the following conditions for P0 and PT (to depict economic policies such as taxes or subsidies on products obtained from extraction), (i.e., XT = Xmax − MEC/β and XT = (1/r)[PT − P0 + r X0 − Qk Log(PT/P0)]). However, formulas are quite large to be detailed and quite uninformative. Indeed, the first derivatives of the stock of XT at the final time T are univocally positive or negative (i.e., if a socially optimal stock of XT exists,

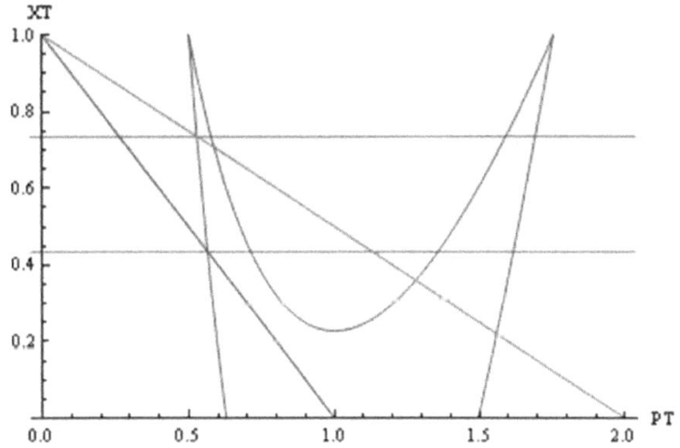

Fig. 4.24 The socially optimal final stock and price values of a non-renewable resource. The decreasing and increasing (blue) curve = MPNB with r = 0.25, the decreasing and increasing (grey) curve = MPNB with r = 0.1, the decreasing (purple) line below = MEC with β = 1, the decreasing (grey) line above = MEC with β = 2. *Notes* If Qk = 1, X0 = 1, P0 = 0.5, r = 0.1, Xmax = 1, XT = 0.43 and PT = 0.56 is an unstable and efficient solution β = 1 and r = 0.1; XT = 0.73 and PT = 0.52 is an unstable and efficient solution with β = 2 and r = 0.25; XT = 0.36 and PT = 1.28 is a stable and efficient solution with β = 2 and r = 0.25; XT = 0.22 and PT = 1.56 is a stable and efficient solution with β = 2 and r = 0.1; XT = 0 and PT = 1 is an unstable and inefficient solution with β = 1 and r = 0.25 (i.e., MPNB > MEC)

then it can be achieved by each policy). In particular:

$$\frac{\partial XT}{\partial P0} = \frac{1}{r}\frac{Qk - P0}{P0} > 0$$

$$\frac{\partial XT}{\partial PT} = \frac{1}{r}\frac{PT - Qk}{PT} < 0$$

To summarise this Section, for renewable resources RES where stocks STO and interactions REL are relevant (i.e., both interaction and independence can prevail), μ ≠ P, apart from the peculiar case of an average opportunity cost at 0 without interaction, where μ = P is unstable; for non renewable resources NON where flows FLO and interactions REL are irrelevant (i.e., only independence prevails), μ ≠ P, apart from the peculiar case of an average extraction cost at 0, where μ = P is unstable.

Remarks on resources with constant flows

If the renewable resource is characterised by a constant flow, then the dynamic model becomes static and the Hamilton function is replaced by the Lagrange equation:

$$LAG = (P - C)H - \lambda(h - H)$$

$$\frac{\partial LAG}{\partial H} = 0 \leftrightarrow P = \lambda$$

$$\lambda(h - H) = 0 \leftrightarrow H = h$$

As a summary of this remark, if flows FLO are relevant, the shadow price equals the equilibrium price (i.e., $\lambda = P^*$), like in contexts where firms are competitive.

4.1.2 Environmental Policy Measures to Achieve Equity

In this Section, I will refer to the two main steps to achieve equity. First, I will measure inequality under the assumption or observation that efficiency is achieved. For this step, I will apply three inequality indexes (i.e., the Gini, Theil, and Shannon indexes) to both individuals (or groups of individuals) and countries (or groups of countries) from the perspective of representative individuals at both national and lower levels. Second, I will combine possible inequality and inefficiency. For this step, I will refer to the main equity equilibria (i.e., Kalai–Smorodinsky, Nash bargaining, Rawls, Harsanyi, sovereignty, responsibility, solidarity) in terms of resources and pollution (i.e., within strong sustainability) or welfare (i.e., within weak sustainability).

Note that I will emphasise the first step for individuals, although it would be possible to discuss alternative laws and regulations to cope with environmental inequity at a national level, but I will emphasise the second step for countries, for which international protocols and agreements are implemented to cope with global environmental issues. Moreover, I will base all equity equilibria on a normative analysis by assuming rationality as well as perfect and complete information, although the equilibria are based on different perspectives (e.g., command-and-control for Kalai–Smorodinsky and Rawls; free negotiation for Nash bargaining and Harsanyi). Finally, for both individuals and countries, I will link the equity equilibria that refer to resources and pollution to mitigation policies or agreements, but I will link equity equilibria that refer to welfare to adaptation policies or agreements.

4.1.2.1 Among Individuals: National Laws and Regulations

Three *main* indexes have been applied to measure vertical inequality (i.e., among individuals or households, to be distinguished from horizontal inequality as related to individuals or groups characteristics such as race or religion): Gini, Theil and Shannon indexes.

The Gini index is given by (see also Sect. 4.2.1.5 on CBA):

$$\mathrm{Gini} = \left[1/\left(2\,n^2 x\right)\right]\Sigma_i \Sigma_j \left|x_i - x_j\right|$$

where x_i is the outcome to people in group i, x_j is the outcome to people in group j, x is the average outcome, and n is the number of groups.

Table 4.5 Demonstrative efficiency and equality for some policies in terms of reduction costs

	Welfare to young people	Welfare to old people	Total welfare
TAX	60	40	100
STA	50	60	110
SUB	30	70	100

Abbreviations: TAX = taxes, STA = standards, SUB = subsidies

The Theil index is given by:

$$\text{Theil} = (1/n)\,\Sigma_i (x_i/x) \text{Log}[x_i/x]$$

where x is the average outcome.

The Shannon index is given by:

$$\text{Shannon} = -\Sigma_i\, p_i \text{Log}[p_i]$$

where $p_i = x_i/(n\ x)$. For example, Table 4.5 presents the welfare costs of some pollution policies in terms of (monetary) reduction costs or (economic) opportunity costs for young and old people.

Note that a TAX is as efficient like a SUB, but it is more equal (e.g., a SUB on green technologies for houses is mostly enjoyed by young people with a larger income); a STA is less efficient than a TAX, but it is more equal; a STA is less efficient than a SUB, but it is more equal. Moreover, taxes TAX and subsidies SUB are consistent with weak sustainability, while standards STA with strong sustainability. Finally, at the same efficiency level, one could prefer a policy PLA characterised by a larger equality (i.e., alternative inequality measures can be applied), while policies PLA characterised by a larger efficiency and a smaller equality can be compared only by introducing an equity criterion (i.e., alternative equilibria can be applied).

In particular, in terms of the three main inequality measures discussed above:

$$\text{Gini (SUB)} = 0.2 > \text{Gini (TAX)} = 0.1 > \text{Gini (STA)} = 0.045$$
$$\text{Theil (SUB)} = 0.082 > \text{Theil (TAX)} = 0.020 > \text{Theil (STA)} = 0.004$$
$$\text{Shannon (STA)} = 0.689 > \text{Shannon (TAX)} = 0.673 > \text{Shannon (SUB)} = 0.610$$

Alternatively, in terms of three inequality aversions applied to the SWF:

$$(\varepsilon = 0)\ \text{STA} > \text{TAX} = \text{SUB}$$
$$(\varepsilon = 0.5)\text{STA(SWF} = 219) > \text{TAX(SWF} = 198) > \text{SUB(SWF} = 192)$$
$$(\varepsilon = 1)\ \text{STA} > \text{TAX} > \text{SUB}$$

Table 4.6 Main articles on equity *among individuals* of national and international policies and agreements

		Quantity (mitigation)	Welfare (adaptation)
Nat PLA	POL	Fullerton and Muehlegger (2019) REEP, Gupta et al. (2020) SS, Zhao et al. (2019) NC, Masri et al. (2020) STE, Dong et al. (2018) RCR	Karydas and Zhang (2019) JEEM, Rozenberg et al. (2020) JEEM, Klenert et al. (2018) ERE, Belfiori (2018) EP, Budolfson et al. (2017) CC, Anthoff et al. (2009) EE, Brink and Wamsler (2017) EPG, Gayer and Viscusi (2016) REEP
	RES	Healy and Barry (2017) EP, Verchere (2017) EE, Boyca et al. (2016) EE, Pérez-Cirera and Lovett (2006) EE, Bai et al. (2020) JEM	Paylor (2017) TWQ, Nahuelhual et al. (2018) STE, Baumgartner et al. (2017) JEEM, Daigneault et al. (2017) EE, Borner et al. (2016) EC
Int PLA	POL	Bohringer et al. (2012) EnEc, Althouse et al. (2020) EE	Rao et al. (2017) NCC, De Cian et al. (2016) ERL
	RES		Almeida-Lenero et al. (2017) EC
Int AGR	POL	Holtz et al. (2018) IEA: PLE, Boran (2017) EPE, Harris (2017) JESS, Huggel et al. (2016) NCC, Okereke and Coventry (2016) WIR: CC, Dooley and Gupta (2017) IEA: PLE, Markkanen and Anger-Kraavi (2019) CP	Roberts and Pelling (2020) CP, Benites and Mello (2019) WD, Yang et al. (2018) NH
	RES	Wang et al. (2017) MASGC	Dawson et al. (2018) ESP, Wong et al. (2019) CP, Isyaku et al. (2017) EC

Abbreviations: Nat = national, Int = international, PLA = policies, POL = pollution, RES = resources: see the list of acronyms for journals

Table 4.6 summarises the main articles on equity among individuals of national and international policies and agreements, by distinguishing articles referring to mitigation measures based on quantities from articles referring to adaptation measures based on welfare.

Note that there are no papers on efficiency of policies to equity. Moreover, most papers are very recent. Finally, there are more papers on pollution than on resources.

To summarise this Section, equity EQU from national and international policies PLA should be measured, whenever some assumptions behind the Economic General Equilibrium Model EGEM (i.e., asymmetric information ASY, uncertainty UNC, imperfect competition IMP) do not hold.

4.1.2.2 Among Countries: Bilateral and Multilateral Agreements

Four *main* criteria have been applied to combine inequality and inefficiency: Nash bargaining, Kalai-Smrodinsky, Rawls, Harshanyi equilibria.

The Nash bargaining solution (short-run) for resources X used by Developed Countries DC and Less Developed Countries LDC in terms of their current status ($U_{DC}(0)$ and $U_{LDC}(0)$) is given by:

$$Max[U_{DC}(X_{DC}) - U_{DC}(0)]^{\gamma}[U_{LDC}(X_{LDC}) - U_{LDC}(0)]^{1-\gamma}$$

where γ represents the bargaining power of DC countries and with $X_{DC} + X_{LDC} = X_{TOT}$.

The Nash bargaining solution (long-run) of DC and LDC countries in terms of their sustainable uses (U_{DC} (X^*) and U_{LDC} (X^*)) (i.e., maximum aversion to inter-generational inequality) is given by:

$$Max\left[U_{DC}(X_{DC}) - U_{DC}(X^*)\right]^{\gamma}\left[U_{LDC}(X_{LDC}) - U_{LDC}(X^*)\right]^{1-\gamma}$$

where X* represents the long-run per capita sustainable resource use.

The Nash bargaining solution (short-run) for emissions Y used by Developed Countries DC and Less Developed Countries LDC in terms of their current status ($U_{DC}(0)$ and $U_{LDC}(0)$) is given by:

$$Max[U_{DC}(Y_{DC}) - U_{DC}(0)]^{\gamma}[U_{LDC}(Y_{LDC}) - U_{LDC}(0)]^{1-\gamma}$$

where γ represents the bargaining power of DC countries and with $Y_{DC} + Y_{LDC} = Y_{TOT}$.

The Nash bargaining solution (long-run) of DC and LDC countries in terms of their sustainable emissions ($U_{DC}(Y^*)$ and $U_{LDC}(Y^*)$) (i.e., maximum aversion to inter-generational inequality) is given by:

$$Max\left[U_{DC}(Y_{DC}) - U_{DC}(Y^*)\right]^{\gamma}\left[U_{LDC}(Y_{LDC}) - U_{LDC}(Y^*)\right]^{1-\gamma}$$

where Y* represents the long-run per capita sustainable pollution emission.

Let us assume that U(X) = X. Let us focus on resources X in terms of current levels. The Nash bargaining equilibrium in the short-run becomes:

$$Max[X_{DC} - YX_{DC}(0)]^{\gamma}[X_{TOT} - X_{DC} - X(0)]^{1-\gamma}$$

Thus:

$$X_{DC} = \gamma X_{TOT} - X_{DC}(0) - \gamma[X_{DC}(0) + X_{LDC}(0)]$$

Or

$$X_{DC} = \frac{\gamma}{1-\gamma}[X_{LDC} - X_{LDC}(0)] + X_{DC}(0)$$

Let us assume that U(Y) = Y. Let us focus on emissions Y in terms of current levels. The Nash bargaining equilibrium in the short-run becomes:

$$Max[Y_{\mathrm{DC}} - Y_{\mathrm{DC}}(0)]^{\gamma}[Y_{\mathrm{TOT}} - Y_{\mathrm{DC}} - Y_{\mathrm{LDC}}(0)]^{1-\gamma}$$

Thus:

$$Y_{\mathrm{DC}} = \gamma Y_{\mathrm{TOT}} - Y_{\mathrm{DC}}(0) - \gamma[Y_{\mathrm{DC}}(0) + Y_{\mathrm{LDC}}(0)]$$

Or

$$Y_{\mathrm{DC}} = \frac{\gamma}{1-\gamma}[Y_{\mathrm{LDC}} - Y_{\mathrm{LDC}}(0)] + Y_{\mathrm{DC}}(0)$$

The Kalai-Smorodinsky solution for resources X is based on:

$$(X_{\mathrm{DC}} - X_{\mathrm{TOT}})/(X_{\mathrm{DC}}(0) - X_{\mathrm{TOT}}) = (X_{\mathrm{LDC}} - X_{\mathrm{TOT}})/(X_{\mathrm{LDC}}(0) - X_{\mathrm{TOT}})$$

Or

$$(X_{\mathrm{TOT}} - X_{\mathrm{DC}})/(X_{\mathrm{TOT}} - X_{\mathrm{DC}}(0)) = (X_{\mathrm{TOT}} - X_{\mathrm{LDC}})/(X_{\mathrm{TOT}} - X_{\mathrm{LDC}}(0))$$

Or

$$(X_{\mathrm{DC}} - X_{\mathrm{DC}}(0))/(X_{\mathrm{TOT}} - X_{\mathrm{DC}}(0)) = (X_{\mathrm{LDC}} - X_{\mathrm{LDC}}(0))/(X_{\mathrm{TOT}} - X_{\mathrm{LDC}}(0))$$

Thus:

$$X_{\mathrm{DC}} = X_{\mathrm{LDC}}\frac{X_{\mathrm{TOT}} - X_{\mathrm{DC}}(0)}{X_{\mathrm{TOT}} - X_{\mathrm{LDC}}(0)} + \frac{X_{\mathrm{TOT}}[X_{\mathrm{DC}}(0) - X_{\mathrm{LDC}}(0)]}{X_{\mathrm{TOT}} - X_{\mathrm{LDC}}(0)}$$

The Kalai-Smorodinsky solution for emissions Y is based on:

$$(Y_{\mathrm{DC}} - Y_{\mathrm{TOT}})/(Y_{\mathrm{DC}}(0) - Y_{\mathrm{TOT}}) = (Y_{\mathrm{LDC}} - Y_{\mathrm{TOT}})/(Y_{\mathrm{LDC}}(0) - Y_{\mathrm{TOT}})$$

Or

$$(Y_{\mathrm{TOT}} - Y_{\mathrm{DC}})/(Y_{\mathrm{TOT}} - Y_{\mathrm{DC}}(0)) = (Y_{\mathrm{TOT}} - Y_{\mathrm{LDC}})/(Y_{\mathrm{TOT}} - Y_{\mathrm{LDC}}(0))$$

Or

$$(Y_{\mathrm{DC}} - Y_{\mathrm{DC}}(0))/(Y_{\mathrm{TOT}} - Y_{\mathrm{DC}}(0)) = (Y_{\mathrm{LDC}} - Y_{\mathrm{LDC}}(0))/(Y_{\mathrm{TOT}} - Y_{\mathrm{LDC}}(0))$$

Thus:

$$Y_{\mathrm{DC}} = Y_{\mathrm{LDC}}\frac{Y_{\mathrm{TOT}} - Y_{\mathrm{DC}}(0)}{Y_{\mathrm{TOT}} - Y_{\mathrm{LDC}}(0)} + \frac{Y_{\mathrm{TOT}}[Y_{\mathrm{DC}}(0) - Y_{\mathrm{LDC}}(0)]}{Y_{\mathrm{TOT}} - Y_{\mathrm{LDC}}(0)}$$

The egualitarian or Ralws solution for resources is based on:

$$Max\ Min[X_{\mathrm{DC}}, X_{\mathrm{LDC}}]$$

With $X_{DC} + X_{LDC} = X_{TOT}$. Thus:

$$X_{\mathrm{DC}} = X_{\mathrm{LDC}}$$

Or

$$X_{\mathrm{DC}} = \frac{1}{2} X_{\mathrm{TOT}}$$

The egualitarian or Ralws solution for emissions Y is based on:

$$Max\ Min[Y_{\mathrm{DC}}, Y_{\mathrm{LDC}}]$$

With $Y_{DC} + Y_{LDC} = Y_{TOT}$. Thus:

$$Y_{\mathrm{DC}} = Y_{\mathrm{LDC}}$$

Or

$$X_{\mathrm{DC}} = \frac{1}{2} X_{\mathrm{TOT}}$$

The Harsanyi solution for resources X if U(X) = X is based on:

$$Max\ SWF = \left(X_{\mathrm{DC}}^{1-\varepsilon} + X_{\mathrm{LDC}}^{1-\varepsilon}\right)^{1/(1-\varepsilon)}$$

With $X_{DC} + X_{LDC} = X_{TOT}$ if $\varepsilon = 0$.
The Harsanyi solution for for emissions Y if U(Y) = Y is based on:

$$Max\ SWF = \left(Y_{\mathrm{DC}}^{1-\varepsilon} + Y_{\mathrm{LDC}}^{1-\varepsilon}\right)^{1/(1-\varepsilon)}$$

With $Y_{DC} + Y_{LDC} = Y_{TOT}$ if $\varepsilon = 0$.

Note that the use of a mitigation perspective (e.g., Kalai-Smorodinsky, Rawls) versus an adaptation perspective (e.g., Nash, Harsanyi) is an ethical decision. Moreover, Nash (i.e., maximisation of the product of surplus utilities), Kalai-Smorodinsky (i.e., equalisation of the ratios of maximal gains or losses) Harsanyi (i.e., maximisation of the total utilities) and Rawls (i.e., maximisation of the minimum of surplus utilities) are Pareto efficient, Rawls is not scale invariant, Kalai-Smorodinsky is not independent on irrelevant alternatives, Nash is not monotonic in total emission or resource. Finally, the reference to free-negotiation agreements (e.g., Nash) versus command-and-control agreements (e.g., Kalai-Smorodinsky) is an ethical decision.

Historical responsibility

Since climate change is caused by cumulative carbon emission, the volume of future emissions could be made dependent on the volume of historical emissions, where the volumes of per capita historical emissions by country i can be a proxy of its responsibility:

$$e_{i,t}^{HR} = \left(\sum_{t=0}^{t-1} e_{\mathrm{world,t}} - \sum_{t=0}^{t-1} e_{\mathrm{i,t}}\right) / e_{\mathrm{world,t}}$$

where $e_{\mathrm{world,t}}$ is the average per capita emission in the world at time t and e_{it} is the per capita emission in country i at time t (Tilton, 2016). For example, if this criterion is applied to DCs and LDCs only, with constant emissions at each time t:

$$e_{DC}^{HR} = ((t-1)Y_{\mathrm{TOT}} - (t-1)Y_{\mathrm{DC}})/(t-1)Y_{\mathrm{TOT}}$$

Thus

$$Y_{\mathrm{DC}} = e_{DC}^{HR} Y_{\mathrm{TOT}} = \frac{Y_{\mathrm{LDC}}(0)}{Y_{\mathrm{DC}}(0) + Y_{\mathrm{LDC}}(0)} Y_{\mathrm{TOT}}$$

If $Y_{\mathrm{DC}} + Y_{\mathrm{LDC}} = Y_{\mathrm{TOT}}$:

$$Y_{\mathrm{DC}} = \frac{Y_{\mathrm{LDC}}(0)}{Y_{\mathrm{DC}}(0)} Y_{\mathrm{LDC}}$$

The HR equilibrium for resources is given by:

$$X_{\mathrm{DC}} = \frac{X_{\mathrm{LDC}}(0)}{X_{\mathrm{DC}}(0)} X_{\mathrm{LDC}}$$

Sovereignty or grandfathering

A larger proportion of emissions could be allowed to countries characterised by larger emissions in the base year t:

$$e_{i,t}^{SG} = \frac{e_{\mathrm{i,t}}}{e_{\mathrm{world,t}}}$$

where $e_{\mathrm{world,t}}$ is the average per capita emission in the world at time t and e_{it} is the per capita emission in country i at time t. For example, if this criterion is applied to DCs and LDCs only:

$$e_{DC}^{SG} = \frac{Y_{\mathrm{DC}}(0)}{Y_{\mathrm{DC}}(0) + Y_{\mathrm{LDC}}(0)}$$

Thus

$$Y_{\mathrm{DC}} = e_{DC}^{SG} Y_{\mathrm{TOT}} = \frac{Y_{\mathrm{DC}}(0)}{Y_{\mathrm{DC}}(0) + Y_{\mathrm{LDC}}(0)} Y_{\mathrm{TOT}}$$

If $Y_{DC} + Y_{LDC} = Y_{TOT}$:

$$Y_{DC} = \frac{Y_{DC}(0)}{Y_{LDC}(0)} Y_{LDC}$$

The SG equilibrium for resources is given by:

$$X_{DC} = \frac{X_{DC}(0)}{X_{LDC}(0)} X_{LDC}$$

Solidarity or capability

The larger (opportunity) cost to reduce carbon emissions could be borne by countries that can afford to a greater extent, where countries' affordability is represented by per capita GDP for each year t:

$$e_{i,t}^{SC} = \frac{gdp_{\mathrm{max,t}} - gdp_{\mathrm{i,t}}}{gdp_{\mathrm{max,t}} \quad gdp_{\mathrm{world,t}}}$$

where $gdp_{max,t}$ is the per capita GDP value for the richest country i and $gdp_{world,t}$ is the average per capita GDP value in the world. For example, if this criterion is applied to DCs and LDCs only, with GDP as a linear function of emissions:

$$Y_{DC} = e_{i,t}^{SC} Y_{TOT} = \frac{Y_{TOT} - Y_{DC}(0)}{Y_{TOT} - (1/2)[Y_{DC}(0) + Y_{LDC}(0)]} Y_{TOT}$$

where $gdp_{DC} = gdp_{max}$ is replaced by $gdp_{TOT} = Y_{TOT}$ to avoid extreme solutions (i.e., $Y_{DC} = 0$ and $Y_{LDC} = 1$). If $Y_{DC} + Y_{LDC} = Y_{TOT}$:

$$Y_{DC} = \frac{Y_{TOT} - Y_{DC}(0)}{Y_{TOT} - Y_{LDC}(0)} Y_{LDC}$$

The SC equilibrium for resources is given by:

$$X_{DC} = \frac{X_{TOT} - X_{DC}(0)}{X_{TOT} - X_{LDC}(0)} X_{LDC}$$

Table 4.7 summarises the main articles on equity among countries of national and international policies and agreements, by distinsguishing articles referring to mitigation measures based on quantities from articles referring to adaptation measures based on welfare. Note that equity from international policies PLA and agreements AGR should be measured, whenever there is interaction between decision-makers (i.e., REL applies).

Figure 4.25 summarises all alternative equity equilibria for Earth resources X between DC and LDC, by a reduction of 20% at global level with respect to the total Ecological Footprint in 2020.

Table 4.7 Main articles on equity *among countries* of international policies and agreements

		Quantity (mitigation)	Welfare (adaptation)
Int PLA	POL	Sun et al. (2019) SUS, Buchholz et al. (2016) ERE, Stuhlmacher et al. (2019) JEM	Aldy et al. (2017) CP, Schenker (2013) ERE, Hallegatte et al. (2007) EE
	RES	Gren (2008) EE (& Welfare), Zagonari and Rossi (2014) JH	Haas et al. (2019) EE
Int AGR	POL	Bauer et al. (2020) NAT, Zimm and Nakicenovic (2020) CP, Winkler (2020) CP, Pauw et al. (2020) CP, Van Zyl et al. (2018) CP, Larkin et al. (2018) CP, Lahn (2018) IEA: PLE, Ritchie and Ready (2017) CP, Liu et al. (2017) CD, Chen et al. (2020) STE, Anderson et al. (2020) CP, Holz et al. (2018) IEA: PLE, Mi et al. (2019) NH, Jabbri et al. (2020) EDS, Duro (2016) EI	Klinsky et al. (2017) CD, Dovie (2019) STE, Gewirtzman et al. (2018) CP, Roberts et al. (2017) EPE, Lange and Schwirplies (2017) ERE, Ueckerdt et al. (2019) ESD, Mejean et al. (2020) CC, Shrivastava and Bhaduri (2019) IEA: PLE
	RES	Zagonari (1998) EDE, Blais-McPherson and Rudiak-Gould (2017) GEC, Huettner et al. (2009) CBM	Aswani et al. (2018) EC

Abbreviations: Int = international, PLA = policies, AGR = agreements, POL = pollution, RES = resources: see the list of acronyms for journals

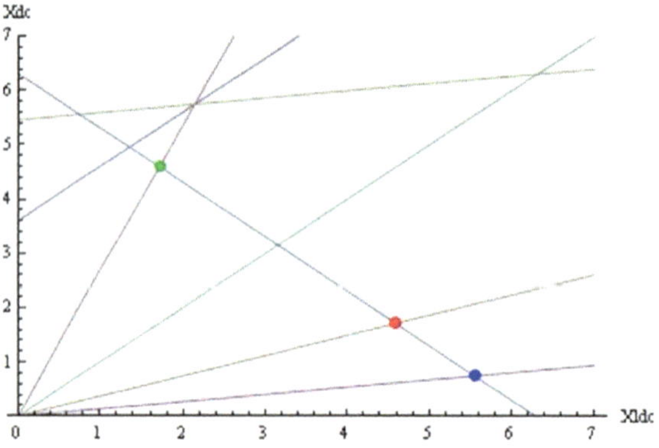

Fig. 4.25 Alternative equity equilibria for a sustainable use of Earth resources by DC and LDC. The decreasing (blue) line = the total sustainable use of Earth resources at 80% of the current total use (i.e., Xdc + Xldc = 5.74 + 2.14 = 7.88). Abbreviations: DC = developed countries, Xdc = per capita use of Earth resources in DC, LDC = less developed countries, Xldc = per capita use of Earth resources in DC. *Notes* If $\gamma = 0.5$, the following equity equilibria are identified from the top on the total sustainable use line: Kalai-Smorodinsky (0.752, 5.552), Nash (1.352, 4.952), Sovereignty (green) (1.712, 4.592), Rawls (3.152, 3.152), Responsibility (red) (4.592, 1.712), Solidarity (blue) (5.552, 0.752). Max SWF with $\varepsilon = 0$ is at Xldc = 4.279, Xdc = 2.024 if Udc = $Xdc^{0.6}$ and Uldc = $Xldc^{0.7}$

Note that the Gini indexes show that, in terms of equity, the international agreements can be ranked as follows: Rawls (Gini = 0), Sovereignty or Grand-fathering SG (Gini = 0.22) = Historical Responsibility HR (Gini = 0.22), although the former in favour of DC and the latter in favour of LDC; Nash (Gini = 0.28); Kalai-Smorodinsky (Gini = 0.38) = Solidarity or Capability SC (Gini = 0.38), although the former in favour of DC and the latter in favour of LDC.

To summarise this Section, equity from international agreements AGR should be measured, whenever there is interaction between decision-makers (i.e., REL applies).

4.2 Environmental Investment Projects

As I noted in the previous Section, the main purpose of this book is to present the simplest theoretical methods to highlight some empirical issues in the literature on environmental sustainability (i.e., it is a textbook with a research outlook). In this Section, I will do this in two steps. First, I will discuss the main features of cost–benefit analysis to achieve efficiency, both in general and in specific contexts (i.e., time, space, uncertainty, interlinkages and inequalities are crucial for decisions). I will perform this step by interpreting cost–benefit analysis as an application of the economic model for weak sustainability (i.e., maximization of total welfare). Similarly, I will discuss the main features of multi-criteria analysis to achieve equity, both in general and in specific contexts (i.e., time, space, uncertainty, interlinkages and inequalities are crucial for decisions). I will do this by interpreting multi-criteria analysis as an application of the ecological model for strong sustainability (i.e., minimization of environmental impacts). Second, I will highlight the main weakness of cost–benefit analysis (i.e., monetary assessment) together with the successful and unsuccessful methods that have been used to cope with this weakness. Similarly, I will highlight the main weakness of multi-criteria analysis (i.e., assessment of relative weights) together with the successful and unsuccessful methods that have been used to cope with this weakness. In particular, I will suggest the development of cost–benefit analysis towards a monetary life-cycle assessment and the development of multi-criteria analysis towards weighted life-cycle assessment.

Note that each specific context is independent of the other specific contexts (i.e., one or more contexts could be relevant in case studies *so* many contexts should be combined). Consequently, students are not expected to read all subsections once they understand the main logic behind this book.

4.2.1 *Cost–Benefit Analysis for Decisions to Efficiency: NPV, BCR, IRR*

Cost–Benefit Analysis CBA can be defined as a systematic set of rules for the economic evaluation (i.e., economic costs and benefits as opportunity costs and gains), with the aim of guiding decisions between alternative potential interventions (i.e., projects), in order to identify the projects that maximise the social welfare (i.e., the net social benefits) (Campbell & Brown, 1983). Note that, within CBA, both a top-down approach (i.e., preferences and welfare of people involved in decisions are better known by experts than by people themselves) or a bottom-up approach (i.e., only people involved in decisions can directly or indirectly express their preferences about alternative projects).

Let us assume that there is a single potential project (i.e., other potential projects are excluded for other reasons) which involves an economic cost C_0 at time t_0 and an economic benefit B_1 at time t_1. Let us assume that there is no a budget problem (i.e., the implementation of the project is not conditional to the satisfaction of a budget constraint). Note that, within CBA, it is possible to sum up benefits and costs in different times, since they are in monetary terms. Moreover, this is *not* a dynamic model, since there is *no* relationship between what happens at time t_0 and what happens at time t_1. Finally, within CBA, it is possible to apply discount factors, since welfare levels (i.e., benefits and costs) in different times are in monetary terms.

The decision to implement or not implement this project can be based on three *main* methodologies.

The Net Present Value NPV suggests the implementation of the project if NPV > 0, where:

$$\mathrm{NPV} = -\mathrm{C}_0 + (1/1 + \mathrm{r})\mathrm{B}_1$$

The Benefit Cost Ratio BCR suggests the implementation of the project if BCR > 1, where:

$$\mathrm{BCR} = [\mathrm{B}_1/(1 + \mathrm{r})]/\mathrm{C}_0$$

The Internal Rate of Return IRR suggests the implementation of the project if IRR > r, where:

$$\text{IRR is r such that NPV} = 0 \text{ or alternatively } \mathrm{C}_0 = \mathrm{B}_1/(1 + \mathrm{IRR})$$

Note that the discount factor $1/(1 + r)$ (i.e., the capitalisation of 1 € from time t_0 to time t_1 equals 1 € $(1 + r)$ versus the actualization of 1 € from time t_1 to time t_0 equals 1 € $(1 + r)^{-1}$) in discrete time applied here (i.e., $1/(1 + r)^t$) is similar to the discount factor in continuous time (i.e., $\exp(-r\ t)$) used in Chap. 3. Indeed, by applying a logarithmic transformation to $\exp(-r\ t) = (1 + r)^{-t}$, one gets $-r\ t = -t\ \mathrm{Log}(1 + r) \sim -t\ r$. Moreover, the discount rate r, to be evaluated from sources external to the project, is used by NPV and BCR methodologies, but not by the IRR methodology. Finally, in this simple context, NPV amounts to BCR which amounts to IRR. Indeed, NPV > 0 can be rephrased as $C_0 < B_1/(1 + r)$, BCR > 1 can be rephrased as $[B_1/(1 + r)] > C_0$, and $IRR = (B_1/C_0 - 1) > r$ can be rephrased as $B_1/(1 + IRR) - C_0 < B_1/(1 + r) - C_0$. However, these three methodologies show different pros and cons in more complicated contexts. In general, if there are many potential projects, NPV compares the differences between benefits and costs characterising these projects, by emphasising absolute gains:

$$NPV = \sum_{t=0}^{T} \frac{1}{(1+r)^t}(B_t - C_t) > 0$$

If there are many potential projects, BCR compares the ratios of benefits and costs characterising these projects, by stressing relative gains from each invested money unit:

$$BCR = \sum_{t=0}^{T} \frac{1}{(1+r)^t}\left(\frac{B_t}{C_t}\right) > 1$$

If there are many future times, IRR could produce ambiguous answers (i.e., many solutions of the equation NPV = 0 with respect to r), by characterising each project by many IRR:

$$IRR > r \text{ with } IRR \text{ from solving } NPV = \sum_{t=0}^{T} \frac{1}{(1+IRR)^t}(B_t - C_t) = 0$$

In summary, NPV is suggested if budget constraints are not relevant, BCR if there are many projects to be compared, IRR if the estimation of the discount rate is unproblematic.

Table 4.8 provides a numerical example of three projects: Project A is chosen in terms of NPV (i.e., A > B > C), Project B is chosen in terms of BCR (i.e., B > A > C) and IRR (i.e., B > A), while IRR cannot be applied to Project C.

Note that the annuity factor $AF_{r,n} = [(1 + r)^n - 1]/[r (1 + r)^n] = C/B$ with r = IRR, and B/C = r with r = IRR if n tends to infinity.

Table 4.8 Net present value, benefit cost ratio, internal rate of return of some demonstrative projects

		C_0	B_1	B_2	NPV at r = 0.5	BCR	IRR
Project	A	−2	10	0	4.66	3.33	4
	B	−0.2	1.1	0	0.53	3.36	4.5
	C	−2	10	−10	0.22	1.11	0.38 and 2.16

Abbreviations: C_0 = costs at time 0, B_1 = benefits at time 1, B_2 = benefits at time 2, r = discount rate

4.2.1.1 Time: DYN, TD

Let us assume that there are two periods in life, 1 and 2, where in period 1 individuals get money M from work and consume goods Q_1, while in period 2 individuals consume goods Q_2 according to the saving produced in period 1: under the assumption that prices are constant (i.e., $P_1 = P_2$), saving equals the amount of money M not used for consumption in period 1 (i.e., $M - Q_1$) capitalized to period 2 in terms of the interest rate r. Under the assumption that the utility from consumption does not change over time (i.e., $U(Q_1) = U(Q_2)$ whenever $Q_1 = Q_2$), the problem to be solved by each individual can be formalized as follows:

$$\text{Maximize } U(Q_1) + 1/(1+\rho)U(Q_2)$$
$$\text{Subject to } Q_2 \leq (M - Q_1)(1 + r)$$

where $\rho \geq 0$ is the inter-temporal marginal preference.

Note that ρ could be positive due to uncertainty about the future. Moreover, this a simple dynamic model (DYN) as opposed to a temporal model, since the outcome in period 2 (i.e., $U(Q_2)$) depends on the action taken in period 1 (i.e., Q_1). Finally, ρ enables the comparison of utility levels (i.e., generally speaking, the combination of welfare or NPV) across time in order to measure the total welfare or NPV to be maximized.

Figure 4.26 depicts a dynamic equilibirium within CBA.

Let us assume that $U(Q) = \text{Log}[Q]$. The first order conditions and the constraint becomes:

$$1/Q_1 = \lambda,\ ((1+r)/(1+\rho))(1/Q_2) = \lambda,\ Q_1 + (1/(1+r))Q_2 = M$$

Thus,

$$\lambda = \frac{2+\rho}{M(1+\rho)},\ Q_1 = \frac{M(1+\rho)}{2+\rho},\ Q_2 = \frac{M(1+r)}{2+\rho}$$

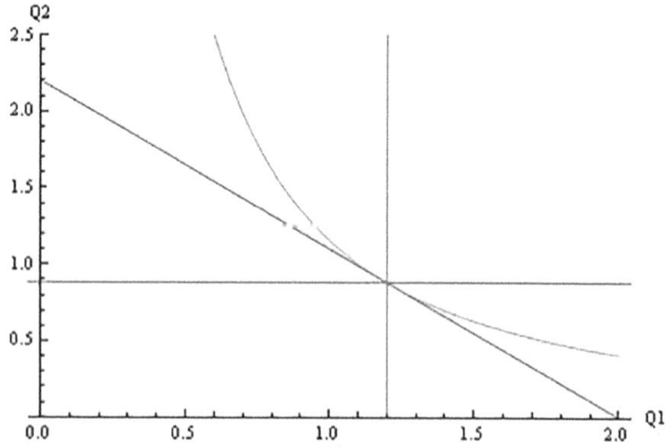

Fig. 4.26 Dynamic equilibrium within CBA. The decreasing (blue) line = the intertemporal budget constraint, the decreasing (grey) curve = the intertemporal indifference curve. Abbreviations: Q1 = consumption at time 1, Q2 = consumption at time 2. *Notes* If M = 2, r = 0.1, ρ = 0.5, the dynamic equilibrium is at Q1 = 1.2 and Q2 = 0.88

Note that $\lambda = 1/M + 1/M$ if $\rho = 0$, where $1/M = \partial U/\partial Q_1(M) = \partial Log/\partial Q_1(M) = \partial U/\partial Q_2(M) = \partial Log/\partial Q_2(M)$ is the marginal utility of consumption in period 1 and period 2 evaluated at M (i.e., the marginal utility of money). In particular:

$$(1/Q_1)/(1+r) = (1/Q_2)[1/(1+\rho)] \text{ or } (1+r) = (1+\rho)(Q_2/Q_1)$$
$$(1+r) = (1+\rho)[(Q_2 - Q_1)/Q_1 + 1] \text{ or, in logarithms, } r \sim \rho + \Delta Q$$

with $\Delta Q = Q_2/Q_1 - 1$ is the consumption growth rate (CGR).

In summary, if a demand side approach is adopted, the theoretical social discount rate r is given by:

$$r = \rho + CGR$$

In particular, r is larger if uncertainty is larger (i.e., a smaller absolute value is attached to future benefits) or consumption growth rate is larger (i.e., a smaller relative value is attached to future benefits). In practice, r is similar to the net return of the governmental long-term bonds. Note that the demand side approach can be used to estimate the social discount rate σ used in Chap. 3. In particular, if a society is assumed to live forever, then $r = \sigma = 0$ for an economically stable society (Scarborough, 2011).

Alternatively, if a supply side approach is adopted, r can be evaluated by referring to the rate of Return on Private Capital Market (RPCM).

Note that, in practice, the demand side and the supply side produce different discount rates, because of uncertainty UNC about the distant future and imperfect competition IMP in capital markets, although there is no uncertainty or imperfect competition in the static theoretical Economic General Equilibrium Model EGEM behind CBA (i.e., $r = \sigma$). Moreover, if public funds are used, their opportunity cost must be considered. In particular, if resources arise from reduced consumption (due to taxes, or bonds or inflation), then r can be evaluated as $r = \rho$. In contrast, if resources come from reduced investment, then r can be evaluated as r = RPCM. Finally, CBA is adequate to maximize the total welfare (i.e., to achieve efficiency EFF) in temporal and dynamic contexts by applying a suitable time discount rate (TD), since welfare levels in different times are in monetary terms. In contrast, inter-temporal inequality aversion at its maximum value (i.e., TIS discussed in Sect. 4.2.2.1 for MCA) is inconsistent with CBA, since it is based on incommensurable time specifications.

4.2.1.2 Space: SPA, SD

Let us assume that there are two places, 1 and 2, where a resource in place 1 X_1 is evaluated to a smaller extent than the same amount of resource in place 2 X_2. For example, population in place 1 POP_1 is smaller than population in place 2 POP_2. Let us assume that it is possible to transfer X_1 from place 1 to place 2, although this implies a loss by r percent. For concreteness, think of a transfer of water resources from place 1 to place 2 (Newburn & Berck, 2011 on alternative polluting technologies in a rural–urban context). Under the assumption that the utility from resource uses does not change in space (i.e., $U(X_1) = U(X_2)$ whenever $X_1 = X_2$), the problem to be solved by policy makers can be formalized as follows:

$$\text{Maximize } U(X_1) + 1/(1-\rho)U(X_2)$$
$$\text{Subject to } X_2 \leq (X - X_1)(1-r)$$

where ρ is the inter-spatial marginal preference.

Figure 4.27 depicts an optimal spatial equilibrium within CBA.

Note that $\rho = (POP_2 - POP_1)/POP_2 < 1$ (i.e., a water loss could compensate a welfare gain from transfering water from a less density populated area 1 to a more density populated area 2). Moreover, this a simple spatial model (SPA) as opposed to a geographical model, since the outcome in space 2 (i.e., $U(X_2)$) depends on the action taken in place 1 (i.e., X_1). Finally, ρ enables the comparison of utility (i.e., generally speaking, the combination of welfare or NPV) across space in order to measure the total welfare or NPV to be maximized.

Let us assume that U(X) = Log[X]. The first order conditions and the constraint becomes:

$$1/X_1 = \lambda,\ ((1-r)/(1-\rho))(1/X_2) = \lambda,\ X_1 + (1/(1-r))X_2 = X$$

Thus,

$$\lambda = \frac{2-\rho}{X(1-\rho)},\ X_1 = \frac{X(1-\rho)}{2-\rho},\ X_2 = \frac{X(1-r)}{2-\rho}$$

In summary, different welfare levels in different places can be combined by applying a spatial discount factor (SD). In other words, CBA is adequate to maximize the total welfare (i.e., to achieve efficiency EFF) in spatial and geographical contexts by applying a suitable spatial discount rate (SD), since welfare levels in different spaces are commensurable *as* measured in monetary terms. In contrast, the

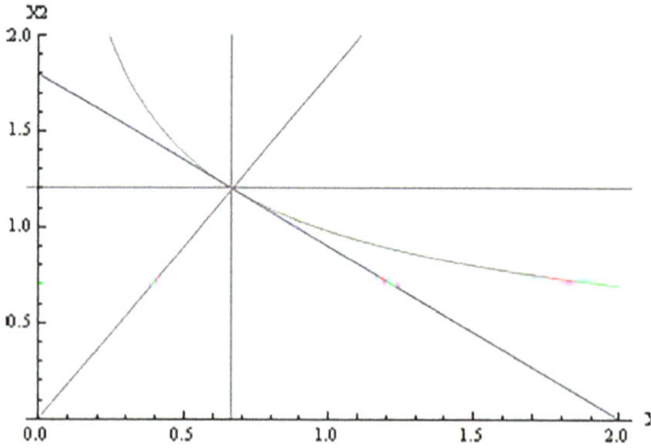

Fig. 4.27 Spatial equilibrium within CBA. The decreasing (blue) line = the interspatial budget constraint, the decreasing (grey) curve = the interspatial indifference curve. Abbreviations: X1 = resource in space 1, X2 = resource in space 2. *Notes* If X = 2, r = 0.1, ρ = 0.5, the spatial equilibrium is at X1 = 0.66 and X2 = 1.2

geographical information systems (i.e., GIS discussed in Sect. 4.2.2.2 for MCA) is inconsistent with CBA, since it is based on incommensurable space specifications.

4.2.1.3 Uncertainty or Risk: EU, CAPM

There is uncertainty whenever it is impossible to identify alternative events *or* to attach a probability to them. In contrast, there is risk whenever it is possible to identify events *and* attach a probability to them.

Let us assume that time is irrelevant for the project under consideration. If alternative possible events or scenarios are known, but it is not possible to attach probabilities to them (i.e., uncertainty applies), then sensitivity analyses can be applied in terms of NPV. Table 4.9 depicts the alternative NPV that can be achieved by implementing 4 actions in 4 stated of the world.

Three *main* methodologies can be applied to compare actions across states of the world. If the Maximin for benefits or Minimax for costs by Wald is applied (i.e., I choose the action that produces the largest minimum NPV) then action B is taken (e.g., Min of A is 0, Min of B is 1, Min of C is 0, Min of D is 0, thus the MaxMin is 1 from B).

Table 4.10 depicts the alternative regret that can be suffered by implementing 4 actions in 4 stated of the world, where each regret is calculated as the difference between what I achieved ex-post (i.e., once uncertainty is solved, since I observed what happened) by the chosen action and what I could have achieved if I had chosen the action that scored best ex-post in that scenario (e.g., if I choose A and the II states of the world happen I get 2, but if I had chosen C instead I could have got 4, thus the regret by choosing A is 2 if the II states of the world happen). If the MinMax Regret by Savage is applied then action D is chosen (e.g., Max Regret of A is 2, Max Regret of A is 2, Max Regret of B is 3, Max Regret of C is 2, Max Regret of D is 1, thus the Min Max Regret is 1 from D).

If the Principle Of Insufficient Reason by Laplace is applied (i.e., I attach the same probabilities to each state of the world since I do not have information to differentiate them) to NPV in Table 4.10, where 4 events imply probability 0.25 attached to each event, then action A is taken, since action A maximizes the weighted sum of outcomes, where weights consist of the same probabilities attached to each

Table 4.9 Net present values in some states of the world

NPV		States of the world			
		I	II	III	IV
Action	A	2	2	0	1
	B	1	1	1	1
	C	0	4	0	0
	D	1	3	0	0

Notes I, II, III, and IV identy four independent states of the world

Table 4.10 Regret of net present values in some states of the world

Regret NPV		States of the world			
		I	II	III	IV
Action	A	0	2	1	0
	B	1	3	0	0
	C	2	0	1	1
	D	1	1	1	1

Notes I, II, III, and IV identy four independent states of the world

state of the world (e.g., 2/4 + 2/4 + 0/4 + 1/4 = 5/4 from A, 1/4 + 1/4 + 1/4 + 1/4 = 1 from B, 0/4 + 4/4 + 0/4 + 0/4 = 1 from C, 1/4 + 3/4 + 0/4 + 1/4 = 1 from D).

Let us assume that probabilities are unknown, but the best and worst possible outcomes are known. Consequently, it is possible to approximate the probability density function by applying the triangular distribution. Figure 4.28 depicts two applications of this procedure to a project A and a project B. Thus, the decision rule seems to be based on calculating the Expected Value $EV = \Sigma_i x_i p_i$, where x_i are the outcomes and p_i are the probabilities attached to the states of the world i with $p_i \geq 0$, and taking the action than maximizes EV.

Note that the Principle Of Insufficient Reason by Laplace amounts to EV with a uniform probability distribution. Moreover, EV can be applied also in a context where probabilities are known (i.e., if risk rather than uncertainty applies). Alternatively, if EV is known but probabilities are unknown, the Fuzzy Set FS methods can be

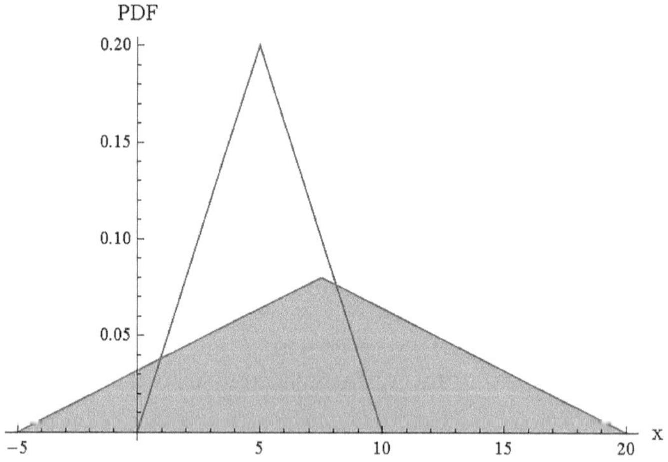

Fig. 4.28 Expected value and triangular distributions. The (purple) area = the probability density function PDF of Project B, based on information about its minimum outcome x = −5, its Expected value at 7.5 and its maximum outcome x = 20; the (white) area = the probability density function PDF of Project A, based on information about its minimum outcome x = 0, its Expected value at 5 and its maximum outcome x = 10

applied: FS are discussed in Sect. 4.2.2.3 for MCA. Finally, a scenario approach is inconsistent with the maximization of the (expected) total welfare within CBA, since a scenario approach assumes that different states are incommensurable and they cannot be combined to obtain an (expected) total NPV.

However, Fig. 4.28 depicts a context where project A is characterized by a smaller expected value EV and a smaller variance VAR, while project B is characterized by a larger EV and a larger VAR. Taking a decision based on EV could be misleading, since the additional information about VAR is disregarded. A decision methodology that considers the different attitudes to risk of different individuals is the Expected Utility Model EUM where $EU = \Sigma_i U(x_i) p_i$ with $U(x_i)$ as the utility U from outcomes x_i.

Note that the EU model is used here as a normative model (i.e., it suggests what to do theoretically) rather than as a positive model (i.e., it explains or predicts what individuals really do). Indeed, the many assumptions on preferences over lotteries behind the EU model (i.e., completeness, transitivity, independence of irrelevant alternatives, continuity) have been scrutinized by the experimental economics literature, although the EU model can also be taken as a positive model in some contexts (Zagonari, 2019).

Figure 4.29 represents the EU model for a fair game (i.e., EV = 0), where an initial wealth W can be reduced or augmented by the same amount h according to the same probabilities p. For concreteness, think of missing a lecture in environmental science (you keep W) versus going to attend that lecture with a bicycle accident (an unlucky event with W − h) or without a bicycle accident (a lucky with W + h). Note that U(EV) > EU (in the example, 3.91 > 3.82) applied to a fair game defines a risk averse attitude (i.e., a risk averse individual prefers to be in a certain context than in a fair game context), which is depicted by any concave utility function on outcomes x (in the example, Log(x)). In particular, the CE (Certainty Equivalent) = U^{-1}(EU) enables the calculation of the CR (Cost of Risk) = EV − CE (e.g., U(x) = Log(x) implies CR = W − Exp(EU).

Let us refer to the Taylor expansion of II degree for a generic function f(x):

$$f(x) \sim f(x_0) + f'(x_0)(x - x_0) + f''(x_0)(x - x_0)^2$$

with $f' = \partial f(x)/\partial x$ and $f'' = \partial f'(x)/\partial x$. For example, the Taylor expansion of I degree for $f(x) = \sqrt{x}$ and $x_0 = 1$ is given by $f(x) \sim \sqrt{1} + (1/2)\,1^{-(1/2)}\,(x - 1) = (1/2) + (1/2)\,x$.

Let f(x) = Log(W) and x_0 = EV:

$$\text{Log}(W) = \text{Log}(EV) + (1/EV)(W - EV) + (-1/EV^2)(W - EV)^2 = K$$

With K a given utility level. If W = EV:

$$K = \text{Log}(EV) - \left(1/EV)^2\right)VAR$$

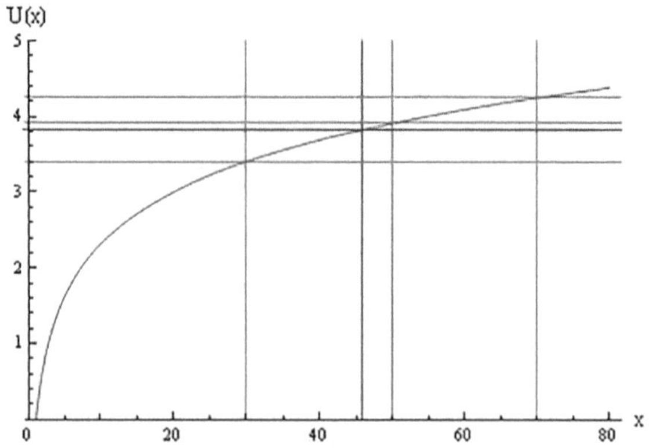

Fig. 4.29 Expected utility of a fair game. The increasing (blue) curve = utility from outcomes U(x) = Log(x), the initial wealth W = 50, h = 20 can be gained with ½ probability so the final wealth is at 60, h = 20 can be lost with ½ probability so the final wealth is at 30. *Notes* EV = 50, U(30) = 3.40, U(80) = 4.38, U(EV) = U(50) = 3.91, EU = 0.5 U(30) + 0.5 U(80) = 3.82, CE = 45.82, CR = 4.18

Thus, the couples of VAR and EV which produce the same level of utility K are given by:

$$VAR = EV^2[Log(EV) - K]$$

The indifference curves for any level of K can be depicted in the [VAR, EV] framework.

Figure 4.30 applies the Mean-Variance approach to the example introduced in Fig. 4.29. Note that projects below the indifference curve are preferred to projects on the indifference curve, since they are characterized by a smaller VAR at the same level of EV (in the example, project B is preferred to project A). Indeed, project B > the project on the indifference curve characterized by the same VAR of B and a smaller EV of B = the project on the *same* indifference curve characterized by the same VAR of A and a larger EV of A > project A: thus B > A.

In summary, CBA is *theoretically* adequate to achieve efficiency EFF (i.e., maximize the expected social welfare) by applying the Mean–Variance approach (i.e., a development of the EUM) whenever risky outcomes are characterized by a normal distribution (i.e., EV and VAR fully represent the outcome distribution) and time is irrelevant. However, decisions depend on risk attitudes which characterize the individuals engaged in decisions or equivalently on the assumed utility function which represent the individuals engaged in decisions (Riddel, 2011). This makes *practically* problematic to apply the Mean–Variance approach, even if a normal distribution can be reasonably assumed. Note that these practical problems vanish if decisions are based on a top-down approach (i.e., experts are assumed to know preferences of people better than people themselves) rather than a bottom-up approach (i.e., people know their preferences better than experts).

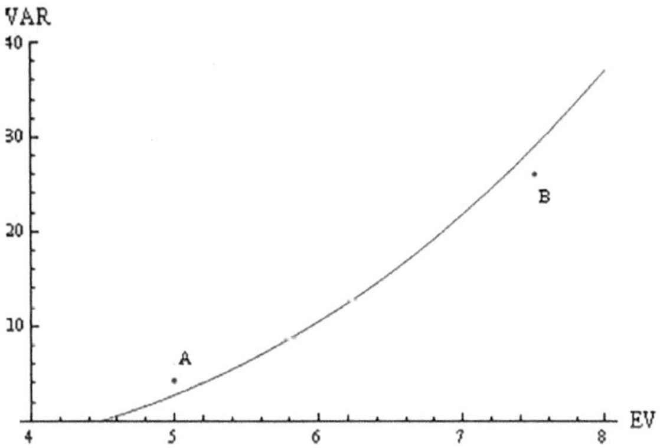

Fig. 4.30 The mean–variance framework. The increasing (blue) curve = the indifference curve between variance VAR and expected value EV at K = 1.5. *Notes* Since Project A (5, 4.16) is above and Project B (7.5, 26.04) is below the indifferent curve, Project B > Project A

Let us assume that time is relevant for the project under consideration. In particular, let us focus on two periods only (Sandsmark & Vennemo, 2007).

Let us assume that the agents maximise the inter-temporal expected utility:

$$\mathrm{Max}_{Q1,x0,xM,xN}\, U(Q_1) + \left(\frac{1}{1+\rho}\right) E\left\{U\left[(W - Q_1)\left(x_0 R_0 + \sum\nolimits_{a\ in\ M,N} x_a R_a\right)\right]\right\}$$

where W is the first period wealth, Q_1 is the first period consumption, x_0 denotes the fraction of wealth invested in the risk free asset, R_0 is the return on the risk free asset, x_a denotes the fraction of wealth invested in the asset a, R_a is the return on the asset a, asset M is the market portfolio and asset N is a climate friendly environmental asset, $\rho \geq 0$ is the inter-temporal marginal preference.

Let us assume that asset N is negatively correlated with asset M (i.e., investing in the climate asset contributes to smoothing consumption, by paying off in states where consumption is low), and that investing in asset N affects ex-ante expected return and variance via a change in the outcome probability (i.e., risk is endogenous). Note that this problems boil downs to the inter-temporal consumption problem if there is no uncertainty: Max $U(Q_1) + (1/(1 + \rho))\, U(W - Q_1)(x_0 R_0 + x_M R_M + x_N R_N)$.

Since

$$x_0 = 1 - \sum\nolimits_{a\ in\ M,N} x_a$$

The uncertain portfolio return is

$$R = R_0 + \sum\nolimits_{a\ in\ M,N} x_a (R_a - R_0)$$

Let $Q_2 = (W - Q_1)/R$ the second-period consumption.

Thus, the maximizing problem becomes:

$$\mathrm{Max}_{Q1,xM,xN}\, U(Q_1) + \left(\frac{1}{1+\rho}\right) E\left\{U\left[(W - Q_1)\left(R_0 + \sum\nolimits_{a\ in\ M,N} x_a (R_a - R_0)\right)\right]\right\}$$

The first-order conditions are for Q_1, x_M and x_N, respectively, are given by:

$$U'(Q_1) = \left(\frac{1}{1+\rho}\right)E\left[U'(Q_2)R\right]$$

$$0 = E\left[U'(Q_2)(x_a(R_M - R_0))\right]$$

$$0 = E\left[U'(Q_2)(R_M - R_0)\right] + \int U(Q_2)\left[\partial F(Q_2, K_M, K_N, \xi)/\partial K_N\right]dQ_2$$

where K_M and K_N are the saving or capital in the market portfolio asset and the climate friendly asset, respectively, and ξ is a random variable.

By using the Hoeffding covariance identity (i.e., COV(x, y) = ∫∫ [F(x,y) − F(x)F(y)] dx dy) and the so-called Stein-Rubinstein lemma (i.e., if y = g(x), COV[g(x), y] = COV(x, y) E[g′(x)]), the third condition becomes:

$$ER_N = R_0 + \beta(ER_M - R_0) - \left\{\int U'(Q_2)\left[\partial F(Q_2, K_M, K_N, \xi)/\partial K_N\right]dQ_2\right\}/E\left[U'(Q_2)\right]$$

where

$$\beta = \frac{COV(Q_2, R_N)}{VAR(Q_2)}$$

However, the Stein-Rubinstein lemma requires that the distribution of Q_2 is normal. Thus:

$$F(Q_2, K_M, K_N, \xi) = \Phi\left[\frac{Q_2 - \mu}{\sigma}\right]$$

where Φ is the cumulative density function for the standardized normal distribution.

Let us assume that:

$$\mu\left(\frac{K_N}{\mu 0}\right) \text{ with } \mu' > 0 \text{ and } \mu 0 = \mu(K_N 0)$$

Thus, the third condition becomes:

$$ER_N = R_0 + \beta(ER_M - R_0) - \frac{E(Q_2')}{E(Q_2)}$$

where β measures the systematic risk within the Capital Asset Pricing Model CAPM.

Let us specify the model as follows:

$$Y^* = AK_M{}^a$$

$$D = \theta_1 T^{\theta 2}$$

$$Y = (1 - D)Y^*$$

$$T = -\eta K_N + T_0$$

$$Y_0 = R_0 K_0$$

$$Q_2 = Y_0 + Y + K_M + K_N$$

Thus:

$$Q_2 = R_0 K_0 + \left[1 - \theta_1 (-\eta K_N + T_0)^{\theta 2}\right] A K_M{}^a + K_M + K_N$$

where Y* is the deterministic production in period 2 based on market portfolio investment, the fixed factor A and the exponential productivity a; D measures the reduced production due to the increase in temperature T according the parameter θ_2, where T starts at T_0 but it negatively depends on the climate friendly investment K_N according the parameter η; θ_1 is normally distributed as $N[\mu_{\theta 1}(K_N), \sigma_{\theta 1}]$ with $\mu_{\theta 1}' < 0$. Thus, the third condition becomes:

$$ER_N \approx R_0 - \frac{\theta_2 \eta}{T}(ER_M - R_0) - ED\frac{-\mu'_{\theta 1}}{\mu_{\theta 1}}$$

Note that $\beta = -(\theta_2\ \eta)/T < 0$ and the expected return of the asset N is larger if θ_2 and η are smaller or if T is larger.

In summary, CBA is theoretically adequate to achieve efficiency EFF (i.e., maximize the expected social welfare) by applying the Capital Asset Pricing Model CAPM (i.e., a development of the EUM) whenever time is relevant and risky outcomes are characterized by a normal distribution. However, the same empirical problems highlighted above where time was irrelevant (i.e., top-down vs. bottom-up approaches) apply also here where time is relevant, unless it is reasonable to assume risk neutral individuals. Indeed, if people engaged in decisions are risk neutral (i.e., experts as decision makers maximize the expected *total* welfare under the assumption that positive and negative outcomes are distributed among individuals with EV = 0), regardless of the probability distribution, EV does not miss any relevant information and can be properly used.

Let us finish this Section with a simple context where the choice at stake is between implement a project today (say, this year) or wait to implement the same project tomorrow (say, next year). If R_H is the high return in the lucky scenario with attached probability q and R_L the low return in the unlucky scenario with attached probability 1 − q, EV = q R_H + (1 − q) R_L. If r is the applied interest to EV from time 0 to infinity, R_0 is the return in period 0 and C_0 is the investment cost in period 0, NPV > 0 if and only if EV/r > $C_0 - R_0$. However, the project can be implemented in period 0, with the following NPV:

$$NPV_0(0) = R_0 + \left[q\, R_H/r + (1-q)R_L/r\right] - C_0$$

In contrast, the project can be implemented in period 1, with the following NPV:

$$NPV_0(1) = q[R_H/r + C_0/(1+r)]$$

Thus, it is better to wait and implement the project in period 1 whenever:

$$NPV_0(1) - NPV_0(0) > 0 \text{ or } EV/r > (C_0 - R_0) + q[R_H/r - C_0/(1+r)]$$

The potential gain from waiting is called the Quasi-Option Value.

4.2.1.4 Linkages: CGEM

Let us assume that there are three sectors. For concreteness, think of agriculture, industry and service sectors. Let us assume that these sectors are linked by inter-sectoral sales and purchases (i.e., sectors purchase goods and services to produce goods or services to be sold to some other sectors). These linkages are measured in terms of monetary values of goods and services purchased and sold. Let us assume that the three sectors also use production factors, together with inputs from other sectors. For concreteness, think of labor and capital. Let us assume that each sector also sell goods and services to a final demand, together with outputs to other sectors. For concreteness, think of private consumption and public consumption combined. Table 4.11 shows a numerical example. Note that the value of production equals the value of demand, since all revenues are used to pay production inputs and production factors, including managers, patent holders, financers.

Table 4.11 A simple input–output matrix for a three sector economy

		Sales			Final demand	Gross output
		1	2	3		
Purchases	1	100	400	300	300	1100
	2	0	400	900	1100	2400
	3	0	0	600	2600	3200
Value added	L	900	1200	1200	4000	
	K	100	400	200		
Gross output		1100	2400	3200		

Abbreviations: Sector 1 = agriculture, Sector 2 = industry, Sector 3 = service, L = labour, K = capital. *Notes* Figures are in any monetary unit

Let x_i the value of production (i.e., 3 values in the last row) and x_{ij} the values in each cell. Thus, the linkages between sectors can be represented by:

$$a_{ij} = x_{ij}/x_i$$

For example, in Table 4.11:

$$a_{12} = x_{12}/x_2 = 400/2400$$

Let y_i the final demand for goods and services produced by the sector i. If the value of production is *totally* distributed to input factors, the following equation applies to all sectors:

$$x_i = \Sigma_j a_{ij} * x_j + y_i$$

In the three sector context, this can be expanded into:

$$x_1 = a_{11}\ x_1 + a_{12}\ x_2 + a_{13}\ x_3 + y_1$$
$$x_2 = a_{21}\ x_1 + a_{22}\ x_2 + a_{23}\ x_3 + y_2$$
$$x_3 = a_{31}\ x_1 + a_{32}\ x_2 + a_{33}\ x_3 + y_3$$

or, by simplifying,

$$(1 - a_{11})x_1 - a_{12}\ x_2 - a_{13}\ x_3 = y_1$$
$$a_{21}\ x_1 - (1 - a_{22})x_2 - a_{23}\ x_3 = y_2$$
$$a_{31}\ x_1 - a_{32}\ x_2 - (1 - a_{33})x_3 = y_3$$

or, by using matrixes,

$$(I - A)X = Y$$

where I is the 3 × 3 identity matrix, A is the coefficient matrix a_{ij}, $X = [x_1\ x_2\ x_3]$, $Y = [y_1\ y_2\ y_3]$.

By solving for X,

$$X = (I - A)^{-1}Y$$

Thus, X measures the direct and indirect productions in each sector to meet the final demand Y for goods and services for each sector, where the direct increase in productions could be measured by $X = I^{-1}\ Y = I\ Y$. Note that $(I - A)^{-1}$ amounts to the multiplicative factor in a single production macro-economic model with fixed prices and no money, where the former accounts for direct and indirect impacts in space, whereas the latter accounts for direct and indirect impacts over time. Indeed, if $C = z + b\ Y$ (i.e., private consumption consists of a fixed amount a plus a proportion b of income Y), if AD = C + I + G (i.e., aggregate demand amounts to private consumption C, private investment I and public consumption and investment G), the equilibrium income Y* (i.e., income such that AD = Y) is Y* = [1/(1 − b)] (z + I + G), where the multiplier [1/(1 − b)] evokes $(I - A)^{-1}$, since b represents a feature of this simplified macro-economy like A represents the structure of the sectoral economy, the scalar 1 represents the neutral element in a one dimension model like the matrix I represents the neutral element in a three dimension model, multiplying by [1/(1 − b)] for scalars is similar to multiplying by $(I - A)^{-1}$ for matrixes, G represents an external increase in aggregate demand like Y represents an external increase in sectoral demand. Moreover, the largest sum of direct and indirect impacts (i.e., $x_1 + x_2 + x_3$) identifies the project to be chosen. Finally, possible preferences on the relative impacts in different sectors can be accounted for by introducing a social welfare function discussed in Sect. 4.2.1.5.

For example, if the sectoral demands from Project F and Project Z are given by $Y_F = [1, 2, 3]$ and $Y_Z = [3, 2, 1]$, respectively, the direct and indirect impacts of these projects are given by $X_F = [2.14, 3.65, 3.70]$ and $X_Z = [3.94, 2.82, 1.23]$, since the matrix A based on Table 4.11 is given by:

$$A = \begin{bmatrix} 0.09 & 0.17 & 0.09 \\ 0.00 & 0.17 & 0.28 \\ 0.00 & 0.00 & 0.19 \end{bmatrix}$$

The value added (labor L and capital K) can be represented by

$$v_L = b_{L1} * x_1 + b_{L2} * x_2 + b_{L3} * x_3$$
$$v_K = b_{K1} * x_1 + b_{K2} * x_2 + b_{K3} * x_3$$

with

$$v_{ij} = b_{ij} * x_j$$

by using matrixes,

$$V = B\,X$$

And

$$V = B(I - A)^{-1}Y$$

Thus, V measures the direct and indirect factors in each sector to meet the final demand Y for goods and services for each sector.

For example, if the sectoral demands from Project F and Project Z are given by $Y_F = [1, 2, 3]$ and $Y_Z = [3, 2, 1]$, respectively, the direct and indirect impacts of these projects are given by $V_F = [4.99, 1.03]$ and $V_Z = [5.11, 0.90]$, since the matrix

B based on Table 4.11 is given by:

$$B = \begin{bmatrix} 0.82\ 0.50\ 0.38 \\ 0.09\ 0.17\ 0.06 \end{bmatrix}$$

The Input–Output model IOM is the complete version of the simplified model discussed in this Section: it is discussed in Sect. 4.2.2.4 about linkages in MCA.

Since IOM assumes fixed prices, it is inadequate to achieve efficiency EFF as pursued by CBA, unless it is reasonable to assume that all projects are small in all markets (i.e., they do not affect equilibrium prices in any market). Note that X depicts outcomes in IOM, whereas CBA compares economic costs and benefits based on utility levels attached to negative and positive outcomes.

The development of the IOM to consider potential changes of prices is the Computable General Equilibrium Model CGEM (i.e., an empirical model based on the Economic General Equilibrium assumptions to depict interactions between individuals and firms).

A *general* CGEM can be presented as follows. The demand side of the model:

$$\mathrm{Max}_{Q_i^m} U^m\left(Q_i^m\right) \mathrm{s.t.} \sum_i^n P_i Q_i^m = Y^m(W, R) = wL^m + RK^m$$

where Q_i^m is the ith commodity demand for consumer m, U^m is an increasing, continuous and concave utility function for consumer m, W is the wage rate, R is the rental rate, P_i is ith commodity price, Y^m is the income level of consumer m, L^m is the labor endowment of consumer m, K^m is the capital endowment of consumer m. The commodity demands for consumers m are the solution of the previous problem:

$$Q_i^m = Q_i^m(P_i, W, R)$$

The supply side of the model:

$$\mathrm{Min}_{L_i, K_i}\, C_i(L_i, K_i) = \mathrm{W}L_i + \mathrm{R}K_i\, \mathrm{s.t.}\, Q_i = \overline{Q_i}$$

where C_i is the direct cost function of each producer in a perfectly competitive industry based on a constant returns to scale production function, L_i is the labor demand, K_i is the capital demand, Q_i bar is a given output level. The derived factor demands for given output levels are the solution of the previous problem:

$$L_i = L_i\left(W, R, \overline{Q_i}\right) \text{ and } K_i = K_i\left(W, R, \overline{Q_i}\right)$$

The excess demand functions for commodities and factors are given by:

$$\sum_m Q_i^m(P_i, W, R) - Q_i \leq 0$$

$$\sum_i L_i(Q_i, W, R) - \sum_m L^m \leq 0$$

$$\sum_i K_i(Q_i, W, R) - \sum_m K^m \leq 0$$

In a competitive industry, the price of each output i is equal to the long-run average cost (i.e., profit is 0 in the long-run):

$$P_i = C_i(L_i, K_i)/Q_i$$

Any set of prices in a CGEM satisfies the Walras' law (i.e., an excess demand in one market must be matched with an excess supply in another market or an examined market must be in equilibrium if all other markets are in equilibrium):

$$\sum_i P_i\left[Q_i - \sum_m Q_i^m(P_i, W, R)\right] + W\left[\sum_i L_i(Q_i, W, R) - \sum_m L^m\right] + R\left[\sum_i K_i(Q_i, W, R) - \sum_m K^m\right] = 0$$

I will present here a *specific* CGEM model to illustrate how to solve it. This model is structurally representative of many other large scale empirical models used for policy analyses. This simplified model has three final goods (agricultural goods a, industrial goods i, services goods s), three factors of production (capital KT in each sector, rural labor RL in each sector, urban labor ULB in each sector), and two classes of consumers (rural households RH, urban households UH). On the production side, production technology in each sector is represented by constant-returns-to-scale Cobb–Douglas (CD) production function. Each factor demand function is derived from a cost minimization problem subject to given technology and given output level. All consumer preferences are represented by a constant-elasticity-of-substitution CD utility function. Each commodity demand is derived from a utility maximization problem subject to the budget constraint faced by each household class.

There are 35 variables for the required equilibrium conditions: 7 exogenous variables (the household endowments of labors and capital) and 28 endogenous variables: 10 prices (Pa, Pi, Ps, RWa, RWi, RWs, UWa, UWi, UWs, r); 6 commodities demanded (RHCa, RHCi, RHCs, UHCa, UHCi, UHCs); 9 factors demanded (DKa, DKi, DKs, RLBa, ULBa, RLBi, ULBi, RLBs, ULBs); 3 commodities supplied (XSa, XSi, XSs).

There are 35 equations: 6 consumer demands, 6 labor supplies, 1 capital supply, 6 labor demands, 3 capital demands, 3 firm supplies, 3 zero profit conditions, 7 no excess factor demand conditions.

Note that labor is assumed to be specific for rural and urban areas (i.e., labor cannot move from area to another), whereas capital is assumed to be shared by rural and urban areas (i.e., households can invest in all areas). Moreover, supply of commodities is assumed to be infinitely elastic (i.e., any quantity is supplied at the equilibrium price). Finally, the supply of labor is assumed to be infinitely elastic (i.e., any quantity is supplied at the equilibrium wage).

There are two households, rural households RH and urban households UH. Both households maximize their own utilities U by solving the following constrained maximization with respect to consumption of agricultural goods Ca, industry goods Ci and service goods Cs:

As for *rural* households:

$$\mathrm{Max}_{\mathrm{RHCa,RHCi,RHCs}}\mathrm{U(RHCa,\ RHCi,\ RHCs)} = \mathrm{RHCa}^{\mathrm{rcsa}}\mathrm{RHCi}^{\mathrm{rcsi}}\mathrm{RHCs}^{\mathrm{rcss}}$$

Subject to

$$\begin{aligned}&\mathrm{Pa\ RHCa + Pi\ RHCi\ + Ps\ RHCs}\\ &\quad \le ((1 - \mathrm{rltr})(\mathrm{RWa\ RLBa + RWi\ RLBi + RWs\ RLBs}) + \mathrm{RHR + URTF + RGTF + RFTF})\\ &\quad (1 - \mathrm{rhtr})\end{aligned}$$

With rcsa, rcsi, rcss are the rural consumption share for agricultural, industry and service goods, respectively; Pa, Pi and Ps are the prices of agricultural, industry and service goods, respectively; rltr is the rural labor tax rate; RWa, RWi, RWs are the rural wages in agriculture, industry and service sectors, RLBa, RLBi, RLBs are the rural labour supply in agriculture, industry and service sectors; RHR are the profits to rural households; URTF are the transfers from urban to rural households; RGTF is the transfer from the government to rural households; RFTF are the transfers from rural households to firms; rhtr is the tax rate on rural income; where

$$\mathrm{RHR} = \mathrm{rhrs} * \mathrm{drs(KTRa + KTRi + KTRs)(1{-}rtr)}$$

$$\begin{aligned}\mathrm{KTRa} = {}& (\mathrm{Pa}(1 - \mathrm{itra}) - \mathrm{Pa\ caa} - \mathrm{Pi\ cia} - \mathrm{Ps\ csa} - \mathrm{cwta} - \mathrm{cwwa})\mathrm{XDa}\\ & - \mathrm{RWa\ RLBa} - \mathrm{UWa\ ULBa} - \mathrm{DEPa}\end{aligned}$$

$$\begin{aligned}\mathrm{KTRi} = {}& (\mathrm{Pi}(1 - \mathrm{itri}) - \mathrm{Pi\ cai} - \mathrm{Pi\ cii} - \mathrm{Ps\ csi} - \mathrm{cwti} - \mathrm{cwwi})\mathrm{XDi}\\ & - \mathrm{RWi\ RLBi} - \mathrm{UWi\ ULBi} - \mathrm{DEPi}\end{aligned}$$

$$\begin{aligned}\mathrm{KTRs} = {}& (\mathrm{Ps}\ (1 - \mathrm{itrs}) - \mathrm{Pa\ cas} - \mathrm{Pi\ cis} - \mathrm{Ps\ css} - \mathrm{cwts} - \mathrm{cwws})\mathrm{XDs}\\ & - \mathrm{RWs\ RLBs} - \mathrm{UWs\ ULBs} - \mathrm{DEPs}\end{aligned}$$

With rhrs is the share of profits distributed to rural households; drs is the distributed profit share, KTRa, KTRi, KTRs are the profits in agriculture, industry and service sectors; rtr is the tax rate on profits; itra, itri, itrs are the tax rates on agriculture agriculture, industry and service revenues; caa, cia, csa are the intersectoral production coefficients for agriculture; cai, cii, csi are the intersectoral production coefficients for industry; cas, cis, css are the intersectoral production coefficients for service; cwta, cwti, cwts are the water fees in agriculture, industry and service sectors; cwwa, cwwi, cwws are the wastewater fees in agriculture, industry and service sectors; DEPa, DEPi, DEPs are depreciations in agriculture, industry and service sectors; XDa, XDi, XDs are demands in agriculture, industry and service sectors.

The derived demand function for *rural* household is:

$$\mathrm{RHCa} * \mathrm{PCa} = (\mathrm{rcsa}/(\mathrm{rcsa} + \mathrm{rcsi} + \mathrm{rcss} + \mathrm{rcswt} + \mathrm{rcsww}))(1 - \mathrm{rhsr})\mathrm{RHY}$$

$$\mathrm{RHCi} * \mathrm{PCi} = (\mathrm{rcsi}/(\mathrm{rcsa} + \mathrm{rcsi} + \mathrm{rcss} + \mathrm{rcswt} + \mathrm{rcsww}))(1 - \mathrm{rhsr})\mathrm{RHY}$$

$$\mathrm{RHCs} * \mathrm{PCs} = (\mathrm{rcss}/(\mathrm{rcsa} + \mathrm{rcsi} + \mathrm{rcss} + \mathrm{rcswt} + \mathrm{rcsww}))(1 - \mathrm{rhsr})\mathrm{RHY}$$

where

$$\mathrm{rcsa} + \mathrm{rcsi} + \mathrm{rcss} + \mathrm{rcswt} + \mathrm{rcsww} = 1$$

With rhsr is the saving rate for rural households.

As for *urban* households:

$$\mathrm{Max}_{\mathrm{RHCa,RHCi,RHCs}}\mathrm{U}(\mathrm{RHCa}, \mathrm{RHCi}, \mathrm{RHCs}) = \mathrm{RHCa}^{\mathrm{rcsa}}\mathrm{RHCi}^{\mathrm{rcsi}}\mathrm{RHCs}^{\mathrm{rcss}}$$

Subject to

$$\begin{aligned}&\mathrm{Pa} * \mathrm{UHCa} + \mathrm{Pi} * \mathrm{UHCi} + \mathrm{Ps}\ \mathrm{UHCs}\\&\quad \le ((1 - \mathrm{rltr})(\mathrm{UWa} * \mathrm{ULBa} + \mathrm{UWi} * \mathrm{ULBi} + \mathrm{UWs} * \mathrm{ULBs}) + \mathrm{UHR} + \mathrm{RUTF} + \mathrm{UGTF} + \mathrm{UFTF})\\&\quad (1 - \mathrm{uhtr})\end{aligned}$$

where

$$\mathrm{UHR} = \mathrm{uhrs} * \mathrm{drs}(\mathrm{KTRa} + \mathrm{KTRi} + \mathrm{KTRs})(1-\mathrm{rtr})$$

With notation for urban is similar to notation for rural households.

The derived demand function for *urban* household is:

$$\mathrm{UHCa} * \mathrm{PCa} = (\mathrm{ucsa}/(\mathrm{ucsa} + \mathrm{ucsi} + \mathrm{ucss} + \mathrm{ucswt} + \mathrm{ucsww}))(1 - \mathrm{uhsr})\mathrm{UHY}$$

$$\mathrm{UHCi} * \mathrm{PCi} = (\mathrm{ucsi}/(\mathrm{ucsa} + \mathrm{ucsi} + \mathrm{ucss} + \mathrm{ucswt} + \mathrm{ucsww}))(1 - \mathrm{uhsr})\mathrm{UHY}$$

$$\mathrm{UHCs} * \mathrm{PCs} = (\mathrm{ucss}/(\mathrm{ucsa} + \mathrm{ucsi} + \mathrm{ucss} + \mathrm{ucswt} + \mathrm{ucsww}))(1 - \mathrm{uhsr})\mathrm{UHY}$$

where

$$ucsa + ucsi + ucss + ucswt + ucsww = 1$$

With notation for urban is similar to notation for rural households. Note that rhrs + uhrs = 1.

Let us assume a Cobb–Douglas production function with constant-return-to-scale for all sectors. Thus, the aggregate producer in each perfectly competitive sector maximizes profits by solving the following constrained cost-minimization problems, subject to the given technology and the given output level.

As for *agricultural* producers:

$$\text{Min}_{KTa,RLBa,UBLa}\ C(KTa, RLBa, UBLa) = PKa\ KTa + RWa\ RLBa + UWa\ UBLa$$

Subject to

$$KTa^{ktsa} RLBa^{rlbsa} UBLa^{ublsa} \geq XSa$$

where

$$ktsa = 1 - rlbsa - ulbsa$$

with KTa is the demand for capital; ktsa, rlbsa, ublsa are the production coefficients in agriculture for capital, rural labor and urban labor.

As for *industrial* producers:

$$\text{Min}_{KTi,RLBi,UBLi} C(KTi, RLBi, UBLi) = PKi\ KTi + RWi\ RLBi + UWi\ UBLi$$

Subject to

$$KTi^{ktsi} RLBi^{rlbsi} UBLi^{ublsi} \geq XSi$$

where

$$ktsi = 1 - rlbsi - ulbsi$$

With notation for the industry is similar to notation for the agriculture sector.

As for *service* producers:

$$\text{Min}_{KTs,RLBs,UBLs}\ C(KTs, RLBs, UBLs) = PKs\ KTs + RWs\ RLBs + UWs\ UBLs$$

Subject to

$$KTs^{ktss} RLBs^{rlbss} UBLs^{ublss} \geq XSs$$

where

$$\text{ktss} = 1 - \text{rlbss} - \text{ulbss}$$

With notation for the industry is similar to notation for the agriculture sector.

The derived labor demands (RLBa, RLBi, RLBs, ULBa, ULBi, ULBs) for given output level XSa, XSi, XSs are:

$$\text{RLBa} = \text{rlbsa}(\text{Pa}(1 - \text{itra}) - \text{Pa caa} - \text{Pi cia} - \text{Ps csa})\text{XSa}/\text{RWa}$$

$$\text{RLBi} = \text{rlbsi}(\text{Pi}(1 - \text{itri}) - \text{Pi cii} - \text{Pi cii} - \text{Ps csi})\text{XSi}/\text{RWi}$$

$$\text{RLBs} = \text{rlbss}\ (\text{Ps}(1 - \text{itrs}) - \text{Ps css} - \text{Pi cis} - \text{Ps css})\text{XSs}/\text{RWs}$$

$$\text{ULBa} = \text{ulbsa}(\text{Pa}(1 - \text{itra}) - \text{Pa caa} - \text{Pi cia} - \text{Ps csa})\text{XSa}/\text{UWa}$$

$$\text{ULBi} = \text{ulbsi}(\text{Pi}(1 - \text{itri}) - \text{Pi cai} - \text{Pi cii} - \text{Ps csi})\text{XSi}/\text{UWi}$$

$$\text{ULBs} = \text{ulbss}(\text{Ps}(1 - \text{itrs}) - \text{Ps cas} - \text{Pi cis} - \text{Ps css})\text{XSs}/\text{UWs}$$

The derived capital demands (DKa, DKi, DKs) for given output level XSa, XSi, XSs are:

$$\text{DKa} = \text{KTa} + \text{DEPa} - (1 - \text{drs})\text{KTRa}$$

$$\text{DKi} = \text{KTi} + \text{DEPi} - (1 - \text{drs})\text{KTRi}$$

$$\text{DKs} = \text{KTs} + \text{DEPs} - (1 - \text{drs})\text{KTRs}$$

The profit conditions (zero in the long-run) are:

$$\begin{aligned}\text{KTRa} = {} & (\text{Pa}(1 - \text{itra}) - \text{Pa caa} - \text{Pi cia} - \text{Ps csa} - \text{cwta} - \text{cwwa})\ \text{XSa} \\ & - \text{RWa RLBa} - \text{UWa ULBa} - \text{DEPa}\end{aligned}$$

$$\begin{aligned}\text{KTRi} = {} & (\text{Pi}(1 - \text{itri}) - \text{Pi cai} - \text{Pi cii} - \text{Ps csi} - \text{cwti} - \text{cwwi})\text{XDi} \\ & - \text{RWi RLBi} - \text{UWi ULBi} - \text{DEPi}\end{aligned}$$

$$\begin{aligned}\text{KTRs} = {} & (\text{Ps}(1 - \text{itrs}) - \text{Pa cas} - \text{Pi cis} - \text{Ps css} - \text{cwts} - \text{cwws})\text{XSs} \\ & - \text{RWs RLBs} - \text{UWs ULBs} - \text{DEPs}\end{aligned}$$

Note that data refer to one year period so it is possible to replace capital remuneration R with interest rate r as follows: $R = P(1 + r)$ (i.e., investment in each sector should remunerate the capital as it is remunerated in the financial market). Moreover, the price of capital is taken as numeraire (i.e., $PK_i = 1$ in all three sectors). Finally, conditions for uniqueness and stability of the equilibrium are based on the excess demand function for capital, since there is no excess demand for labor, due to the assumed infinitely elastic supply (i.e., the excess demand function for capital must be decreasing and continuous).

Excess demand conditions for goods EDa, EDi, EDs are:

$$XSa \leq caa\ XDa + cai\ XDi + cas\ XDs + RHCa + UHCa + LGCa + DKa + DNEa + FNEa$$

$$XSi \leq cia\ XDa + cii\ XDi + cis\ XDs + RHCi + UHCi + LGCi + DKi + DNEi + FNEi$$

$$XSs \leq csa\ XDa + csi\ XDi + css\ XDs + RHCs + UHCs + LGCs + DKs + DNEs + FNEs$$

where LGCa, LGCi, LGCs are the local government demands for agriculture, industry, service; DNEa, DNEi, DNEs are the domestic next export demands for agriculture, industry, service; FNEa, FNEi, FNEs are the foreign next export demands for agriculture, industry, service.

The excess demands for capital is given by:

$$DKa + DKi + DKs \leq SKa + SKi + SKs$$

The Walras' law is given by:

$$DKa + DKi + DKs = SKa + SKi + SKs$$

As for numerical solutions, I will apply the single equation approach with factor price revision to the Social Accounting Matrix for the Chinese Province of Suqian (Table 4.12).

The Mathematica code for the numerical solutions is provided below:

```
KTa=ktsa    (Pca    (1-itra)-Pca    caa-PCi    cia-PCs    csa-cwta-
cwwa) Xsa/Pka
Kti=ktsi    (PCi    (1-itri)-PCi    cai-PCi    cii-PCs    csi-cwti-
cwwi) Xsi/Pki
KTs=ktss    (PCs    (1-itrs)-PCs    cas-PCi    cis-PCs    css-cwts-
cwws) XSs/PKs

RLBa=rlbsa (Pca (1-itra)-Pca caa-PCi cia-PCs csa) Xsa/Rwa
RLBi=rlbsi (PCi (1-itri)-PCi cai-PCi cii-PCs csi) Xsi/Rwi
RLBs=rlbss (PCs (1-itrs)-PCs cas-PCi cis-PCs css) XSs/RWs

ULBa=ulbsa (Pca (1-itra)-Pca caa-PCi cia-PCs csa) Xsa/Uwa
ULBi=ulbsi (PCi (1-itri)-PCi cii-PCi cii-PCs csi) Xsi/Uwi
ULBs=ulbss (PCs (1-itrs)-PCs css-PCi cis-PCs css) XSs/Uws
```

Table 4.12 The social accounting matrix for the Chinese Province of Suqian

		ACT			COM			FAC					INS					INV	EXPORT		TOT
		agr	ind	ser	agr	ind	ser	rlb	ulb	kt	wt	ww	firm	rh	uh	lg	cg	INV	dom	for	TOT
ACTIVITIES	agr				15915																15915
	ind					39744															39744
	ser						12550														12550
COMMODITIE	agr	3936	1952	434										1209	476	58		335	7513	3	15918
	ind	3123	24683	2355										3818	2167	0	1	12162	0	444	48754
	ser	352	3894	2777										3247	1827	2081		507	0	3	14688
FACTORS	rlb	6732	3788	1282																	11802
	ulb	1106	816	2589																	4511
	kt	67	2502	853																	3422
	wt	33	592	195										113	25						958
	ww	0	30	5											1						36
INSTITUTIONS	firm									3423											3423
	rh							11802	785				140			0					12727
	uh							0	3650				1339			169					5158
	lg	223	758	327				0	76		958	036	388								2766
	cg				0	0	0			-1				-2152	-1424						-3577
SAV		343	728	1733									1556	6492	2085	458				-391	13004
IMPORT	dom				0	8954	2138										-3579				7513
	for				3	56	0														59
TOT	TOT	15915	39744	12550	15918	48754	14688	11802	4511	3422	958	36	3423	12727	5159	2766	-3578	13004	7513	59	

Abbreviations: agr = agriculture, ind = industry, ser = service, rlb = rural labour, ulb = urban labour, kt = capital, wt = water, ww = wastewater, rh = rural households, uh = urban households, lg = local government, cg = central government, INV = investments, dom = domestic, for = foreign, SAV = savings. *Notes* Figures are in Million Yuan

```
KTRa=343;
KTRi=728;
KTRs=1733;

DEPa=depa Xsa;
DEPi=depi Xsi;
DEPs=deps XSs;
```

```
Solve[{343==(Pca (1-itra)-Pca caa-PCi cia-PCs csa-cwta-
cwwa) Xsa-Rwa RLBa-Uwa ULBa-DEPa,728==(PCi (1-itri)-
Pca cai-PCi cii-PCs csi-cwti-cwwi) Xsi-Rwi RLBi-Uwi ULBi-
DEPi,1733==(PCs (1-itrs)-Pca cas-PCi cis-PCs css-cwts-
cwws) XSs-RWs RLBs-Uws ULBs-DEPs},{Pca,PCi,PCs}]

Pca =5.35001
PCi =1.66332
PCs =1.8362

RHR=rhrs drs (KTRa+KTRi+KTRs) (1-rtr);
UHR=uhrs drs (KTRa+KTRi+KTRs) (1-rtr);

RHY=((1-rltr) (Rwa*RLBa+Rwi*RLBi+RWs*RLBs)+RHR+URTF+RGTF+RFTF)
(1-rhtr); UHY=((1-ultr) (Uwa*ULBa+Uwi*ULBi+Uws*ULBs)+UHR+RUTF+
RGTF+RFTF)(1-uhtr);

RHCa=(rcsa/(rcsa+rcsi+rcss+rcswt+rcsww))(1-rhsr) RHY/Pca
RHCi=(rcsi/(rcsa+rcsi+rcss+rcswt+rcsww))(1-rhsr) RHY/PCi
RHCs=(rcss/(rcsa+rcsi+rcss+rcswt+rcsww))(1-rhsr) RHY/PCs

UHCa=(ucsa/(ucsa+ucsi+ucss+ucswt+ucsww))(1-uhsr) UHY/Pca
UHCi=(ucsi/(ucsa+ucsi+ucss+ucswt+ucsww))(1-uhsr) UHY/PCi
UHCs=(ucss/(ucsa+ucsi+ucss+ucswt+ucsww))(1-uhsr) UHY/PCs

LGCa=slgca Xsa;
DNEa=sdnea Xsa;
FNEa=sfnea Xsa;

LGCi=slgci Xsi;
DNEi=sdnei Xsi;
FNEi=sfnei Xsi;

LGCs=slgcs XSs;
DNEs=sdnes XSs;
FNEs=sfnes XSs;

Dka=335/15918 Xsa
Dki=12162/48754 Xsi
DKs=507/14688 XSs

Xda=caa Xsa+cai Xsi+cas XSs+RHCa+UHCa+LGCa+Dka+DNEa+FNEa
Xdi=cia Xsa+cii Xsi+cis XSs+RHCi+UHCi+LGCi+Dki+DNEi+FNEi
XDs=csa Xsa+csi Xsi+css XSs+RHCs+UHCs+LGCs+DKs+DNEs+FNEs
RLBae=rlbsa (Pca (1-itra)-Pca caa-PCi cia-
PCs csa) Xsae/Rwa /. {Xsae→15915,Xsie→48754,Xsse→14688}
RLBie=rlbsi (PCi (1-itri)-PCi cai-PCi cii-
PCs csi) Xsie/Rwi /. {Xsae→15915,Xsie→48754,Xsse→14688}
RLBse=rlbss (PCs (1-itrs)-PCs cas-PCi cis-
PCs css) Xsse/RWs /. {Xsae→15915,Xsie→48754,Xsse→14688}

ULBae=ulbsa (Pca (1-itra)-Pca caa-PCi cia-
PCs csa) Xsae/Uwa /. {Xsae→15915,Xsie→48754,Xsse→14688}
ULBie=ulbsi (PCi (1-itri)-PCi cai-PCi cii-
PCs csi) Xsie/Uwi /. {Xsae→15915,Xsie→48754,Xsse→14688}
ULBse=ulbss (PCs (1-itrs)-PCs css-PCi cis
PCs css) Xsse/Uws /. {Xsae→15915,Xsie→48754,Xsse→14688}
```

```
DKTa=Kta+DEPa-(1-drs) KTRa;
DKTi=Kti+DEPi-(1-drs) KTRi;
DKTs=KTs+DEPs-(1-drs) KTRs;

DKTae=DKTa /. {Xsae→15915,Xsie→39744,Xsse→12550}
DKTie =DKTi/.{Xsae→,Xsie→39744,Xsse→12550}
DKTse =DKTs/.{Xsae→15915,Xsie→39744,Xsse→12550}

Solve[RLBae==6732,Rwa]
Solve[RLBie==3788,Rwi]
Solve[RLBse==1282,RWs]

Solve[ULBae==1106,Uwa]
Solve[ULBie==816,Uwi]
Solve[ULBse==2589,Uws]

Rwa=7.187
Rwi=2.126
RWs=3.002
Uwa=7.187
Uwi=2.126
Uws=1.979

Solve[-(313159/1426) + (343 Xsae)/15915 + (0.045` Xsae)/PK -
332332/713 + (91 Xsie)/4968 + (0.127 Xsie)/PK - 1582229/1426 +
(1733 Xsse)/12550 + (0.211 Xsse)/ PK == (335 + 12162 + 507 -
536 - 391)] /. {Xsae→15915,Xsie→48754,Xsse→14688}

1640+ 9279/PK == 12077

PK = 0.89
```

Figure 4.31 shows the numerical capital excess demand function.

Note that the capital excess demand function is decreasing and convex.

Let us assume that water used in agriculture increases by 50% (i.e., 1030 MCM added to 2060 MCM). Moreover, let us remind that rural population is 4.808 Million people, urban population is 0.244 Million people. Finally, let us assume that water was the only constraint to agriculture production (i.e., agriculture production increases by 50%). Table 4.13 summarizes the main results obtained by CGEM in CBA.

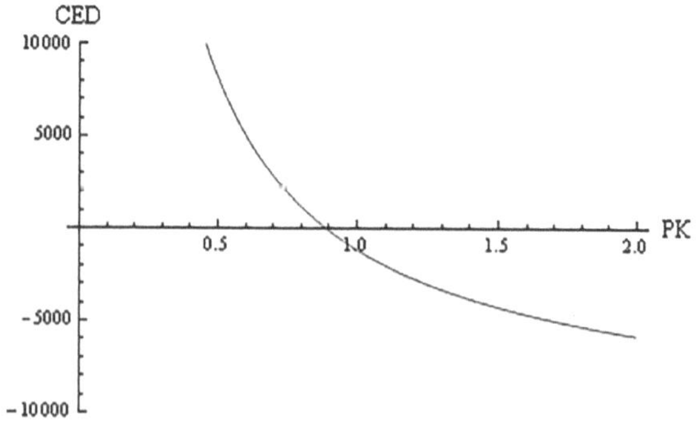

Fig. 4.31 The capital excess demand and its equilibrium price in a computable general equilibrium model. The decreasing (blue) curve — capital excess demand CED in terms of the price of capital PK. *Notes* The equilibrium PK* = 0.89

Table 4.13 Consequences of a 50% increase in water availability within CBA

The old prices were Pa = 5.350, Pi = 1.663, Ps = 1.836
The new prices are Pa = 4.522, Pi = 1.469, Ps = 1.709
The old total incomes were RHY = 49,452, URY = 7271
The old per capita incomes were RHY = 10,285, URY = 29,799
The new total incomes were RHY = 59,504, URY = 8663
The new per capita incomes were RHY = 12,376, URY = 35,504
The old productions were XSae = 15,918, XSie = 48,754, XSse = 14,688
The new productions are XSae = 23,877, XSie = 185,307, XSsc = 42,673
The old total wages were RWa = 7.187, RWi = 2.126, RWs = 3.002, UWa = 7.187, UWi = 2.126, UWs = 1.979
The old per capita wages were RWa = 1.494, RWi = 0.442, RWs = 0.624, UWa = 29.454, UWi = 8.713, UWs = 8.110
The new total wages are RWa = 9.066, RWi = 1.823, RWs = 2.83, UWa = 9.066, UWi = 1.823, UWs = 1.886
The new per capita wages are RWa = 1.885, RWi = 0.379, RWs = 0.590, UWa = 37.157, UWi = 7.474, UWs = 7.732
The old price of capital was PK = 0.89
The new price of capital was PK = 0.87

Abbreviations: Pa = price in agriculture, Pi = price in industry, Ps = price in service, RHY = income of rural households, URY = income of urban households, XSae = equilibrium product supply in agriculture, XSie = equilibrium product supply in industry, XSse = equilibrium product supply in service, RWa = rural wage in agriculture, RWi = rural wage in industry, RWs = rural wage in service, UWa = urban wage in agriculture, UWi = urban wage in industry, UWs = urban wage in service, PK = price of capital

In other words, inequality between rural and urban wages is reduced.

In summary, CGEM enables CBA to maximize the social welfare. In contrast, the social accounting matrix (i.e., SAM discussed in Sect. 4.2.2.4 for MCA) would be inadequate for CBA to achieve efficiency EFF.

4.2.1.5 Inequalities: IM, SFW

Let us assume that there two groups of people (i.e., n = 2). For concreteness, think of poor and rich people. Let us assume that different projects produce different levels of NPV for poor and rich people. For simplicity, let us equate NPV and welfare (i.e., a linear utility function so welfare is measures in terms of NPV). Table 4.14 provides an example.

Table 4.14 Distributions of net present values between rich and poor people in demonstrative projects

		Net present values		
		Rich	Poor	Total
Project	A	60	40	100
	B	50	30	80
	C	20	80	100
	D	40	50	90

Three *main* criteria of efficiency can be evaluated. A project is Harsanyi efficient if it produces the largest total welfare. A project is Pareto efficient if it is not possible to find an alternative project where somebody is better off and nobody is worse off (i.e., there are no losers in moving from the reference project to another). A project is Kaldor-Hicks efficient if it is not possible to find an alternative project where better off people are in a position to compensate worse off people (i.e., gainers cannot compensate losers in moving from the reference project to another).

In Table 4.14, projects A and C are Harsanyi efficient, projects A, C and D are Pareto efficient, projects A and C are Kaldor-Hicks efficient.

Let us depict NPV to rich and poor people from each project as points in the $[NPV_{rich}, NPV_{poor}]$ framework (Fig. 4.32). Let us represent the couples of NPV to rich and NPV to poor people which produce the same total NPV:

$$NPV_{tot} = NPV_{rich} + NPV_{poor}$$

Let us call it the Harsanyi efficient curve:

$$NPV_{rich} = NPV_{tot} - NPV_{poor}$$

Note that projects A and C are on the same Harsanyi efficient curve: in terms of efficiency, A = C > D and A = C > B.

If some projects achieve the same level of efficiency, they can be compared by evaluating two *main* measures of inequality. The Gini index is given by:

$$\text{Gini} = \left[1/\left(2\,n^2x\right)\right]\Sigma_i\,\Sigma_j\left|x_i - x_j\right|$$

where x_i is the outcome to people in group i, x_j is the outcome to people in group j, x is the average outcome, and n is the number of groups. The Gini index is in [0, (n − 1)/n], where Gini Index = 0 represents the maximum equality. For project A:

$$\text{Gini} = \left[1/\left(2\,2^2 50\right)\right][|60 - 40| + |40 - 60|] = 0.1$$

The Atkinson index is given by:

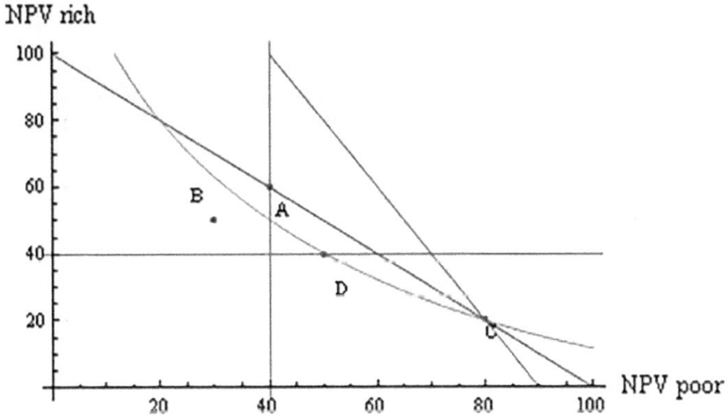

Fig. 4.32 Alternative efficiency and equity concepts in demonstrative projects. The decreasing (grey) curve = indifference curve about the distribution of Net Present Values NPV to rich and poor people based on the Social Welfare Function SWF with ε = 0.5, the decreasing (blue) flatter line = indifference curve about the distribution of NPV to rich and poor people based on the SWF with ε = 0 (Harsanyi efficiency), the decreasing (purple) steeper line = indifference curve about the distribution of NPV to rich and poor people based on relative weights (weighted efficiency discussed in Sect. 4.2.2.5 for MCA), vertical and horizontal (grey) lines = indifference curve about the distribution of NPV to rich and poor people based on the SWF with ε = 0 (Rawls equity). *Notes* Projects A, B, C, D (red points) refer to Table 4.14; A > D = C for Harsanyi efficiency, A = D > C for Rawls equity

$$\text{Atkinson} = 1 - \left[(1/n)\Sigma_i(x_i/x)^{(1-\varepsilon)}\right]^{1/(1-\varepsilon))}$$

With ε the inequality aversion in [0, 1[depicts the concern for inequality from its minimum with ε = 0 to its maximum with ε = 1. The Atkinson index in [0, 1- $n^{(-\varepsilon/(1-\varepsilon))}$], where Atkinson index = 0 represents the maximum equality For projcct A with if ε = 0.5:

$$\text{Atkinson} = 1 - (1/2)[\surd(60/50) + \surd(40/50)]^2 = 0.01$$

Note that the Atkinson index requires additional information from external sources, while the Gini index does not. Moreover, the inequality indexes can be compared in terms of the following properties:

- Weak principle of transfers: it falls (raises) with an income transfer from richer (poorer) to poorer (richer) individuals
- Strong principle of transfers: its change due to income transfers depends only on the distance between individual ranks, not on their location in the income distribution (NO Atkinson)
- Scale invariance: it is invariant to proportional changes of the original incomes
- Transaction invariance: it is invariant to uniform addition or subtractions of the original incomes
- Principle of population: it is invariant to replications of the original population

- Decomposability: if the original population is split into n groups, the total inequality is the sum of the n group inequalities (NO Gini)

Finally, the Atkinson index can characterize individuals in terms of inequality aversions, while the Gini index does not.

In project A and C, the Gini indexes are 0.1 and 0.3, respectively, while the Atkinson indexes if $\varepsilon = 0.5$ in Project A and C are 0.01 and 0.1, respectively. Thus, in terms of inequality, A > C. Note that the Gini index for Project B and D are 0.125 and 0.055, respectively, whereas the Atkinson index for Project B and D are 0.015 and 0.003, respectively.

Whenever the ranking of projects in terms of efficiency is different from the ranking of projects in term of equality, unless it is assumed no concern for equality, an equity criterion is required to compare and possibly combine levels of efficiency and levels of equality.

To do so, two *main* methods have been used: inequality weights IW are discussed in Sect. 4.2.2.5 for MCA; the Social Welfare Function SWF evaluates NPV to maximize the social welfare SW:

$$\mathrm{SWF} = \left[\mathrm{NPV_{rich}}^{(1-\varepsilon)} + \mathrm{NPV_{poor}}^{(1-\varepsilon)}\right]^{(1/(1-\varepsilon))}$$

Note that in the $[\mathrm{NPV_{rich}}, \mathrm{NPV_{poor}}]$ framework, this amounts to reach the highest indifference curve. If $\varepsilon = 0$, then linear indifference curves apply:

$$\mathrm{NPV_{rich}} = \mathrm{NPV_{tot}} - \mathrm{NPV_{poor}}$$

This amounts to the Harsanyi efficiency: thus, A = C > D.

If $\varepsilon = 1$, then Leontiev indifference curves apply:

$$\mathrm{SWF} = \mathrm{Min}\left[\mathrm{NPV_{rich}}, \mathrm{NPV_{poor}}\right]$$

This amounts to the Rawls equity: thus, A = D > C.

If the inequality aversion ε is]0, 1[, the SWF amounts to apply the Atkinson inequality index (Adler, 2016). For example, if $\varepsilon = 0.5$, then A > C = D.

Note that strong sustainability is close to Rawls in terms of resources, while weak sustainability is close to Harsanyi in terms of welfare.

In summary, SWF enables CBA to achieve efficiency EFF, where SWF includes all inequality aversions as specific cases, by spanning from Harsanyi efficiency to Rawls equity. In contrast, inequality weights (i.e., IW discussed for MCA in Sect. 4.2.2.5) would be inadequate for CBA to maximize the total welfare.

4.2.1.6 Monetary Assessment: PA (DR, RC, OP, PC) and UA (HP, TC, CV, CE)

In contexts discussed in previous Sections, all benefits and costs are assumed to be properly evaluated in monetary terms. Indeed, if there is a competitive market of the flow resource X or the flow pollution Y to be estimated in monetary terms, then P* (equilibrium price) is at WTP (willingness to pay from the Marshall demand) – MEC (marginal external costs) = MC (the opportunity cost from the marginal production cost), where WTP represents consumers, MC depicts producers, and MEC represents people affected by an uncompensated externality.

Note that shadows prices for pollution stocks with or without interactions in static and dynamic contexts (i.e., λ and μ, respectively) were obtained by comparing the demand for goods and services producing pollutions Q with the marginal external costs, by assuming MC = 0 for each Q or Y: the MEC for pollution flows without interaction was obtained by comparing MPNB = P – MC. Moreover, shadows prices for pollution flows without interaction in case of taxes or subsidies were obtained by a distortion applied to equilibrium market prices (e.g., $P_{tax} = P\,(1 + tax)$ where P is the shadow price; $P_{sub} = P\,(1 - sub)$ where P is the shadow price). Finally, shadows prices for resource stocks with or without interactions in static and dynamic contexts (i.e., λ and μ, respectively) were obtained by comparing the supply for goods and services using resources with the marginal external costs, by assuming MC = 0 for non-renewable resources. In contrast, in this Section, we refer to a demand for environmental goods X or Y (i.e., a demand for goods and services X or Y net of external costs from producing pollution Y or using resources X) and a supply of environmental goods X or Y (i.e., marginal production costs to reduce pollutions or preserve resources).

Let us assume that the project does not affect the equilibrium price (i.e., the project is marginal).

Thus, the observed equilibrium price P* is an adequate evaluation of X and Y, where P* combines the economic value of X and Y in terms of WTP on the demand curve (e.g., MEC for renewable and non-renewable resources) with economic value of X and Y in terms of opportunity cost on the supply curve (e.g., MC).

Note that the equilibrium quantities X and Y maximize the social welfare (i.e., the sum of producer surplus and consumer surplus net of external costs included in the demand for X or Y), where monetary transfers from consumers to producers are disregarded, since the social welfare is based on the producer surplus (i.e., what producers obtain as the equilibrium price above the minimum they require to produce) and on the consumer surplus (i.e., what consumers are willing to pay WTP above what they pay as the equilibrium price so 0 for the marginal unit).

Let us assume that the project is still marginal, but some distortions make the market non-competitive. Note that goods and services can be defined as non-tradable if the following condition holds:

$$\text{FOB} < \text{domestic price} < \text{CIF}$$

where FOB = Free On Board for export and CIF = Cost, Insurance and Freight for import. In particular, the first inequality (i.e., FOB < domestic price) excludes export, whereas the second inequality (i.e., domestic price < CIF) excludes imports.

Let us focus on non-tradable environmental goods and services. Many market distortions can be considered.

Figure 4.33 depicts the distortion arising from a subsidy on fossil fuel used in agriculture. In particular, WTP for is at 10€ per liter, while farmers pay 5€ per liter due to a subsidy of 5€ per liter: the proper value of fossil fuel as an input in a project is 10€ per liter, since a subsidy is a monetary transfer (i.e., subsidies to farmers are financed by additional taxes on other firms) or the social benefit is represented by the WTP at 10€ per liter on the demand curve or the social cost is represented by the opportunity cost at 10€ per liter on the supply curve without subsidy. Note that a subsidy implies that units from 15 to 20 are produced, although their marginal cost is larger than the willingness to pay for them (i.e., the social loss is 12.5). Thus, this example (i.e., an excess of demand due to a reduction of the equilibirium price) is represented by the forth cell in Table 4.15 (i.e., an input from an additional supply so ignore distortion) and the distortion amounts to 37.5€ due to excess demand (i.e., the area of the trapezoid below the demand curve from the quantity used 20 L at the subsidized price of fuel 5€ to the quantity used 15 L at the price of fuel without subsidy 15€ to depict the added demand).

Note that the price that consumers are willing to pay includes transfers such as taxes and subsidies, while the production cost measures the sacrificed resources so that it does not include transfers such as taxes and subsidies. Moreover, transport or insurance costs must be considered as production costs, while profits should be ignored as monetary transfers. Finally, additional distortions should be applied to public funds (e.g., money comes from taxes or bonds).

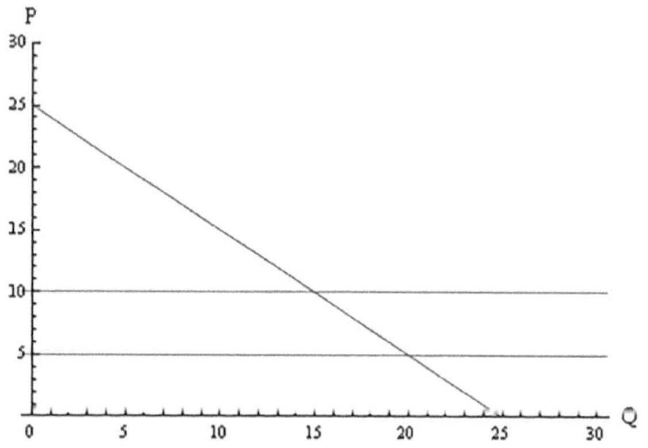

Fig. 4.33 Distortions arising from a subsidy on a fossil fuel. The decreasing (blue) line = demand curve as willingness to pay WTP, horizontal line above = supply curve as Marginal Cost MC, horizontal line below = subsidized supply curve = P – SUB. *Notes* The distortion amounts to (10 + 5) * 5/2 = 37.5 below the demand curve. The social economic value is 10, although the price paid is 5

Table 4.15 Reference to demand or supply curves to evaluate non-tradable inputs or outputs

	Demand curve	Supply curve
Output-benefit	It satisfies additional demand KEEP DISTORTIONS BUT (−tax + subsidy if corrective)	It satisfies existent demand IGNORE DISTORTIONS AND (−tax + subsidy if corrective)
Input-cost	Originating from an alternative market use KEEP DISTORTIONS AND (−tax + subsidy if corrective)	Originating from additional supply IGNORE DISTORTIONS BUT (−tax + subsidy if corrective)

Figure 4.34 depicts the distortion arising from a quota on bio fuel used in industry. In particular, firms using bio fuel pay 20€ per ton, while farmers producing bio fuel could supply it at 5€ per ton: the proper value of fossil fuel as an input in a project is 20€ per ton, since the social benefit is represented by the WTP at 20€ per ton on the demand curve, although the social cost is represented by the opportunity cost at 5€ per ton on the supply curve without quota. Note that a quota implies that units from 5 to 12.5 are *not* produced, although their marginal cost is smaller than the willingness to pay for them (i.e., the social loss is 56.25). Thus, this example (i.e., a lack of supply due to a reduction of the equilibirium quantity) is represented by the third cell in Table 4.15 (i.e., an input from an alternative market use so keep distortion) and the distortion amounts to 121.875€ due to shortage demand (i.e., the area of the trapezoid below the demand curve from the quantity used 5 ton at the quote price of bio fuel 20€ to the quantity used 12.5 ton at the price of bio fuel without quota 12.5€ to depict the subtracted demand).

Table 4.15 summarizes the solutions suggested in the literature to cope with market distortions in non-tradable goods and services for marginal projects.

In summary, the theoretical and practical solution is the appropriate reference to the demand curve (i.e., WTP) *or* the supply curve (i.e., MC): indeed, they are different due to the market distortions.

Let us focus on tradable environmental goods and services.

A distortion can be due to devaluation in case of fixed exchange rates (to increase exports and reduce imports).

Figure 4.35 graphically calculates this distortion in terms of the area to the left of the curves of demand for foreign currency (i.e., money used to buy imports) and supply for foreign currency (i.e., money obtained from selling exports). In particular, in terms of domestic consumption (UNIDO), the official exchange rate is OER = 2 = 1/0.5 with 0.5 on the supply curve (i.e., an increase in supply of \$/year is due to a reduction of domestic consumption and an increase in export); in terms of money demand (OECD), the shadow exchange rate is SER = 0.66 = 1/1.5 with 1.5 on the demand curve (i.e., a reduction in demand for \$/year is due to a reduction of import). Note that the distortion amounts to 0.625€ (OECD) (i.e., the area of the trapezoid

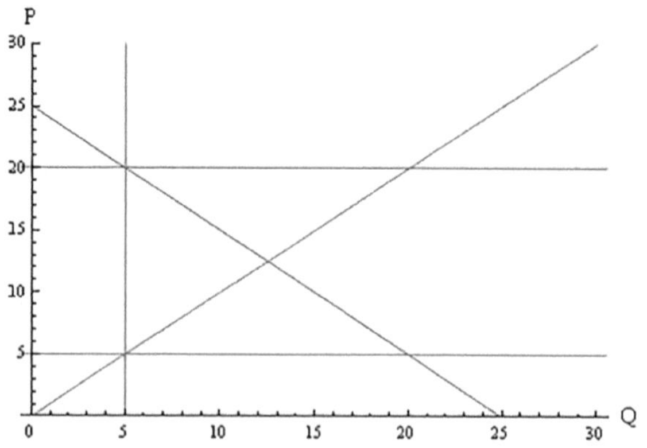

Fig. 4.34 Distortions arising from a quota on bio fuel. The decreasing (blue) line = demand curve as willingness to pay WTP, the increasing (purple) line = supply curve as Marginal Cost MC, the supply curve with a quota becomes fixed at Q = 5 for Q ≥ 5. *Notes* The distortion amounts to (20 + 12.5) * 7.5/2 = 121.875 below the demand curve. The social economic value is 5, although the price paid is 20

right to the demand curve from the quantity demanded for imports 0.5$/year at the devaluated exchange rate 0.5€/$ to the quantity demanded for imports 1$/year at the exchange rate without devaluation 1€/$ to depict the reduced imports). Alternatively, the distortion amounts to 0.625€ (UNIDO) (i.e., the area of the trapezoid right to the supply curve from the quantity supplied for exports 1.5$/year at the devaluated exchange rate 0.5€/$ to the quantity supplied for exports 1$/year at the exchange rate without devaluation 1€/$ to depict the increase exports). In other words, an imported input paid in $ should be evaluated in terms of SER = 0.66€/$.

A distortion can be due to tariffs (to reduce import) and subsidies (to increase exports) in the context of flexible exchange rates. Figure 4.36 graphically calculates this distortion in terms of the area to the left of the curves of demand and supply for foreign currency. In particular, if SU is the supply of foreign currency in terms of the OER, if DE are the demand of foreign currency in terms of the OER, if SUt is the supply of foreign currency affected by the export subsidy, and DEt is the demand of foreign currency affected by the import tariff:

$$\mathrm{SU} = \mathrm{aa} + \mathrm{bb} * (1 * \mathrm{OER})$$
$$\mathrm{DE} = \mathrm{cc} - \mathrm{dd}(1 * \mathrm{OER})$$
$$\mathrm{DEt} = \mathrm{DE}/(1 + \mathrm{nt})$$
$$\mathrm{SUt} = \mathrm{SU} * (1 + \mathrm{ns})$$

then, the international market prices as shadow-prices

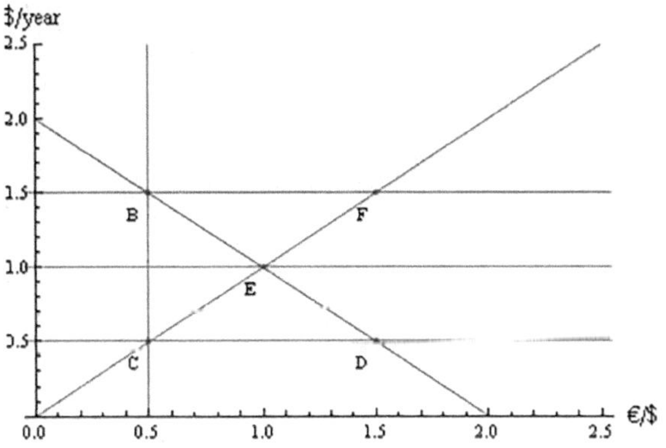

Fig. 4.35 Distortions arising from a devaluation of a fixed exchange rate. The decreasing (blue) line = demand for foreign currency, the increasing (purple) line = supply of foreign currency. *Notes* OER = 0.5 instead of the equilibrium exchange rate at 1. Distortion = area to the left of the demand curve (blue) (i.e., 0.5, C, D, E, 1) = 0.625 *or* area to the left of the supply curve (purple) (i.e., 1.5, B, F, E, 1) = 0.625. SER = 0.66

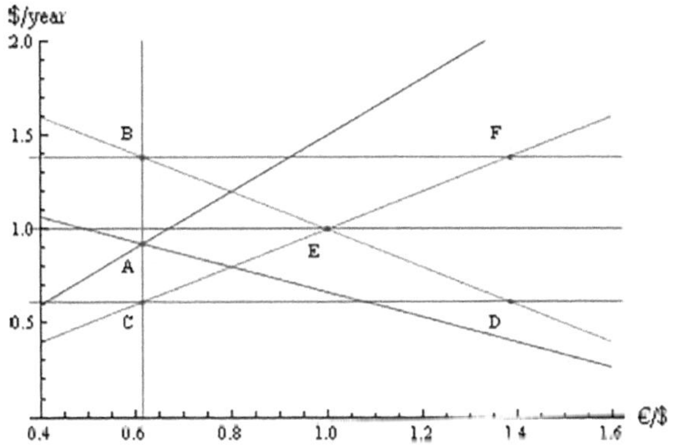

Fig. 4.36 Distortions arising from tariffs t and subsidies s with flexible exchange rates. The decreasing (grey) line above = demand for foreign currency without tariff, the decreasing (purple) line below = demand for foreign currency with tariff, the increasing (green) line below = supply of foreign currency without subsidy, the increasing (blue) line above = supply of foreign currency with subsidy. *Notes* OER = 1. A tariff t of 50% on imports and a subsidy s of 50% of exports: aa = 0, bb = 1, cc = 2, dd = 1. If import requires both foreign currency diverted from other imports (at a rate M/(M + X)) and foreign currency from additional exports (at a rate X/(M + X)), total distortion = (1/2) area to the left of the demand curve (grey) (i.e., 0.615, C, D, E, 1) + (1/2) area to the left of the supply curve (green) (i.e., 1.385, B, F, E, 1) = 0.5 * 0.49 + 0.5 * 0.495 = 0.918. SER = 0.66

$$\text{SER} = \text{OER}/[(\text{M}(1+\text{nt}))/(\text{X}+\text{M}) + (\text{X}(1+\text{ns}))/(\text{X}+\text{M})]$$
$$\text{OER} = \text{SER}(1+\text{FEP})$$
$$\text{OER} < \text{SER iff FEP} = \text{OER}/\text{SER} - 1 < 0$$

Note that, if import requires both foreign currency diverted from other imports (at a rate M/(M + X)) and foreign currency from additional exports (at a rate X/(M + X)), the distortion amounts to 0.918€ as a sum of 0.459€ (i.e., the area of the trapezoid right to the demand curve from the quantity demanded for imports 1.385$/year at the devaluated exchange rate 0.615€/$ to the quantity demanded for imports 1$/year at the exchange rate without devaluation 1€/$ to depict the reduced imports) together with 0.459€ (i.e., the area of the trapezoid right to the supply curve from the quantity supplied for exports 1.385$/year at the devaluated exchange rate 0.615€/$ to the quantity supplied for exports 1$/year at the exchange rate without devaluation 1€/$ to depict the increase exports). In other words, an imported input paid in $ should be evaluated in terms of SER = 1/(0.5 * 1.5 + 0.5 * 1.5) = 0.66€/$.

In summary, the theoretical and practical solution is the appropriate application of SER: indeed, SER is different from OER due to the market distortions.

Let us assume that the project is *non-marginal* (i.e., the project affects the equilibrium price). This implies that the change in surplus is not measured by the equilibrium price (i.e., Willingness To Accept WTA ≠ Willingness to Pay WTP ≠ the equilibrium price).

Note that projects involving tradable goods and services (e.g., they import inputs and export outputs) can be assumed to be marginal (i.e., they are unlikely to affect exchange rates).

Let us calculate the Equivalence Variance and the Compensation Variance for the following consumer problem:

$$\text{Marshall demand from Max } U = Q_1^{\,2} Q_2 \text{ subject to } P_1\, Q_1 + P_2\, Q_2 \le M$$
$$\text{Hicks demand from Min } P_1\, Q_1 + P_2\, Q_2 \text{ subject to } U \ge \bar{U}$$

Let us assume that P_1 and M are fixed with $P_1 = 1$ and $M = 150$. Let us call $Q_2 = Q$. Let us assume that P_2 can be $P_2 = 2.5$ (solutions A and C) or $P'_2 = 1$ (solutions B and D). Figure 4.37 shows all four solutions. Thus, in A (Marshall demand), the solution is $Q_1 = 100$ and $Q_2 = 50$, U(A) = 200,000; the indifference curve in A is $M = 200\sqrt{5/Q}$; the budget constraint in A is $M = 150 - 2.5\,Q$. The solution in B (Hicks demand) is $Q_1 = 36.84$ and $Q_2 = 73.68$; the budget constraint in B is $M = 110.52 - Q$. In D (Marshall demand), the solution is $Q_1 = 100$ and $Q_2 = 20$, U(D) = 500,000; the indifference curve in D is $M = 500\sqrt{5/Q}$; the budget constraint in D is $M = 150 - Q$. The solution in C (Hicks demand) is $Q_1 = 27.14$ and $Q_2 = 135.72$; the budget constraint in C is $M = 203.57 - 2.5\,Q$.

Let us focus on a damage due to an increase in price P_2 (from D to A through B) (i.e., case 2) and on a benefit due to a decrease in price P_2 (from A to D through B). In particular, the equivalence value is measured by the change in M from D to B; the compensation value is measured by the change in M from A to B. In the example,

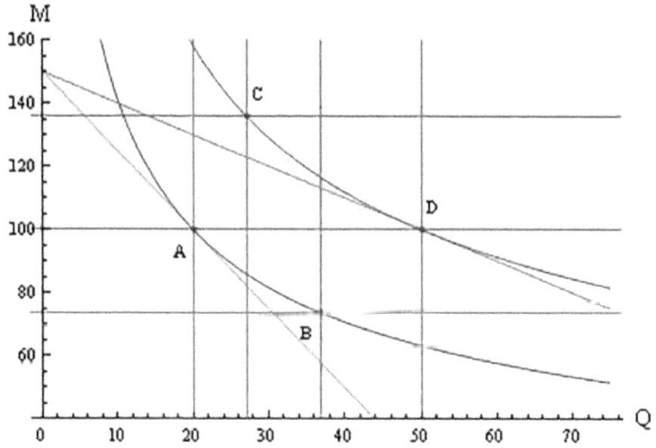

Fig. 4.37 The Hicks compensation value and equivalence value. The decreasing (green) line above = budget constraint at the smaller P_2, the decreasing (grey) line below = budget constraint at the larger P_2, the decreasing (purple) curve = indifference curve at the smaller P_2, the decreasing (blue) curve = indifference curve at the larger P_2. Abbreviations: M = available money, Q = quantity demanded. *Notes* $P_1 = 1$, $P_2 = 1$ and $P_2' = 2.5$, M = 150. The Hicks compensation value means new prices and old utility; the Hicks equivalence value means old prices and new utility. In case 2 in Table 4.16 from D to B (i.e., citizens pay to avoid the construction of a waste landfill), equivalence value = 40; in case 3 in Table 4.16 from A to B (i.e., citizens pay to construct a waste incinerator), compensation value = 40

150 in D − 110.52 in B (i.e., WTP as an equivalence value); 150 in A − 110.52 in B (i.e., WTP as a compensation value).

Figure 4.38 shows the underestimation and overestimation of surplus by using the Marshal demand instead of the Hicks demand. In the example, from A to D, Marshal demand overestimation is (50 − 36.84)(5 − 2)/2 = 19.74; from D to A, Marshal demand underestimation is (27.14 − 20)(5 − 2)/2 = 10.71.

Note that the Marshall demand is observable, whereas the Hicks demand is unobservable. Moreover, if income effect = 0, then VC = VE. Finally, WTA might be larger than WTP due also to loss aversion.

Table 4.16 provides the four theoretical cases where to apply which method. For concreteness, think of citizens who have the right to rely on waste incinerator nearby (and to the related reduction in energy price), but it is not constructed by the municipality (case 1): how much money are they willing to accept as compensation? Think of citizens who do *not* have the right to refute a waste landfill nearby (case 2): how much money are they willing to pay for it not be constructed? Think of citizens who do *not* have the right to rely on waste incinerator nearby (case 3): how much money are they willing to pay for it to be constructed (and enjoy the related reduction in energy price)? Think of citizens who have the right to refute a waste landfill nearby, but it is constructed by the municipality (case 4): how much money are they willing to accept as compensation?

Note that monetary assessments in the theoretical cases 1 and 4 can be linked to changes in prices in a related market (i.e., they can be practically estimated by both revealed and stated preferences within utility approaches discussed below),

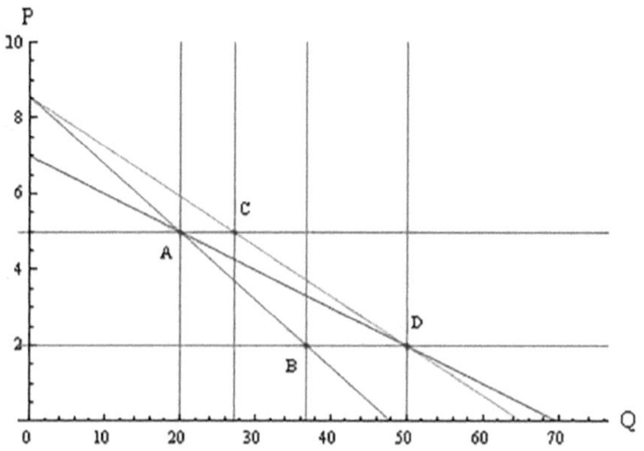

Fig. 4.38 The Hicks and Marshall demand functions. The decreasing (blue) line through A and D = Marshal demand, the decreasing (grey) line through C and D = the Hicks demand based on compensation value, the decreasing (purple) line through A and B = the Hicks demand (equivalence value). *Notes* Marshal demand overestimates the change in surplus in case 2 (from D to A) with respect to the (equivalence value) Hicks demand; Marshal demand underestimates the change in surplus in case 3 (from A to D) with respect to the (compensation value) Hicks demand

Table 4.16 Assessment methods in alternative circumstances

	Cases	Circumstances	Assessment methods	Compared solutions	Monetary assessments
Equivalence variance No change occurs (Old prices)	1	Benefit Right to a change	WTA	C–A	203–150
	2	Damage No right to reject a change	WTP	D–B	150–110
Compensation variance A change occurs (New prices)	3	Benefit No right to have a change	WTP	A–B	150–110
	4	Damage Right to avoid a change	WTA	D–C	203–150

Abbreviations: WTP = willingness to pay, WTA = willingness to accept. *Notes* Compared solutions and monetary assessments refer to points in Fig. 4.38

whereas monetary assessments in cases 2 and 3 refer to presence or absence of a bad (i.e., waste landfill) or a good (e.g., waste incinerator) (i.e., they can be practically estimated only by stated preferences within utility approaches discussed below).

In summary, the *theoretical* solution is the appropriate estimation of the Hicks demand curve for resources X or pollution Y on in a related market: indeed, the Marshal demand is different from the Hicks demand due to the income effect arising from a change in the equilibrium price. However, the Hicks demand is *practically* more complicated to be estimated than the Marshal demand curve, since it is not observed. Note that Contingent Valuation CV and Choice Experiment CE discussed below deal with cases where there are no related markets.

Let us assume that there *no* is a market for the resource X or the pollution Y, where the project can be both marginal or not and the related market can both exist or not. Two main approaches have been adopted in the literature: either a market supply is simulated or a market demand is simulated.

Let us focus on *production approaches* PA, by referring to a simulated supply curve (Diaz-Balteiro & Romero, 2008; Farnsworth et al., 2015; Fitter, 2013; Kuosmanen & Kortelainen, 2007; Scrieciu et al., 2013). For concreteness, let us refer to a case where pollution from agricultural fertilizers and pesticides (flow) negatively affects water quality in rivers (flow) and damages coral reefs (stock) in the sea that negatively impacts on tourism revenues (flow). Note that this example (i.e., tourism activity {coral reef [water quality (agricultural activity)]} to depict it in a formula) refers to non marginal projects with related markets.

Four main methods have been applied in the literature:

- Dose response DR (after the damage): monetary impact of an increase in pollution or a reduction of resources on related economic activities. In the previous case, the value of the coral reef amounts to the loss of tourism activity in GDP
- Replacement costs RC (after the damage): the cost to repair the damage (e.g., reduce pollution or increase resources) to achieve the previous economic activities. In the previous case, the value of the coral reef amounts to the cost to repair it or to replace it with an artificial coral reef or to clean water quality to achieve the same tourism activity in GDP
- Opportunity cost OC (before the damage): resources in terms of missed production to preserve them. In the previous case, the value of the coral reef amounts to the loss of agricultural activity (i.e., the reduce production in monetary terms) if agricultural fertilizers and pesticides are not used
- Preventive cost PC (before the damage): resources in terms of production cost to preserve them. In the previous case, the value of the coral reef amounts to the cost to move from traditional to biological agriculture to achieve to achieve the same agriculture activity in GDP

In summary, within PA, OC and PC are adequate to estimate X and Y if either flows or stocks are relevant: indeed, they estimate the cost to avoid a damage. In contrast, DR and RC are inadequate to estimate X and Y if stocks are relevant: indeed, they estimate the short-run market losses of the experienced damage.

Let us focus on *utility approaches* UA, by referring to a simulated demand curve (Hajkowicz, 2007; Kneese, 1986). Two main methods have been applied in the literature: revealed preferences (i.e., X and Y are estimated by referring to related markets where real consumption decisions can be observed); stated preferences (i.e., X and Y are estimated by referring to a simulated market where hypothetical consumption decisions can be observed).

Within the revealed preferences, let us refer to the two main methods: hedonic prices and travel costs.

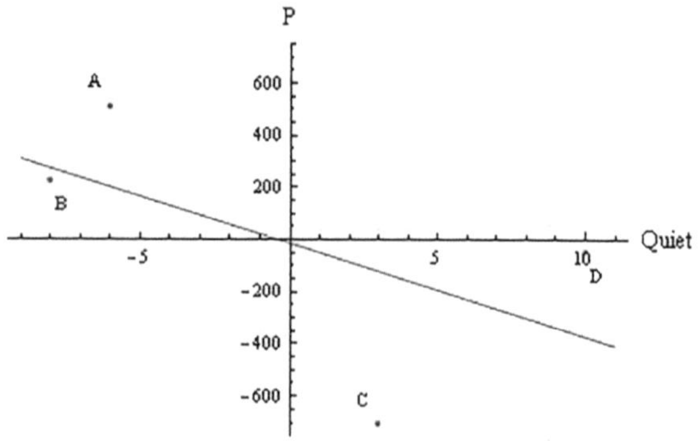

Fig. 4.39 The hedonic price framework. The decreasing (blue) line = linear interpolation of observed couples of quiet levels in dB and flat prices in €/m² observed in points A, B, C, D. *Notes* The evaluated willingness to pay for quiet is bb = 35.758

Hedonic prices HP, based on the Marshall demand, estimate the relevant environmental part of the equilibrium price observed in a related competitive market (e.g., the value of the quiet characteristic of a flat is assumed to be included in its market price). Note that this example refers to marginal projects.

Figure 4.39 shows the statistical estimation of the demand curve for quiet from four observations of prices P in €/m² of flats in different areas characterised by different quiet levels in dB:

$$P = aa - bb\ Quiet + ee$$
$$|dP/dQuiet| = bb$$

Or, by assuming a unit change in quiet,

$$dP = bb\ dQuiet = bb$$

Thus, bb is the estimated value of one unit of quiet: it must be multiplied by the average size of flats in the area to measure the value of quiet in the area.

Travel costs TC, based on the Marshall demand, estimate the complementary expenditures (e.g., costs of fuels or hotels) that individuals are willing to pay to enjoy the environmental goods or services to be evaluated (i.e., it is assumed that the value that consumers attach to X and Y is at least equal to the money they pay to enjoy it). Note that this example refers to non-marginal projects.

Figure 4.40 shows the expenditures per visit of two different consumers that enjoy a natural park different times per year N by paying different amounts of money per visit C:

$$C = aa - bb\ N + ee$$
$$|dC/dN| = bb$$

Or, by assuming a single visit,

$$dC = bb\ dN = bb$$

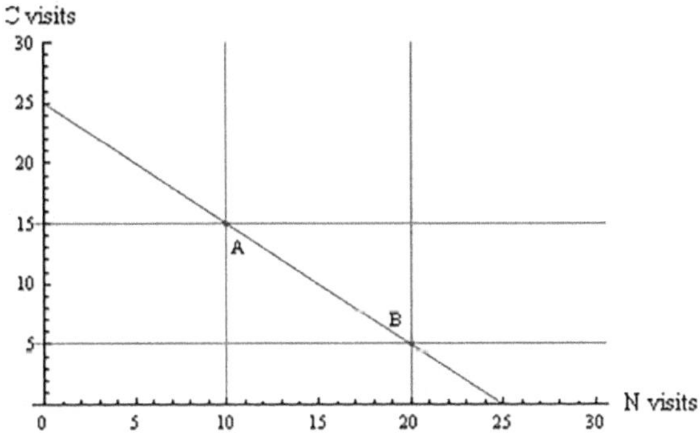

Fig. 4.40 The travel cost framework. The decreasing (blue) line = linear interpolations of observed couples of number of visits and costs per visit for two individuals in points A and B. *Notes* Surplus A = 50 = (10 * 10)/2, surplus B = 200 = (20 * 20)/2, AC = 10, average N = 15, number of people = 2, total surplus [(25 – 10) * 15)/2] * 2 = 225: bb = −1

Thus, bb is the estimated value of one visit to the natural park: it must be multiplied by the expected total visits to measure the value of the natural park.

In summary, within the revealed preferences, HP is adequate to estimate X and Y if either flows or stocks are relevant: indeed, it estimates a long-run use value. In contrast, TC is inadequate to estimate X and Y if stocks are relevant: indeed, it estimates a short-run use value. Note that there are no theoretical issues and practical problems to be solved.

Within the stated preferences, let us refer to the two main methods (Faccioli et al., 2016): Contingent valuation and Choice experiment.

Contingent valuation CV, based on the Hicks demand, estimates the WTP or WTA of involved individuals for the introduction or the elimination of an environmental good or service (i.e., it is assumed that the value interviewees attach to resource X and pollution Y is at least equal to the money they are willing to pay or accept) (Venkatachalam, 2004). The Contingent Valuation for single-bounded dichotomous choices (i.e., I refer to yes or no answers: alternatively, one could use open-ended answers) in its general model can be formulated as follows. Individuals deterministically maximise their known utility, by comparing the utility level with the presence of some additional environmental good Q_1 (i.e., the case of disutility associated to the environmental bad is neglected) together with the payment of C, on one side, and, on the other side, the utility level with the absence of some additional environmental good Q_0 together with the whole original income M (i.e., the utility from money is considered):

$$U(Q_1, M - C) \geq U(Q_0, M)$$

However, researchers do not know some features of the individuals' utility and the previous inequality based on the Hicks compensation value becomes stochastic (i.e., I use the random utility model: alternatively, one could use the random expenditure function):

$$V(Q_1, M - C, \varepsilon_1) \geq V(Q_0, M, \varepsilon_0)$$

where ε_0 and ε_1 are stochastic variables.

Thus, the probability that the individual is willingness to pay C is given by:

$$\text{Prob(yes to pay C)} = \text{Prob}\left[V(Q_1, M - C, \varepsilon_1) \geq V(Q_0, M, \varepsilon_0)\right]$$

Let us refer to a Box-Cox utility function with $\lambda = 1$ (i.e., a linear indirect utility function):

$$V_1 = \text{aa}\, Q_1 + \text{bb}(M - C) + \varepsilon_1 \geq V_0 = \text{aa}\, Q_0 + \text{bb}\, M + \varepsilon_0$$

where aa and bb are positive parameters. Let $Q_1 - Q_0 = 1$ (i.e., the individual compares the presence with the absence of the environmental good). In other words, the reference is to non-marginal projects. Let us assume that there is no a tight budget constraint (i.e., $C < M$). Thus, the *maximum* amount that the individual is willingness to pay for this environmental good is given by max $C = (\text{aa} + (\varepsilon 1 - \varepsilon 0))/\text{bb}$. Note that the linear model assumes that the elasticity of C with respect to M is 0, while bb eliminates the currency dimension from the estimations of WTP (i.e., it scales the estimations vertically).

Let $\varepsilon = \varepsilon_1 - \varepsilon_0$ (i.e., the stochastic difference in utility with and without the environmental good) has a standardised normal distribution (e.g., $N[\mu, \sigma]$) (Fig. 4.41) (alternatively, one could use the Turnbull estimator within non-parametric models). Note that $E(\varepsilon)$ is the mean value that individuals attach to the environmental good (i.e., it scales the estimations horizontally).

Figure 4.42 shows the stochastic cumulative difference of utility with and without the environmental good, while Fig. 4.43 shows the probability individuals are willingness to pay C.

Thus, the probability that individuals are willingness to pay C is given by:

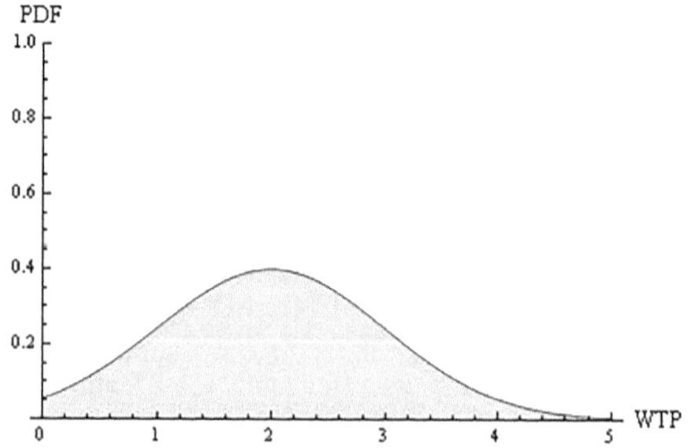

Fig. 4.41 The stochastic difference in utility. The probability density function PDF of the willingness to pay for one unit of an environmental good if aa = 2, bb = 1 and $\varepsilon \sim$ Normal[2, 1]

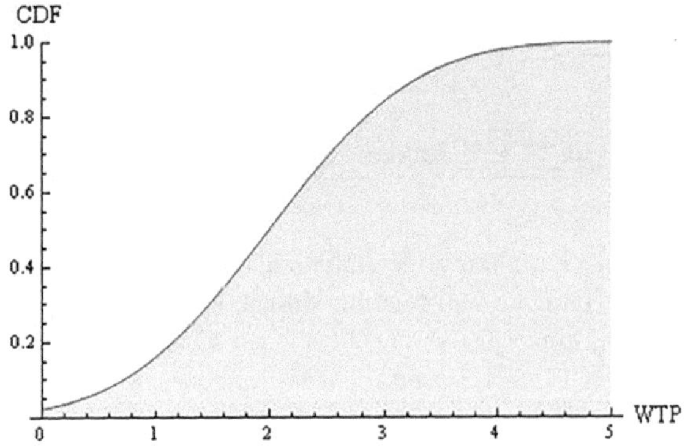

Fig. 4.42 The stochastic cumulative difference in utility. The cumulative density function CDF of the willingness to pay for one unit of an environmental good if aa = 2, bb = 1 and ε ~ Normal[2, 1]

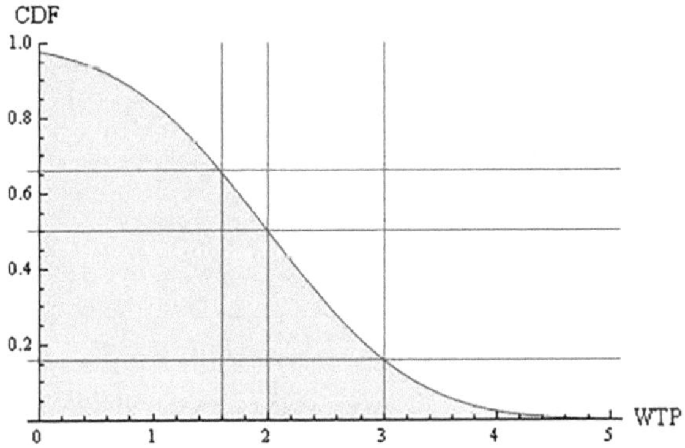

Fig. 4.43 Mean and median willingness to pay within contingent valuation. The cumulative density function CDF of the willingness to pay for one unit of an environmental good if aa = 2, bb = 1 and ε ~ Normal[2, 1]. *Notes* Mean WTP = 2, median 50% WTP = 2, median 66% WTP = 1.587, Prob(yes to C = 3) = 1 − Φ(C = 3) = 0.287

$$\text{Prob(yes)} = \Phi\left(\frac{\mu - C}{\sigma}\right) - 1 - \Phi\left(\frac{C - \mu}{\sigma}\right)$$

with $C \leq \mu$, or, equivalently,

$$\text{Prob(yes)} = 1 - \Phi(\text{aa} - \text{bb C})$$

where aa = μ/σ and bb = $1/\sigma$ are the parameters values estimated by the probit model, with the estimated coefficient bb as negative, since it represents the marginal utility of M.

Note that Cameron (1988) suggests to use a different G_ε for each individual instead of introducing a linear indirect utility function as Hanemann (1984), Moreover, more general Box-Cox utility functions make more difficult the estimation of G_ε, although the linear utility function assumes an income elasticity at 0. Finally, a Weibull distribution together with a logit model could be used instead of a normal distribution together with a probit model.

Thus, the mean and median maximum amount that the agent is willing to pay are:

$$\text{Mean WTP} = \int_0^\infty (1 - G_\varepsilon(\mathrm{C}))\mathrm{dC} = -\frac{\mathrm{aa}}{\mathrm{bb}}$$

$$\frac{\mathrm{aa} + 50\%\ \text{Median WTP}}{\mathrm{bb}} = 0.5 \text{ or } \frac{\mathrm{aa} + 66\%\ \text{Median WTP}}{\mathrm{bb}} = 0.66$$

Note that alternative statistical distributions (e.g., Weibull) do not significantly affect the mean and median values. Moreover, the Mean WTP is smaller if E(ε) is smaller, since G_ε is lower for each C, where the maximum C is given by [aa + E(ε)]/bb. Finally, I used the non-truncated distribution.

The Choice Experiment in its general model can be formulated as follows. Individuals are asked to choose between alternative goods which are described in terms of their attributes, one of which could be its price (or some proxy of its price). The underlying utility function of individual i is given by:

$$U_{ij} = \mathrm{U}(x_j, p_j)$$

$$U_{ik} = \mathrm{U}(x_k, p_k)$$

where x_j and x_k are vectors of attributes describing j and k, while p_j and p_k are the prices or costs associated with each of the alternatives. Individual i will choose alternative j over alternative k if and only if $U_{ij} > U_{ik}$. If $U_{ij} = U_{ik}$ then the individual is indifferent between the two alternatives (i.e., if $U_{ij} = U_{ik} = 0$ then the individual receives no satisfaction form either alternative).

Let the utility functions be partitioned into two components:

$$U_{ij} = \mathrm{V}(x_j, p_j) + \varepsilon(x_j, p_j)$$

$$U_{ik} = \mathrm{V}(x_k, p_k) + \varepsilon(x_k, p_k)$$

where V(.) is deterministic and observable (i.e., it is the so-called indirect utility function), while ε(.) is random and unobservable. Thus, the probability that individual i will choose alternative j over alternative k is given by:

$$\mathrm{Prob}_i(\mathrm{j}|\mathrm{x}) = \mathrm{Prob}(V_{ij} + \varepsilon_{ij} > V_{ik} + \varepsilon_{ik})$$

where x is the complete set of alternatives.

Let assume that the error terms are independently and identically distributed (e.g., Gumbel-distributed): this assumptions entails the property of independence from irrelevant alternatives. Thus:

$$\mathrm{Prob}_i(\mathrm{j}|\mathrm{x}) = \exp(\omega V_{ij}) / \sum_x \exp(\omega V_{ix})$$

where ω is a scale parameter which is inversely proportional to the standard deviation of the error distribution: it is typically assumed $\omega = 1$, implying a constant error variance. Let

$$V_{ij} = \varphi + \text{bb}\, p_j + \text{aa}'\, x_{ij}$$

$$V_{ik} = \varphi + \text{bb}\, p_k + \text{aa}'\, x_{ik}$$

Therefore

$$\text{Prob}_i(\text{j}|\text{x}) = \exp\left(\varphi + \text{bb}\, P_j + \text{aa}'\, x_{ij}\right) / \sum_x \exp\left(\varphi + \text{bb}\, P_x + \text{aa}'\, x_{ix}\right)$$

Let us assume that the error terms are distributed with an extreme-value distribution (i.e., a standardised logistic distribution): this implies that the probability of any particular alternative being chosen as the most preferred can be expressed in terms of the logistic distribution, which results in the so-called conditional logit model.

The implicit WTP for any attribute is

$$WTP_x = -\frac{aa\ x}{bb}$$

where bb is the parameter estimate of the price variable P and aa is the parameter estimate of the specific attribute x. Note that CE estimates the marginal contribution of each attribute to the overall value of the environmental good (i.e., the estimated constant matters). Moreover, CV is simpler than CE, since it focuses on a single attribute (i.e., presence vs. absence of the environmental good). Finally, CE estimates the implicit price of an attribute in monetary terms, under the assumption that attributes are discrete variables and the marginal utility of income is constant.

Figure 4.44 depicts mean and median WTP within CE.

However, within the stated preferences, apart from some practical warnings which can be properly tackled (e.g., strategic bias, starting point bias, budget constraint bias,

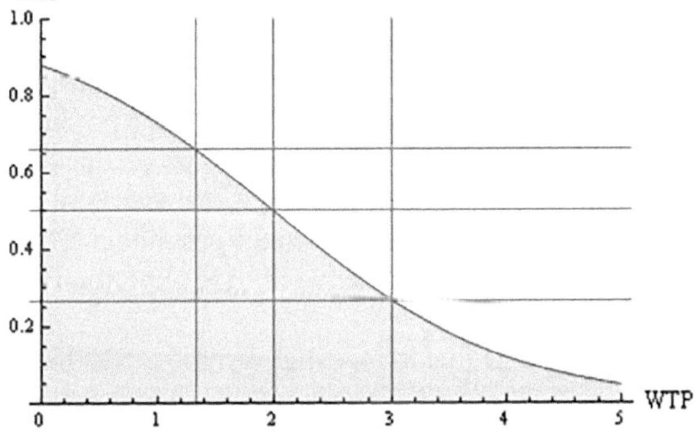

Fig. 4.44 Mean and median willingness to pay within choice experiment. The cumulative density function CDF of the willingness to pay for one unit of an environmental good if aa = 2, bb = 1 and ε ~ Logistic[2, 1]. *Notes* Mean WTP = 2.126, median 50% WTP = 2, median 66% WTP = 1.336, Prob(yes to C = 3) = 1 − Φ(C = 3) = 0.268

frame effect, embedding effect, scope effect incentive compatibility such as payment vehicle, uncertainty about actual costs, decisiveness of responses) (Ahlheim, 1998; Carson et al., 2001; Schläpfer & Brauer, 2007; Schläpfer & Getzner, 2020; Schläpfer et al., 2004; Whitehead, 2016), there are both theoretical issues (i.e., what is vs. what should be estimated) and practical problems (i.e., assumed vs. verified hypotheses). In particular, some assumptions of the estimated model could not hold and data might not be available to check them (e.g., as for CV, commensurability of values in Beckerman & Pasek, 1997; endogeneity of values in Lyssenko and Martinez-Espineira, 2012; non-constant marginal utility of income in Medin et al., 2001; cognitive limitations such as anchor effect or scale effect in Schläpfer & Schmitt, 2007; as for CE, ignored attributes in Carlsson et al., 2010; additivity bias in Dachary-Bernard & Rambonilaza, 2012; heterogeneity in price sensitivity in Giergiczny et al., 2012; non-constant values in Scheufele & Bennett, 2012; overevalauation of economic and underevalaution of ecological impacts in Prato, 1999). Similarly, some misunderstandings with respondents could arise and time might not enough to clarify them (e.g., as for CV, instead of estimating the economic values of the environment, it is estimated responsibility in Blamey, 1998; altruism in Johansson, 1992; moral satisfaction in Kahneman & Knetsch, 1992; duty in Walsh et al., 1990; goods other than natural resources in Madariaga & McConnell, 1987).

Thus, even if all these theoretical and practical problems are solved, CV and CE estimate direct and indirect use values, by missing non-use values (i.e., existence, bequest, option values related to stocks STO) (Common et al., 1997; Plottu & Plottu, 2007; Stoeckl et al., 2018) and social values (i.e., values arising from interactions REL between humans and non-humans) (Chee, 2004; Groeneveld, 2020), by deliberatively disregarding distributional issues if the total WTP is the sum of individual WTP (Price, 2000), and by neglecting the institutional impacts on value formation over time, if preferences are assumed to be fixed (Vatn, 2004). For example, WTP is inadequate for climate change mitigation (Krupnik & McLaughlin, 2012).

In summary, within the stated preferences, CV and CE are inadequate to estimate X and Y if either interactions or stocks are relevant: indeed, they estimate a short-run use value. In other words, if stocks STO and interactions REL are relevant, the focus on the market as the only institution which provides the compose value as the equilibrium price might be reductive (i.e., non-use and social values are missed), where REL refer also to interactions between humans and non-humans.

Let us finish this Section with a simple context where all benefits and costs characterizing a single project are properly evaluated except for two items. For concreteness, let us refer to a local development project that produces a NPV to the reference group of stakeholders of 40€. However, this project also involves a wetland degradation that could be replaced elsewhere at the total cost of K€, while the same amount of money invested elsewhere would generate a NPV to another reference group of stakeholders of 15€. In other words, the opportunity cost of the wetland preservation is 25€).

Let us assume that the local government has three alternative options:

(i) To preserve the wetland intact, by stopping the project implementation

Fig. 4.45 The threshold analysis framework. Minimum Benefits B and Costs C which identify alternative scenarios. *Notes* Choose option (i) if B > 25 and C > 25; choose option (ii) if B < 25 and B < C (trapezoid to the right of the blue line); choose option (iii) if C < 25 and B > C (trapezoid above the blue line)

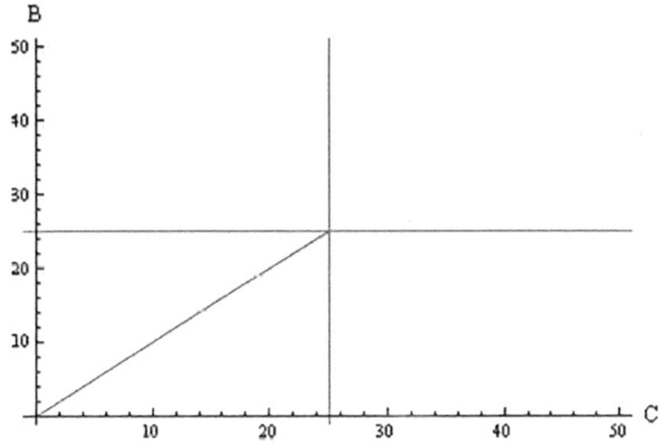

(ii) To implement the project, without replacing the wetland elsewhere
(iii) To implement the project, by replacing the wetland elsewhere

The choice depends on the value attached to the wetland ecosystems B.

Figure 4.45 summarises the relevant scenarios for choosing to implement or not to implement this project (i.e., identifying the values to attach to the two items to justify it implementation in terms of a positive NPV) in the framework [C, B].

$$\text{Thus, Choose (i) if (i)} > \text{(ii) B} > 25 \text{ (i.e., B} > 40 - 15) \text{ and (i)} > \text{(iii) C} > 25 \text{ (i.e., B} > \text{B} + 40 - 15 - \text{C)}$$
$$\text{Choose (ii) if (ii)} > \text{(i) B} < 25 \text{ and (ii)} > \text{(iii)B} < \text{C (i.e.,} 40 - 15 > 40 - 15 + \text{B} - \text{C)}$$
$$\text{Choose (iii) if (iii)} > \text{(i) C} < 25 \text{ and (iii)} > \text{(ii) B} > \text{C}$$

This method is called Threshold Analysis.

To summarise this Section, if stocks STO are relevant, within UA, HP is adequate, provided the related market refers to long-run estimates (e.g., prices of flats): indeed, HP estimates the expected benefits or costs from the environment in the long-run. In contrast, TC, CV, CE are inadequate to estimate X and Y: indeed, they estimate the WTP or WTA for the environment in the short-run. Within PA, OC and PC are adequate, while DR and RC are inadequate. The same reasoning applies if interactions REL are relevant. Thus, within CBA, stocks STO or interactions REL should be combined with OC or PC in a dynamic model.

4.2.1.7 Monetary Life-Cycle Assessment from LCA = EM or EX or RCM or ERM

Life-cycle assessment LCA can be defined as an holistic method that systematically analyses the full range of effects associated with all stages of a product's life (from creation to disposal) in terms of energy used, GHG emissions, materials used, to support the development of business strategies and to improve product and process designs (Fauzi et al., 2019).

In particular, Emergy EM measures the energy directly or indirectly used to produce and consume a given good or service (Pollok et al., 2021); Exergy EX measures the energy left after the production process and the consumption life (Traverso et al., 2018); the recycle content materials RCM measures the amount of materials, used to produce and consume a given good or service, that arise from recycling (Costa et al., 2019); the end-of-life recycling materials ELRM measures the amount of materials that can be recycled after the production process and the consumption life (Suhariyanto et al., 2017).

Table 4.17 characterise LCA as a combination of EM, EX, RCM and ELRM.

Note that LCA adopts a methods close to CBA, by translating different indexes into a single unit (i.e. this is not often a monetary index, but can be easily transformed into a monetary value by using the competitive market price of energy, emissions or materials), although its logics is close to strong sustainability, by minimising impacts

Table 4.17 The main features of life-cycle assessment

		EM	EX	RCM	ELRM	LCA
Focus	Product/technology/user		V		V	V
	Producer/organisation/donor	V		V		V
	Sector					
	Society					
Goal	Rank	V	V	V	V	V
	Score	V	V	V	V	V
Impacts	Environmental	V	V	V	V	V
	Social					
	Economic					
	Environment/society					
	Environment/economy	V	V	V	V	V
	Society/economy					
	Time	X	X	X	X	X
	Space					
	Uncertainty					
	Linkages IOM	V	V	V	V	V
	Inequality					
Standardization	Monetary assessment	V	V	V	V	V
	Relative weights					
Aggregation		V	V	V	V	V
Participation		X	X	X	X	X

Abbreviations: EM = emergy, EX = exergy, RCM = the recycle content materials, ELRM = the end-of-life recycling materials, IOM = Input–Output Model. V = fine, X = never, blank = maybe, but it is still a methodological problem. LCA focused on user is also called consequential LCA, LCA focused on donor is also called attributional LCA in terms of energy use, GHG emissions, materials use (i.e., its implicit goal is the preservation of the status quo, without discounting impacts over time, in terms of human health, ecosystem quality and resource availability) (Sureau et al., 2018). Moreover, Life Cycle Inventory Assessment and Life Cycle Impact Assessment can be considered to be preliminary to LCA (i.e., they guide the data collection) (Fontes et al., 2018). Finally, standardisation can rely on internal statistics (e.g., dividing by the difference between max and min values, dividing by the mean value in the sample under consideration) or on external sources (e.g., dividing by the absolute maximum, dividing by the target value), where the latter method is preferred due to cases of rank reversal (Petti et al., 2018).

In summary, the main weaknesses (on feasibility) of LCA as a tool to achieve sustainability can be summarised as follows:

1. Considering participation is essential to measure sustainability (McCabe & Halog, 2018)
2. Aggregating different forms of energy might be misleading (Bonilla-Alicea & Fu, 2019)
3. Including social impacts could be essential (Pizzol et al., 2015)

However, LCA can be developed to MLCA if all issues from time 0 (i.e., creation) to time T (i.e., disposal) are expressed in consistent units (e.g., money, water, or energy consumption, and use of materials) for all dynamic variables:

$$MLCA = \sum\nolimits_{k=0}^{T} \left[\frac{1}{(1+r)^k} \right] [x_{\text{eco}}(k) + x_{\text{soc}}(k) + x_{\text{env}}(k)]$$

where $x_{\text{eco}}(k)$, $x_{\text{soc}}(k)$, and $x_{\text{env}}(k)$ represent the dynamic *net* economic, social, and environmental impacts, respectively, from time 0 to time t in *monetary* terms. Thus, a decision is taken if MLCA > 0.

4.2.2 *Multi-criteria Analysis for Decisions to Equity: MAUT, TOPSIS, VIKOR, ELECTRE, PROMETHEE*

Multi-Criteria Analysis MCA can be defined as a systematic set of methods for structuring decision problems that involve more than one criterion and unit to find non-dominated alternative solutions by incorporating preference information (Etxano & Villalba Eguiluz, 2021; Lindfors, 2021).

Note that Cost-Effectiveness Analysis (if many projects) identifies which project to implement at a smaller cost for a specified objective (impact index/cost to achieve it), whereas the Environmental Impact Analysis (a single project) evaluates many physical indexes, without economic indexes (different performances with respect to different criteria).

For simplicity, let us refer to the *main* five methodologies suggested in the literature to combine different impacts and criteria.

The Multi-Attribute Utility Theory MAUT for the option a can be generally formulated as follows:

$$\mathrm{MAUT}(a) = \sum\nolimits_{i=1}^{n} w_{\mathrm{i}} \Delta f_{\mathrm{i}}(\mathrm{t})/f_{\mathrm{i}}(0)$$

where $\Delta f_i(t)$ is the change in criterion i in period t (i.e., the time span of the project), $f_i(0)$ is the criterion level i at time 0 (i.e., the reference time of the project), and w_i is the relative weight attached to criterion i (i.e., $\sum w_i = 1$). Note that impacts (i.e., changes) are in percentages, since they are divided by the initial levels (i.e., they are standardised).

For concreteness, let us refer to three criteria only (i.e., economic, social and environmental indicators for economic, social and environmental impacts). Thus, MAUT becomes:

$$\mathrm{MAUT}(a) = w_{\mathrm{eco}}\left[\frac{f_{\mathrm{eco}}(t)}{\mathrm{Max}\, f_{\mathrm{eco}}}\right] + w_{\mathrm{soc}}\left[\frac{f_{\mathrm{soc}}(t)}{\mathrm{Max}\, f_{\mathrm{soc}}}\right] + w_{\mathrm{env}}\left[\frac{f_{\mathrm{env}}(t)}{\mathrm{Max}\, f_{\mathrm{env}}}\right]$$

where w_{eco}, w_{soc}, w_{env} represent the economic, social and environmental relative weights, $f_{\mathrm{eco}}(t), f_{\mathrm{soc}}(t), f_{\mathrm{env}}(t)$ represent the *net* economic, social and environmental impacts expected at time t, and Max $f_{\mathrm{eco}}(t)$, Max $f_{\mathrm{soc}}(t)$, Max $f_{\mathrm{env}}(t)$ represent the best economic, social and environmental impacts that can be expected. The option a is chosen if it scores the largest MAUT. Note that impacts are again in percentages, although a different standardisation procedure is applied.

The Technique for Order of Preference by Similarity to Ideal Solution TOPSIS procedure consists of the following steps:

(a) Calculate the normalised decision matrix. The normalised value of impact i in option j r_{ij} is calculated as:

$$\mathrm{r_{ij}} = \mathrm{f_{ij}} / \sqrt{\sum_{\mathrm{j=1}}^{\mathrm{J}} \mathrm{f_{ij}^2}}$$

(b) Calculate the weighted normalised decision matrix. The weighted normalised value v_{ij} is calculated as

$$\mathrm{v_{ij}} = \mathrm{w_i}\ \mathrm{r_{ij}}$$

where w_i is the relative weight of the ith attribute or criterion.

(c) Determine the positive and negative ideal solution

$$A^{+} = \left\{v_1^{+}, \ldots, v_n^{+}\right\} = \left\{(\max j\ vij | i\epsilon I'), (\min j\ vij | i\epsilon I'')\right\}$$

$$A^- = \{v_1^-, \ldots, v_n^-\} = \{(\min j\ vij|i\epsilon I'), (\max j\ vij|i\epsilon I'')\}$$

where I′ is associated with benefit criteria and I″ is associated with cost criteria.

(d) Calculate the separation measures, using the n-dimensional Euclidian distance. The separation of each alternative from the positive ideal solution is given as:

$$D_j^* = \sqrt{\sum_{i=1}^{n} (v_{ij} - v_i^+)^2}$$

Similarly, the separation from the negative ideal solution is given as

$$D_j^- = \sqrt{\sum_{i=1}^{n} (v_{ij} - v_i^-)^2}$$

(e) Calculate the relative closeness to the ideal solution: the relative closeness of the alternative aj with respect to A + is defined as

$$C_j^+ = D_j^- / \left(D_j^+ + D_j^-\right)$$

(f) Rank the preference order

For concreteness, let us refer to three criteria only (i.e., 1 = economic, 2 = social and 3 = environmental indicators for economic, social and environmental impacts) and two options only (i.e., a and b). Thus, TOPSIS becomes:

$$\mathrm{r_{1a}} = \mathrm{f_{1a}} / \sqrt{\mathrm{f_{1a}^2} + \mathrm{f_{1b}^2}}$$

$$\mathrm{r_{2a}} = \mathrm{f_{2a}} / \sqrt{\mathrm{f_{2a}^2} + \mathrm{f_{2b}^2}}$$

$$\mathrm{r_{3a}} = \mathrm{f_{3a}} / \sqrt{\mathrm{f_{3a}^2} + \mathrm{f_{3b}^2}}$$

Then

$$\mathrm{v_{1a}} = \mathrm{w_1\ r_{1a}},\ \mathrm{v_{2a}} = \mathrm{w_2\ r_{2a}},\ \mathrm{v_{3a}} = \mathrm{w_3\ r_{3a}}$$

Similarly for Project B. Then

$$A^+ = \{v_1^+, v_2^+, v_3^+\} = \{(\max a, b\ \mathrm{v_{ij}}|i\epsilon I'), (\min j\ \mathrm{v_{ij}}|i\epsilon I'')\}$$

$$A^{-} = \{v_1^-, v_2^-, v_3^-\} = \{(\min j\ v_{ij}|i\epsilon I'), (\max j\ v_{ij}|i\epsilon I'')\}$$

Then

$$D_a^+ = \sqrt{\sum_{i=1}^{3} \left(v_{ia} - v_i^+\right)^2}$$

$$D_a^- = \sqrt{\sum_{i=1}^{3} \left(v_{ia} - v_i^-\right)^2}$$

Finally

$$C_a^+ = D_a^- / \left(D_a^+ + D_a^-\right)$$

Similarly for Project B. The option a is preferred to option b if $C_a^+ > C_b^+$.

The Multi-Criteria Optimisation and Compromise Solutions (Serbian) VIKOR is based on the following form of Lp-metric:

$$L_{p,j} = \left\{ \sum_{i=1}^{n} \left[w_i \left(f_i^+ - f_{ij}\right) / \left(f_i^+ - f_i^-\right) \right]^p \right\}^{1/p}$$

With p in [1, ∞], where n is the number of criteria, j denotes options, and f_{ij} is the value of ith criterion function for the alternative j, and where f_i^+ is the best value and f_i^- is the worst value.

The compromise ranking algorithm VIKOR is based on the following steps:

(a) Determine the best fi^+ and the worst fi^- values of all criterion functions, i = 1, 2,..., n. If the ith function represents a benefit then

$$f_i^+ = \max \text{j} f_{ij} \text{ and } f_i^- = \min \text{j} f_{ij}$$

(b) Compute the values S_j and R_j, j = 1,2, ...J by the relations:

$$S_j = \sum_{i=1}^{n} w_i \left(f_i^+ - f_{ij}\right) / \left(f_i^+ - f_i^-\right)$$

$$R_j = \max_i w_i \left(f_i^+ - f_{ij}\right) / \left(f_i^+ - f_i^-\right)$$

where w_i are the weights of criteria, expressing their relative importance.

(c) Compute the values Q_j, j, 1,2, ...,J by the relation:

$$Q_j = v\frac{(S_j - S^+)}{(S^- - S^+)} + (1 - v)\frac{(R_j - R^+)}{(R^- - R^+)}$$

where

$$S^+ = \min j\ S_j,\ S^- = \max j\ S_j,\ R^+ = \min j\ R_j,\ R^- = \max j\ R_j$$

And v is the weight attached to the strategy "maximizing the group utility of the majority of criteria" and 1 − v is the weight attached to the strategy "minimising the individual regret of the opponent criterion".

(d) Rank the alternatives, sorting by the values of S, R, and Q in decreasing order: three ranking lists are obtained.
(e) Propose as a compromise solution the option "a" which is ranked the best by the measure Q (minimum) if the following two conditions are satisfied:

C1. "Acceptable advantage"

$$Q(b) - Q(a) \geq DQ$$

where "b" is the alternative with second position in the ranking list by Q, DQ = 1/(J − 1), and J is the number of alternatives.

C2. "Acceptable stability of decision making"

Option "a" must be the best ranked by S or/and R. This compromise solution is stable within a decision making process which could be: voting by majority rule (v > 0.5), voting by consensus (v = 0.5) or voting with veto (v < 0.5).

If one of the conditions is not satisfied, then a set of compromise solutions could be proposed: alternative "a" and "b" if only condition C2 is not satisfied; alternative "a", "b", … "z" if condition C1 is not satisfied and "z" is determined by the relation Q(z) − Q(a) < DQ for maximum z.

For concreteness, let us refer to three criteria only (i.e., 1 = economic, 2 = social and 3 = environmental indicators for economic, social and environmental impacts) and two options only (i.e., a and b). Thus, VIKOR becomes:

$$L_{p,j} = \{[w_1(f_1^+ - f_{1a})/(f_1^+ - f_1^-)]^p + [w_2(f_2^+ - f_{2a})/(f_2^+ - f_2^-)]^p$$
$$+ [w_3(f_3^+ - f_{3a})/(f_3^+ - f_3^-)]^p\}^{1/p}$$

Then

$$S_a = \frac{w_1(f_1^+ - f_{1a})}{(f_1^+ - f_1^-)} + \frac{w_2(f_2^+ - f_{2a})}{(f_2^+ - f_2^-)} + \frac{w_3(f_3^+ - f_{3a})}{(f_3^+ - f_3^-)}$$

$$R_a = \max\left[\frac{w_1(f_1^+ - f_{1a})}{(f_1^+ - f_1^-)}, \frac{w_2(f_2^+ - f_{2a})}{(f_2^+ - f_2^-)}, \frac{w_3(f_3^+ - f_{3a})}{(f_3^+ - f_3^-)}\right]$$

Then

$$Q_a = v\frac{(S_a - S^+)}{(S^- - S^+)} + (1 - v)\frac{(R_a - R^+)}{(R^- - R^+)}$$

Similarly for Project B. The option a is chosen if it is better ranked in terms of S, R or Q.

The ELimination Et Choix Traduisant La REalite ELECTRE procedure consists of the following steps. Let $f_i(a)$ the score of criterion i in option a. For each pair of option a and b:

$$N_{ab}^+ = \{i \in N | f_i(a) > f_i(b)\}$$

$$N_{ab}^= = \{i \in N | f_i(a) = f_i(b)\}$$

$$N_{ab}^- = \{i \in N | f_i(a) < f_i(b)\}$$

The concordance index is defined as:

$$c(a, b) = \sum_{i \in N_{ab}^+}^{n} w_j + \sum_{i \in N_{ab}^=}^{n} w_j$$

Let α a given concordance level, usually within the range [0.5, 1 − min w_i]. If c(a, b) ≥ α then the assertion a outranks b is validated by accordance.

Let the discordance index defined as:

$$d(a, b) = max_{i \in N}\left\{\frac{f_i(a) > f_i(b)}{\beta}\right\}$$

where β represents a given veto threshold. If d(a, b) ≥ 1 then the assertion a outranks b is no longer valid. If this assertion passes both concordance and discordance validation (i.e., c(a, b) ≥ α and d(a, b) < 1) then a outranks b.

The credibility matrix indicates the reliability of the outranking hypothesis.

For concreteness, let us refer to three criteria only (i.e., 1 = economic, 2 = social and 3 = environmental indicators for economic, social and environmental impacts) and two options only (i.e., a and b). Thus, ELECTRE becomes:

$$N_{ab}^+ = \{i \in N | f_1(a) > f_1(b), f_2(a) > f_2(b), f_3(a) > f_3(b)\}$$

$$N_{ab}^= = \{i \in N | f_1(a) = f_1(b), f_2(a) = f_2(b), f_3(a) = f_3(b)\}$$

$$N_{ab}^{-} = \{i \in N | f_1(a) < f_1(b), f_2(a) < f_2(b), f_3(a) < f_3(b)\}$$

For example

$$N_{ab}^{+} = \{i \in N | f_1(a) > f_1(b)\}$$

$$N_{ab}^{=} = \{i \in N | f_2(a) = f_2(b)\}$$

$$N_{ab}^{-} = \{i \in N | f_3(a) < f_3(b)\}$$

Then

$$c(a, b) = w_1 + w_2$$

Then

$$d(a, b) = \max\left[\frac{f_1(a) > f_1(b)}{\beta}, \frac{f_2(a) > f_2(b)}{\beta}, \frac{f_3(a) > f_3(b)}{\beta}\right]$$

The option a is chosen if c(a, b) $\geq \alpha$ and d(a, b) < 1.

The Preference Ranking Organisation METHod for Enrichment Evaluation PROMETHEE procedure is based on the following steps.

Let strict preference thresholds (i.e., positive ideal solution) represented by:

$$f_i^{+} = \max_i \{f_{ij} | i = 1, 2, \ldots, n; j = 1, 2, \ldots J\}$$

Let indifferent thresholds (i.e., negative ideal solution) represented by:

$$f_i^{-} = \min_i \{f_{ij} | i = 1, 2, \ldots, n; j = 1, 2, \ldots J\}$$

where f_{ij} is the score of the jth alternative for the ith criterion.

Let $p_i(a, b)$ the preference function for alternative a outranking alternative b for the ith criterion represented by:

$$p_i(a, b) = \left\{ \begin{array}{cc} 0 & if\ f_{ia} - f_{ib} \leq f_i^{-} \\ \frac{(f_{ia} - f_{bi}) - f_i^{-}}{f_i^{+} - f_i^{-}} & if\ f_i^{-} \leq f_{ia} - f_{ib} \leq f_i^{+} \\ 1 & f_{ia} - f_{ib} > f_i^{+} \end{array} \right\}$$

A multi-criteria preference index indicating the advantage of alternative u over alternative v can be given by:

$$\pi(a,b) = \sum_{i}^{n} w_i p_i(a,b)$$

where w_i is the influence weight for the ith criterion.

Let ϕ^+(a) the leaving flow representing the extent to which alternative a outranks the other alternatives, and ϕ^-(a) the entering flow representing the extent to which other alternatives outrank the alternative a:

$$\phi^+(a) = \sum_{b}^{z} \pi(a,b)$$

$$\phi^-(a) = \sum_{b}^{z} \pi(a,b)$$

where both terms could be multiplied by 1/(n − 1).

The net flow ϕ(a) is the final score of the alternative a:

$$\phi(a) = \phi^+(a) - \phi^-(a)$$

where both terms could be calculated in absolute value.

For concreteness, let us refer to three criteria only (i.e., 1 = economic, 2 = social and 3 = environmental indicators for economic, social and environmental impacts) and two options only (i.e., a and b). Thus, PROMETHEE becomes:

$$f_1^+ = \max[f_{1a}, f_{1b}], f_2^+ = \max[f_{2a}, f_{2b}], f_3^+ = \max[f_{3a}, f_{3b}],$$

$$f_1^- = \min[f_{1a}, f_{1b}], f_2^- = \min[f_{2a}, f_{2b}], f_3^- = \min[f_{3a}, f_{3b}],$$

Then

$$p_1(a,b) = \left\{\begin{array}{cc} 0 & \text{if } f_{1a} - f_{1b} \le f_1^- \\ \frac{(f_{1a}-f_{1b})-f_1^-}{f_1^+-f_1^-} & \text{if } f_1^- \le f_{1a} - f_{1b} \le f_1^+ \\ 1 & f_{1a} - f_{1b} > f_1^+ \end{array}\right\}$$

$$p_2(a,b) = \left\{\begin{array}{cc} 0 & \text{if } f_{2a} - f_{2b} \le f_2^- \\ \frac{(f_{2a}-f_{2b})-f_2^-}{f_2^+-f_2^-} & \text{if } f_2^- \le f_{2a} - f_{2b} \le f_2^+ \\ 1 & f_{2a} - f_{2b} > f_2^+ \end{array}\right\}$$

$$p_3(a,b) = \left\{\begin{array}{cc} 0 & \text{if } f_{3a} - f_{3b} \le f_3^- \\ \frac{(f_{3a}-f_{3b})-f_3^-}{f_3^+-f_3^-} & \text{if } f_3^- \le f_{3a} - f_{3b} \le f_3^+ \\ 1 & f_{3a} - f_{3b} > f_3^+ \end{array}\right\}$$

Then

$$\pi(a, b) = w_1 p_1(a, b) + w_2 p_2(a, b) + w_3 p_3(a, b)$$

Then

$$\phi^+(a) = \pi(a, b) > 0$$

$$\phi^-(a) = 0$$

Finally

$$\phi(a) = \phi^+(a) - \phi^-(a) > 0$$

The option a is chosen if ϕ(a) > 0. For example, if economic, social and environmental impacts for Project A are 4, 2, and 1, and 1, 3, and 5 for Project B, with relative weights are all at 1/3, then MAUT and ELECTRE suggest to implement Project B, VIKOR suggest to implement Project A, while PROMETHEE does not reach any suggestion (Table 4.18).

In summary, MAUT is suggested if each solution must evaluated in terms of weighted scores; TOPSIS if each solution must be close to the theoretical best and away from the theoretical worst option, in terms of weighted scores; VIKOR if each solution must minimise regret from avoiding alternative options, in terms of weighted scores; ELECTRE if some solutions must be compared according to majority and veto rules; PROMETHEE if each solution must be compared with many alternative options, in terms of weighted scores. Note that all five methodologies maximise positive impacts and minimise negative impacts: this represent strong sustainability if the former impacts increase resilience and the latter impacts reduce resilience.

Table 4.18 MCA ranking of Project A and Project B with equal relative weights

	MAUT	TOPSIS	VIKOR			ELECTRE		PROMETHEE
			S	R	Q	c(i,j)	d(i,j)	Φ(i)
Project A	6.22	*0.224*	**0.66**	0.33	na	0.33	0.5	*0.25*
Project B	**7.33**	*0.222*	0.33	0.33	na	**0.66**	0.5	*0.25*

Abbreviations: MAUT = Multi-Attribute Utility Theory, TOPSIS = The Technique for Order of Preference by Similarity to Ideal Solution, VIKOR = The Multi-Criteria Optimisation and Compromise Solutions (Serbian), ELECTRE = The ELimination Et Choix Traduisant La REalite, PROMETHEE = The Preference Ranking Organisation METHod for Enrichment Evaluation. *Notes* S, R, Q, c(i,j), d(i,j) and Φ(i) are detailed in text. Bold text identifies the chosen project (in rows) according to the specified methodology (in columns).

4.2.2.1 Time: DYN, TIS

Let us assume that there are two periods in life, 1 and 2, where in period 1 individuals get money M from work and consume goods Q_1, while in period 2 individuals consume goods Q_2 according to the saving produced in period 1: under the assumption that prices are constant (i.e., $P_1 = P_2$), saving equals the amount of money M not used for consumption in period 1 (i.e., $M - Q_1$) capitalized to period 2 in terms of the interest rate r. Under the assumption that the utility from consumption does not change over time (i.e., $U(Q_1) = U(Q_2)$ whenever $Q_1 = Q_2$), the problem to be solved by individuals can be formalized as follows:

$$\text{Maximize Min}\left[w_1\, Q_1,\, w_2\, Q_2\right]$$
$$\text{Subject to } Q_2 \leq (M - Q_1)(1 + r)$$

where w_1 and w_2 are the relative weights attached to consumption in period 1 and 2, respectively.

Figure 4.46 depicts a dynamic equilibrium within MCA.

In summary, the temporal information system TIS enables MCA to take decisions in temporal and dynamic contexts, where impacts in different times can be compared by possibly attaching different weights, although they cannot be combined by a using a compensation procedure. Note that the function Min represents the largest degree of temporal inequality aversion. In other words, although MCA avoids ex-ante assumptions to compare alternative temporal scenarios, it might be forced to introduce ex-post criteria whenever a combination is required.

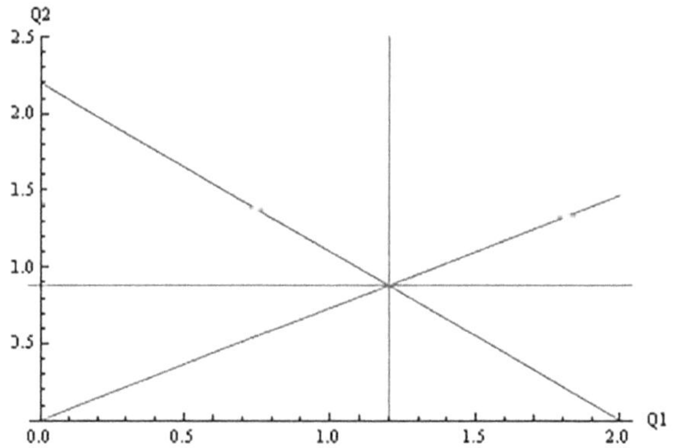

Fig. 4.46 Dynamic equilibrium within MCA. The decreasing (blue) line = the intertemporal budget constraint, the horizontal and vertical (grey) lines = the intertemporal indifference curve, the increasing (purple) line = couples of Q1 and Q2 such that w1 Q1 = w2 Q2. Abbreviations: Q1 = consumption at time 1, Q2 = consumption at time 2. *Notes* If M = 2, r = 0.1, w2 = 22, w1 = 30, the dynamic equilibrium is at Q1 = 1.2 and Q2 = 0.88

4.2.2.2 Space: SPA, GIS

Let us assume that there are two places, 1 and 2, where a resource in place 1 X_1 is differently evaluated than the same resource in place 2 X_2. It is possible to transfer this resource from place 1 to place 2, although this implies a loss by r percent. For concreteness, think of a transfer of water from place 1 to place 2. Under the assumption that the utility from resource uses does not change in space (i.e., $U(X_1) = U(X_2)$ whenever $X_1 = X_2$), the problem to be solved by policy makers can be formalized as follows:

$$\text{Maximize Min}[w_1 X_1, w_2 X_2]$$
$$\text{Subject to } X_2 \leq (X - X_1)(1 - r)$$

where w_1 and w_2 are the relative weights attached to resources in places 1 and 2, respectively.

Figure 4.47 depicts a spatial equilibrium with MCA.

In summary, geographical information systems GIS enables MCA to take decisions in spatial and geographical contexts, where impacts in different places can be compared by possibly attaching different weights, although they cannot be combined by a using a compensation procedure. Note that the function Min represents the largest degree of spatial inequality aversion. In other words, although MCA avoids ex-ante assumptions to compare alternative spatial scenarios, it might be forced to introduce ex-post criteria whenever a combination is required.

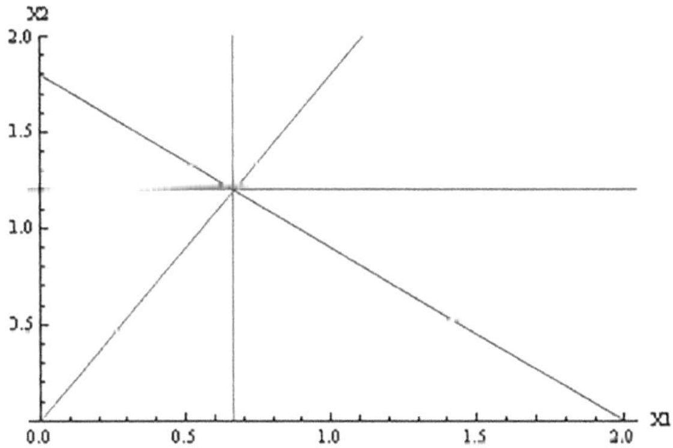

Fig. 4.47 Spatial equilibrium within MCA. The decreasing (blue) line = the interspatial budget constraint, the horizontal and vertical (grey) lines = the interspatial indifference curve, the increasing (purple) line = couples of X1 and X2 such that w1 X1 = w2 X2. Abbreviations: Abbreviations: X1 = resource in space 1, X2 = resource in space 2. *Notes* If X = 2, r = 0.1, w2 = 44, w1 = 80, the spatial equilibrium is at X1 = 0.66 and X2 = 1.2

4.2.2.3 Uncertainty or Risk: EV, FS

Let us assume to have uncertainty (i.e., to face a non-probabilistic problem) due to incomplete or inaccurate information.

Four *main* methods of FS theories have been applied to sustainability science: ordinary, type 2, intuitionistic, and hesitant FS (Kahraman et al., 2017). Let us refer to the most recent hesitant FS. In particular, let us focus on preferences or indices (xij) in crisp real numbers or interval numbers within [0,1] or linguistic labels in {N, L, M, H, P}, where N represents Null, L represents Low, M represents Medium, H represents High, and P represents Perfect.

Let us assume that expert j has clear ideas about values for each category i so that triangular fuzzy numbers can be used to aggregate heterogeneous preferences. Thus, information in different forms (xij) can be transformed into fuzzy sets γhij for reference index h, by using the BLTS method suggested by Herrera et al. (2005). In particular, if h = 1, 2, 3, 4, and 5 are the reference indices and θ is the value specified by experts, then in the case of real numbers within [0,1]:

$$\gamma 1ij = 1, \gamma 2ij = 0, \gamma 3ij = 0, \gamma 4ij = 0, \text{ and } \gamma 5ij = 0 \text{ if } \theta = 0$$

$$\gamma 1ij = (\theta - 0)/(0.25 - 0), \gamma 2ij = (\theta - 0)/(0.5 - 0), \gamma 3ij = 0, \gamma 4ij = 0, \text{ and } \gamma 5ij = 0 \text{ if } \theta \text{ is within } (0, 0.25]$$

$$\gamma 1ij = 0, \gamma 2ij = (\theta - 0)/(0.5 - 0), \gamma 3ij = (\theta - 0.25)/(0.75 - 0.25), \gamma 4ij = 0, \text{ and } \gamma 5ij = 0 \text{ if } \theta \text{ is within } (0.25, 0.5]$$

$$\gamma 1ij = 0, \gamma 2ij = 0, \gamma 3ij = (\theta - 0.25)/(0.75 - 0.25), \gamma 4ij = (\theta - 0.5)/(1 - 0.5), \text{ and } \gamma 5ij = 0 \text{ if } \theta \text{ is within} (0.5, 0.75]$$

$$\gamma 1ij = 0, \gamma 2ij = 0, \gamma 3ij = 0, \gamma 4ij = (\theta - 0.5)/(1 - 0.5), \text{ and } \gamma 5ij = (1 - \theta)/(1 - 0.75) \text{ if } \theta \text{ is within } (0.75, 1)$$

$$\gamma 1ij = 0, \gamma 2ij = 0, \gamma 3ij = 0, \gamma 4ij = 0, \text{ and } \gamma 5ij = 1 \text{ if } \theta = 1$$

If h = 1, 2, 3, 4, and 5 are the reference indices and [θL, θU] are the lower and upper bounds of the intervals specified by experts, respectively, then in the case of interval numbers within [0,1]:

$$\gamma 1ij = 0.5, \gamma 2ij = 0.5, \gamma 3ij = 0, \gamma 4ij = 0, \text{ and } \gamma 5ij = 0 \text{ if } [\theta L, \theta U] = [0, 0.25]$$
$$\gamma 1ij = 0, \gamma 2ij = 0.5, \gamma 3ij = 0.5, \gamma 4ij = 0, \text{ and } \gamma 5ij = 0 \text{ if } [\theta L, \theta U] = [0.25, 0.5]$$
$$\gamma 1ij = 0, \gamma 2ij = 0, \gamma 3ij = 0.5, \gamma 4ij = 0.5, \text{ and } \gamma 5ij = 0 \text{ if } [\theta L, \theta U] = [0.5, 0.75]$$
$$\gamma 1ij = 0, \gamma 2ij = 0, \gamma 3ij = 0, \gamma 4ij = 0.5, \text{ and } \gamma 5ij = 0.5 \text{ if } [\theta L, \theta U] = [0.5, 1]$$

If h = N, L, M, H, and P are the reference indices and θ is the label specified by the experts, then in the case of linguistic labels:

$$\gamma 1ij = 1, \gamma 2ij = 0, \gamma 3ij = 0, \gamma 4ij = 0, \text{ and } \gamma 5ij = 0 \text{ if } \theta = N$$
$$\gamma 1ij = 0, \gamma 2ij = 1, \gamma 3ij = 0, \gamma 4ij = 0, \text{ and } \gamma 5ij = 0 \text{ if } \theta = L$$
$$\gamma 1ij = 0, \gamma 2ij = 0, \gamma 3ij = 1, \gamma 4ij = 0, \text{ and } \gamma 5ij = 0 \text{ if } \theta = M$$
$$\gamma 1ij = 0, \gamma 2ij = 0, \gamma 3ij = 0, \gamma 4ij = 1, \text{ and } \gamma 5ij = 0 \text{ if } \theta = H$$
$$\gamma 1ij = 0, \gamma 2ij = 0, \gamma 3ij = 0, \gamma 4ij = 0, \text{ and } \gamma 5ij = 1 \text{ if } \theta = P$$

For example, if evaluations of economic, social and environmental impacts are characterised by high, medium and low levels of reliability (i.e., there are no Null and Perfect labels and indexes to be applied to H, M, L are 0.75, 0.5, 0.25, respectively), the deterministic values at 4, 2 and 1 become the stochastic values at 3, 1 and 0.25.

Note that these relationships are called membership degree in the FS literature. Moreover, these data are transformed into values within [0,1] by applying $Xij = \sum_{j=0}^{H} h\, \gamma hij / \sum_{j=0}^{H} \gamma hij$ (see Herrera et al., 2005, for general formulations): here, no semantic overlapping is considered. Finally, the interval numbers expressed by the experts are assumed to coincide with the specified intervals [θL, θU] (see Herrera et al., 2005, for general cases).

In summary, FS enables MCA to take decisions under uncertainty. In contrast, FS would be inadequate within CBA (i.e., the EUM without time or CAPM with time discussed in Sect. 2.1.3).

4.2.2.4 Linkages: SAM

Let us refer to the Social Accounting Matrix SAM used in Sect. 2.1.4 to estimate a CGEM.

Let us apply the SAM at fixed price to estimate impacts of inequalities within MCA.

The Mathematica code for the numerical solutions is presented below:

```
KTa=ktsa (PCa (1-itra)-PCa caa-PCi cia-PCs csa-cwta-cwwa) XSa/PKa
KTi=ktsi (PCi (1-itri)-PCi cai-PCi cii-PCs csi-cwti-cwwi) XSi/PKi
KTs=ktss (PCs (1-itrs)-PCs cas-PCi cis-PCs css-cwts-cwws) XSs/PKs

RLBa=rlbsa (PCa (1-itra)-PCa caa-PCi cia-PCs csa) XSa/RWa
RLBi=rlbsi (PCi (1-itri)-PCi cai-PCi cii-PCs csi) XSi/RWi
RLBs=rlbss (PCs (1-itrs)-PCs cas-PCi cis-PCs css) XSs/RWs

ULBa=ulbsa (PCa (1-itra)-PCa caa-PCi cia-PCs csa) XSa/UWa
ULBi=ulbsi (PCi (1-itri)-PCi cii-PCi cii-PCs csi) XSi/UWi
ULBs=ulbss (PCs (1-itrs)-PCs css-PCi cis-PCs css) XSs/UWs

KTRa=343;
KTRi=728;
KTRs=1733;

DEPa=depa XSa;
DEPi=depi XSi;
DEPs=deps XSs;
```

```
Solve[{343==(PCa (1-itra)-PCa caa-PCi cia-PCs csa-cwta-cwwa) XSa-
RWa RLBa-UWa ULBa-DEPa,728==(PCi (1-itri)-PCa cai-PCi cii-
PCs csi-cwti-cwwi) XSi-RWi RLBi-UWi ULBi-DEPi,1733==(PCs (1-
itrs)-PCa cas-PCi cis-PCs css-cwts-cwws) XSs-RWs RLBs-UWs ULBs-
DEPs},{PCa,PCi,PCs}]

PCa =5.35001
PCi =1.66332
PCs =1.8362

RHR=rhrs drs (KTRa+KTRi+KTRs) (1-rtr);
UHR=uhrs drs (KTRa+KTRi+KTRs) (1-rtr);
RHY=((1-rltr) (RWa*RLBa+RWi*RLBi+RWs*RLBs)+RHR+URTF+RGTF+RFTF)(1-
rhtr);
UHY=((1-ultr) (UWa*ULBa+UWi*ULBi+UWs*ULBs)+UHR+RUTF+RGTF+RFTF)(1-
uhtr);

RHCa=(rcsa/(rcsa+rcsi+rcss+rcswt+rcsww))(1-rhsr) RHY/PCa
RHCi=(rcsi/(rcsa+rcsi+rcss+rcswt+rcsww))(1-rhsr) RHY/PCi
RHCs=(rcss/(rcsa+rcsi+rcss+rcswt+rcsww))(1-rhsr) RHY/PCs
UHCa=(ucsa/(ucsa+ucsi+ucss+ucswt+ucsww))(1-uhsr) UHY/PCa
UHCi=(ucsi/(ucsa+ucsi+ucss+ucswt+ucsww))(1-uhsr) UHY/PCi
UHCs=(ucss/(ucsa+ucsi+ucss+ucswt+ucsww))(1-uhsr) UHY/PCs

LGCa=slgca XSa;
DNEa=sdnea XSa;
FNEa=sfnea XSa;

LGCi=slgci XSi;
DNEi=sdnei XSi;
FNEi=sfnei XSi;

LGCs=slgcs XSs;
DNEs=sdnes XSs;
FNEs=sfnes XSs;

DKa=335/15918 XSa
DKi=12162/48754 XSi
DKs=507/14688 XSs

XDa=caa XSa+cai XSi+cas XSs+RHCa+UHCa+LGCa+DKa+DNEa+FNEa
XDi=cia XSa+cii XSi+cis XSs+RHCi+UHCi+LGCi+DKi+DNEi+FNEi
XDs=csa XSa+csi XSi+css XSs+RHCs+UHCs+LGCs+DKs+DNEs+FNEs

Solve[{caa XSa+cai XSi+cas XSs+RHCa+UHCa+LGCa+DKa+DNEa+FNEa==XSa},
{XSa}]
Solve[{cia XSa+cii XSi+cis XSs+RHCi+UHCi+LGCi+DKi+DNEi+FNEi==XSi},
{XSi}]
Solve[{csa XSa+csi XSi+css XSs+RHCs+UHCs+LGCs+DKs+DNEs+FNEs==XSs},
{XSs}]

XSa=10.103+0.215 XSi+0.168 XSs
XSi=287.832+4.056 XSa+2.108 XSs
XSs=45.274+0.412 XSa+0.182 XSi
```

```
Solve[{csa XSa + csi XSi + css XSs + RHCs + UHCs + LGCs + DKs + DNEs
+ FNEs == XSs, cia XSa + cii XSi + cis XSs + RHCi + UHCi + LGCi + DKi
+ DNEi + FNEi == XSi, csa XSa + csi XSi + css XSs + RHCs + UHCs + LGCs
+ DKs + DNEs + FNEs == XSs}, {XSi, XSs}]

XSi = 622.173+ 7.996 XSa, XSs = 158.585+ 1.868 XSa
```

In particular, let us assume that water used in agriculture increases by 50% (i.e., 1030 MCM added to 2060 MCM). Moreover, let us remind that rural population is 4.808 Million people, urban population is 0.244 Million people. Finally, let us assume that water was the only constraint to agriculture production (i.e., agriculture production increases by 50%). Table 4.19 summarizes the main results obtained by SAM in MCA.

In other words, inequality between rural and urban incomes is reduced.

Table 4.19 Consequences of a 50% increase in water availability within MCA

The old prices were Pa = 5.350, Pi = 1.663, Ps = 1.836
The new prices are Pa = 4.522, Pi = 1.469, Ps = 1.709
The old total incomes were RHY = 49,452, URY = 7271
The old per capita incomes were RHY = 10,285, URY = 29,799
The new total incomes were RHY = 59,504, URY = 8663
The new per capita incomes were RHY = 12,376, URY = 35,504
The old productions were XSab = 15,918, XSib = 48,754, XSsb = 14,688
The new productions are XSab = 23,877, XSib = 185,307, XSsb = 42,673
The old total wages were RWa = 7.187, RWi = 2.126, RWs = 3.002, UWa = 7.187, UWi = 2.126, UWs = 1.979
The old per capita wages were RWa = 1.494, RWi = 0.442, RWs = 0.624, UWa = 29.454, UWi = 8.713, UWs = 8.110
The new total wages are RWa = 9.066, RWi = 1.823, RWs = 2.83, UWa = 9.066, UWi = 1.823, UWs = 1.886
The new per capita wages are RWa = 1.885, RWi = 0.379, RWs = 0.590, UWa = 37.157, UWi = 7.474, UWs = 7.732
The old price of capital is PK = 0.89
The new price of capital is PK = 0.87

Abbreviations: Pa = price in agriculture, Pi = price in industry, Ps = price in service, RHY = income of rural households, URY = income of urban households, XSae = equilibrium product supply in agriculture, XSie = equilibrium product supply in industry, XSse = equilibrium product supply in service, RWa = rural wage in agriculture, RWi = rural wage in industry, RWs = rural wage in service, UWa = urban wage in agriculture, UWi = urban wage in industry, UWs = urban wage in service, PK = price of capital

In summary, MCA is adequate to take decisions where linkages are relevant, by applying SAM.

4.2.2.5 Inequalities: IM, IW

Two main criteria has been suggested within MCA to deal with inequalities. Let us refer to the example introduced in Sect. 4.2.1.5 to discuss inequalities within CBA.

The MaxMin criterion by Rawls ensures a minimum to each individual i:

$$\text{Max Min}\left[NPV_{poor}, NPV_{rich}\right]$$

The relative inequality weights IW attach a larger NPV to less advantaged individuals i (Roseland and Spiliotopoulou 2016) apply a triple-bottom line to urban sustainability). In particular, the IW are calculated as follow:

$$U = x^a \text{ with } 0 < a \leq 1$$
$$\partial U/\partial x = a\, x^{a-1}$$
$$\partial U/\partial \bar{x} = a\, \bar{x}^{a-1}$$
$$(\partial U/\partial x)/(\partial U/\partial \bar{x}) = (x/\bar{x})^{a-1}$$

where U is the utility from outcome x and $\bar{x}$ is the average outcome observed Thus, $w_i = (x_i/\bar{x})^{-m}$ with m = 1-a. Note that in Harsanyi efficiency m = 0.

Graphically, the IW can be applied to produce couples of NPV to rich and poor people which produce the same total weighted NVP:

$$WNPV_{tot} = NPV_{rich}(INC_{rich}/INC_{mean})\hat{}(-m) + NPV_{poor}\left(INC_{poor}/INC_{mean}\right)\hat{}(-m)$$

where INC_{rich} and INC_{poor} are the income levels of rich and poor people, respectively. Or indifference curve for equity in the framework [NPV poor, NPV rich]

$$WNPV_{rich} = NPV_{tot}/(INC_{rich}/INC_{mean})\hat{}(m) - NPV_{poor}\left(INC_{poor}/INC_{mean}\right)\hat{}(-m)$$

For example, if $INC_{rich} = 2000$, $INC_{poor} = 500$, a = 0.5, then:

$$WNPV_{tot} = NPV_{rich}(2000/1250)^{-1/2} + NPV_{poor}(500/1250)^{-1/2}$$

Or

$$WNPV_{rich} = NPV_{tot}\sqrt{1.6} - NPV_{poor}(\sqrt{1.6}/\sqrt{0.4})$$

Thus, in terms of relative IW (i.e., $w_{poor}/w_{rich} = 2$), C > A = D (i.e., Project A = D in Fig. 4.32 by applying these inequality weights).

In summary, MCA is adequate to take decisions where inequalities are relevant, by applying IW.

4.2.2.6 Relative Weights: AHP, RSP, FA, LR

Four *main* methods have been used to estimate relative criterion weights. Within the statistical methods, let us refer to the conditional mean analysis or linear regression (LR) and the factor analysis (FA). Within the preference-elicitation methods, let us refer to the revised Simos Procedure (RSP) and the Analytical Hierarchy Process (AHP). Zagonari (2016) summarizes the pros and cons of these methods. In particular, let us focus on the determination of the perceived importance of three categories (i.e., economic, social, environmental).

The LR estimation is based on the following equation:

$$W_{ij} = \beta_{eco} D_{eco} + \beta_{soc} D_{soc} + \beta_{env} D_{env} + \varepsilon_{ij}$$

where W_{ij} is the preference value (from 1 to 5) expressed in question i by stakeholder j and D_k is a dummy variable for the 3 categories (i.e., $D_{eco} = 1$ if the answer is linked to economic issues, $D_{soc} = 1$ if linked to social issues, and $D_{env} = 1$ if linked to environmental issues). See Zagonari (2016) for a numerical example.

The FA estimation is based on the following two steps. Firstly, find λ_{ij} (i.e., factor loadings) that minimises the mean square error in the off-diagonal residuals (Θ) of the correlation matrix (Ψ):

$$W_i = \Lambda\ \eta_i + \varepsilon_i$$
$$\Sigma = \Lambda\ \Psi\ \Lambda' + \Theta$$

With

$$W_{ij} = \lambda_{i1}\ \eta_{i1} + \lambda_{i2}\ \eta_{i2} + \lambda_{i3}\ \eta_{i3} + \varepsilon_{ij}$$

where W_{ij} is the preference value (from 1 to 5) expressed in question i by stakeholder j, η_i are the factor values, ε_i are the residuals with zero means and no correlations with the factors. Note that I assumed that there is no intercept and that the number of relevant factors is 3. This step identifies the three factors (i.e., η_1, η_2, η_3) that better explain the observed variability of scores. Secondly, calculate the correlation matrix of the factor values η_i and the assumed categories (i.e., Eco, Soc, Env). This step matches a factor to a category. See Zagonari (2016) for e numerical example.

As for AHP, each stakeholder is expected to express the preference of issue i over issue j in a 1 to 7 scale (e.g., $a_{i,j}$ compares each couple of economic, social and environmental issues). Then, the AHP estimation is based on the following two steps. Firstly, construct a matrix A with dimension 3 × 3 (Table 4.20):

Table 4.20 The general analytical hierarchy process matrix

	Eco	Soc	Env
Eco	1	$a_{Eco,Soc}$	$a_{Eco,Env}$
Soc	$1/a_{Eco,Soc}$	1	$a_{Soc,Eco}$
Env	$1/a_{Eco,Env}$	$1/a_{Soc,Eco}$	1

Abbreviations: Eco = economic features, Soc = social features, Env = environmental features, $a_{I,J}$ = relative importance of feature I versus feature J

Table 4.21 A specific analytical hierarchy process matrix

	Eco	Soc	Env
Eco	1	2	0.5
Soc	0.5	1	0.25
Env	2	4	1

Abbreviations: Eco = economic features, Soc = social features, Env = environmental features

Note that the lower diagonal is filled by analysts, while stakeholders express their preferences for 3 couples only. Secondly, calculate eigenvectors and eigenvalues of A (i.e., $A X = \lambda X$). Note that relative weights are obtained by normalising eigenvalues to 100. For example, if the matrix A for a stakeholder is given by (Table 4.21):

Then the non-normalised eigenvector attached to the eigenvalue $\lambda = 3$ is given by [0.436, 0.218, 0.872]. Note that $w_{env} = 2\ w_{eco} = 2\ w_{soc}$.

As for RSP, each stakeholder is expected to rank three cards which represent the main economic, social and environmental features of the project. For example, a stakeholder could rank environmental more important than economic issues, and economic more important than social issues (i.e., EnvEcoSoc). Next, each stakeholder is allowed to introduce blank cards in order to stress the greater importance attached to the feature which is ranked better. For example, a stakeholder could introduce a blank card between environmental and economic issues and no blank cards between economic and social issues to express that environmental issues are *much* more important than economic ones, and economic issues are more important than social ones (i.e., EnvBEcoSoc). Then, the RSP estimation is based on the following equations:

$$P(r) = 1 + u(s_0 + \ldots s_{r-1}) \text{ with } s_0 = 0 \text{ and } P(1) = 1$$
$$u = (z - 1)/s$$
$$s = \Sigma_{r=1}{}^{n-1} s_r$$
$$s_r = s_{r-1}{}^{b} + 1$$

where P(r) is the non-normalised weight, z represents the answer to the following question "how many times more important is the most important compared to the least important issue?", $s_{r-1}{}^{b}$ is the number of blank cards between issue ranked r − 1 and the issue ranked r. For example, figures for a stakeholder with no blank cards (i.e., EnvEcoSoc) are given by:

$$P(Env) = 1 + [(3 - 1)/3](0 + 1 + 2) = 3$$
$$P(Eco) = 1 + [(3 - 1)/3](0 + 1) = 1.66$$
$$P(Soc) = 1 + [(3 - 1)/3](0) = 1$$

With z = 3. Again, figures for a stakeholder with a blank card between environmental and economic issues and no blank cards between economic and social issues (i.e., EnvBEcoSoc) are given by:

$$P(Env) = 1 + [(4 - 1)/4](0 + 1 + 3) = 4$$
$$P(Eco) = 1 + [(4 - 1)/4](0 + 1) = 1.75$$
$$P(Soc) = 1 + [(4 - 1)/4](0) = 1$$

With z = 4. Note that some Bayesian has been recently developed (Laurila-Pant et al., 2019; Salliou et al., 2017).

To summarise this Section, if stocks STO are irrelevant, AHP, RSP, FA, LR are all adequate methods to estimate relative weights for average decisions under the assumption of a representative stakeholder with unimodal preferences if flows are relevant. However, within MCA, if stocks STO are relevant, AHP, RSP, FA or LR should be combined with a dynamic model.

4.2.2.7 Weighted Life-Cycle Assessment from LCSA = LCA and LCC and SLCA

Life-Cycle *Sustainability* Assessment LCSA can be defined as an holistic method that incorporates the three pillars of sustainable development (i.e., economic, social, environmental impacts) into a single formulation, by retaining the life cycle perspective (from creation to disposal) to support the development of a whole sector or economy (Wulf et al., 2019).

In particular, Life Cycle Costing LCC measures all costs directly covered by participants in the product system through its life cycle to assess its economic sustainability (França et al., 2021); Social Life Cycle Assessment SLCA measures the real and potential, positive and negative, socio-economic impacts of a product system through its life cycle on 5 stakeholder groups (i.e., workers, local community, society, consumers value chain actors) (Ramos Huarachi et al., 2020): LCA was discussed in Sect. 4.2.1.7 within CBA.

Table 4.22 characterise LCSA as a combination of LCA, LCC and SLCA.

Note that LCSA adopts a methods close to MCA, by aggregating incommensurable features by using relative weights attached to standardised impacts, although its logics is close to weak sustainability, by monetising environmental externalities in terms of willingness to pay (i.e., its implicit goal is the maximisation of human well-being in terms of health and happiness) (Jorgensen et al., 2013; Toniolo

Table 4.22 The main features of life-cycle sustainability assessment

		LCA	LCC	SLCA	LCSA
Focus	Product/technology/user	V	V		V
	Producer/organisation/donor	V		V	V
	Sector				V
	Society			V	V
Goal	Rank			V	V
	Score	V	V		
Impacts	Environmental	V		V	V
	Social			V	V
	Economic		V	V	V
	Environment/society				
	Environment/economy	V			V
	Society/economy			V	V
	Time TD	X	V	V	
	Space				
	Uncertainty FS			V	
	Linkages CGEM	V	V	V	V
	Inequality				
Standardization	Monetary assessment				X
	Relative weights			V	
Aggregation		V	V	V	
Participation		X		V	V

Abbreviations: LCA = Life-Cycle Assessment, LCC = Life-Cycle Costing, SLCA = Social Life-Cycle Assessment, LCSA = Life-Cycle Sustainability Assessment, TD = Time Discount, FS = Fuzzy Set, CGEM = Computable General Equilibrium Model. V = fine, X = never, blank = maybe, but it is still a methodological problem

et al., 2019). Moreover, monetisation of environmental impacts in SLCA shows the same problems of CBA if stocks STO or interactions REL are relevant, while weighing different impact categories shows the same problems of MCA in terms of top-down versus bottom-up approaches if properly sensitivity analyses are not properly performed (i.e., simplex method) (Hoogmartens et al., 2014; Sharma & Gupta, 2020). Finally, linear aggregation is the most common method, although this implies the full compensation between the different impact categories, where outranking methods (e.g., TOPSIS, VIKOR, PROMETHEE) are rarely applied (Miah et al., 2017; Onat et al., 2017).

In summary, the main weaknesses (on feasibility) of LCSA as a tool to achieve sustainability can be summarised as follows:

1. All 3 dimensions should have the same spatial and temporal system (Van der Wilde & Newel, 2021)

2. All 5 stakeholder groups should be equally involved (Subramanian et al., 2018)
3. Using a low discount rate is unrealistic for the perspective of financial investors (i.e., economic sustainability in LCC), while using a high discount rate is unfair in terms of inter-generational equity (i.e., social sustainability in SLCA) (Tokede & Traverso, 2020)
4. SLCA is hampered by a lack of focus on materiality impacts and a lack of understanding of the main impact pathways for social and economic impacts (Weidema, 2018)
5. Uncertain data (e.g., lack of knowledge, overestimation of quantitative data over qualitative data) and uncertain results (e.g., representation of reality linked to the modelling technique, technical parameters with little or no specific meaning) (Macombe et al., 2018)
6. SLCA combines two different incommensurable scientific paradigms in terms of ontology, epistemology and methodology, since LCA refers to disciplines such as biology, chemistry and physics, while SLCA refers to disciplines such as economics, sociology and management (Iofrida et al., 2018)

However, LCSA can be developed to WLCA if many issues from time 0 (i.e., creation) to time T (i.e., disposal) are expressed in different units by dynamic variables (e.g., income effects in terms of revenues, social effects in terms of employment, environmental effects in terms of biodiversity), standardised and combined with relative weights:

$$WLCA = \sum_{k=0}^{T} w_{\text{eco}}[\Delta x_{\text{eco}}(k)]/x_{\text{eco}}(0) + w_{\text{soc}}[\Delta x_{\text{soc}}(k)]/x_{\text{soc}}(0) + w_{\text{env}}[\Delta x_{\text{env}}(k)]/x_{\text{env}}(0)$$

where $x_{\text{eco}}(k)$, $x_{\text{soc}}(k)$, and $x_{\text{env}}(k)$ represent the *net* economic, social, and environmental impacts, respectively, from time 0 to time T in *different* units and combined with relative weights (Zagonari, 2021). Thus, a decision is taken if WLCA > 0.

4.3 Main Insights of Chap. 4

In this chapter, I identified the main policies that can be implemented to pursue efficiency and equity. In particular, I discussed policy measures to achieve efficiency for pollution (i.e., taxes, standards, subsidies, permits) and policy measures to achieve efficiency for renewable and non-renewable resources (i.e., taxes, standards, subsidies, exploitation rights, protected areas, technological support). For equity, I discussed inequality measures (i.e., the Gini, Theil, and Shannon indexes) and equity criteria (i.e., Kalai–Smorodinsky, Nash bargaining, Rawls, Harsanyi, sovereignty, responsibility, solidarity). I also discussed the main methodologies that are applied by cost–benefit analysis for efficiency and by multi-criteria analysis for equity, to guide decisions about the implementation of investment projects for both easy contexts and problematic contexts. The methodologies applied by cost–benefit analysis in easy contexts include the net present value, benefit to cost ratio, and internal rate of return;

within multi-criteria analysis, the methodologies applied in easy contexts include the multi-attribute utility theory (MAUT), Technique for Order of Preference by Similarity to Ideal Solution (TOPSIS), multi-criteria optimisation and compromise solutions VIKOR (the Serbian acronym), ELimination Et Choix Traduisant La REalite (ELECTRE), and the Preference Ranking Organisation METHod for Enrichment Evaluation (PROMETHEE). The problematic contexts include the relevant time, space, uncertainty, linkages, and inequalities. In particular, for the easy contexts within cost–benefit analysis, net present value is suggested if budget constraints are not relevant, whereas the benefit to cost ratio is suggested if there are many projects to be compared and the internal rate of return is suggested if estimation of the discount rate is problematic. For the easy contexts within multi-criteria analysis, MAUT is suggested if each solution must evaluated in terms of weighted scores; TOPSIS is suggested if each solution must be close to the theoretical best solution and distant from the theoretical worst option in terms of weighted scores; VIKOR is suggested if each solution must minimise the regret from avoiding alternative options in terms of weighted scores; ELECTRE is suggested if some solutions must be compared according to majority and veto rules; and PROMETHEE is suggested if each solution must be compared with many alternative options in terms of weighted scores.

Similarly, in problematic contexts within cost–benefit analysis, options include time discounting when time is relevant, spatial discounting when space is relevant, expected utility when uncertainty is relevant, a computable general equilibrium model when linkages are relevant, and a social welfare function when inequalities are relevant. In problematic contexts within multi-criteria analysis, options include a temporal information system when time is relevant, a geographical information system when space is relevant, fuzzy sets when uncertainty is relevant, a social accounting matrix when linkages are relevant, and inequality weights when inequalities are relevant. Note that if there is a market for resources or pollution *and* the project is marginal (i.e., the project does not affect the equilibrium market price), then willingness to pay or marginal costs are suitable references to account for potential market distortions for non-tradable goods and services, whereas the shadow exchange rate is a suitable reference for tradable goods and services. Similarly, if there is no a market for resources or pollution *or* if the project is non-marginal, then the Hicks or Marshall demand functions are suitable references to evaluate willingness to pay or willingness to accept, with or without a related market.

However, I highlighted some mistakes for policy measures and investment projects. For the policy measures, efficiency is *not* achieved if asymmetric information or uncertainty or imperfect competition exist. Efficiency is also *not* achieved if an environmental decision interaction exists (i.e., if pollution produced or resources used by one decision-maker affects welfare or options of pollution production and resource use for another decision-maker), both with flows and with stocks, unless cooperative equilibria are implemented, and shadow prices differ from equilibrium market prices, which measure the marginal utilities for the representative individual. For investment projects, if stock environmental problems exist, both for resources and pollution, then utility-based approaches to simulate markets for resources or pollution such as hedonic pricing, travel costs, contingent valuation, and choice experiments as well as dose responses and replacement costs within production-based approaches in cost–benefit analysis do not achieve efficiency. This is because monetary assessments might not properly measure the shadow prices. If stock environmental problems exist, then subjective methods to assess relative weights such as the analytical hierarchy process, linear regression, the revised Simos procedure, and factor analysis in multi-criteria analysis do not achieve equity. This is because the net beneficial and detrimental impacts might not be properly weighted. Similarly, I highlighted some concerns about policy measures: standards are unsuitable for efficiency; and taxes, subsidies, permits are unsuitable for equity. There are also concerns for investment projects: use of temporal and geographical information systems, fuzzy sets, a social accounting matrix, and inequality weights are improper in cost–benefit analysis; time discounting, spatial discounting, expected utility, a computable general equilibrium model, and social welfare functions are improper in multi-criteria analysis.

Note that I discussed the potential of monetary life-cycle assessment to account for the main weaknesses of life cycle assessment (i.e., participation, aggregation of different forms of energy, inclusion of social impacts) as a tool to achieve sustainability. Similarly, I discussed the potential of weighted life-cycle assessment to account for the main weaknesses of life cycle sustainability assessment (i.e., consider the three main dimensions, involve the five main stakeholder groups, combine material impacts with social and economic impacts, cope with uncertain data and results, and combine incommensurable scientific paradigms in terms of ontology, epistemology and methodology) as a tool to achieve sustainability.

4.4 Remarks on Participation

Three main roles evaluate pollutions and resources in this Chapter: people who use goods and services obtained by using resources and producing pollutions (i.e., this value is summarised in the willingness to pay WTP); firms that produce these goods and services (i.e., this value is summarised in the marginal cost MC); people negatively affected by the use of resources and the production of pollutions (i.e., this value is summarised in the marginal external cost MEC). Note that the equilibrium price represents a reasonable compromise or average between WTP and MC if the market of the good and service under consideration is perfectly competitive. Participation refers to what extent and how these roles are involved in sustainability evaluations and decisions.

However, any institution can be defined as a set of (ethical) values and a set of (behavioral) rules. Thus, participation refers to what extent and how these roles are involved in (new) institutions to cope with environmental issues.

Table 4.23 summarizes the main features of the collective-action problems in the set of environmental situations under consideration (Zagonari, 2020).

These problems represent situations in which an individual's action (e.g., conserving energy, recycling wastes, purchasing organic food) leads to a clearly

Table 4.23 Main characteristics of some collective action problems in environmental contexts

	Energy conservation	Waste recycling	Purchasing organic food
Benefits to others only	Y/N	N	N
Repeated actions	Y	Y	Y
Lack of governmental rules	Y	Y	Y
Lack of social sanctions	Y	Y	Y
Lack of alternative technologies	Y	Y/N	Y
Overdetermined benefits	N	N	N
International benefits	Y	Y/N	N

Abbreviations: Y = yes, N = no, Y/N = sometimes yes, sometimes no. *Notes* "Overdetermined benefits" refers to an individual expected benefit of 0

better collective status but has such a low potential benefit that its consciously or unconsciously expected utility cannot explain any obligation to perform this action. In other words, an individual action (i.e., to do or refrain from doing something) that is characterized by an opportunity cost (e.g., in terms of time, money, or both) produces benefits for others and for the individual too (Talbot, 2018).

Note that institutions at a large scale such as markets in Smith and societies in Hobbes should make self-interest individuals (i.e., cooperation is mediated) achieve solutions of collective-action problems (based on averages), whereas institutions at a small scale could make relational individuals achieve solutions of collective-action problems (based on majorities). Moreover, collective-action problems are relevant whenever there are interactions (i.e., $\lambda \neq P$) or stocks (i.e., $\mu \neq P$) or both. Finally, the literature on solving environmental collective-action problems by relying on ethics is still in its infancy. For example, Bolis et al. (2017) found that only 3 out of 151 papers in their review supported a deep, disruptive transition based on value-oriented rational action (i.e., substantive rationality, which focuses on the ethics of an action, rather than instrumental rationality, which focuses on whether the means can achieve the desired end, irrespective of the ethics of that means), in which the resulting collective or individual decisions tend to be aligned with personal values and beliefs.

Two main approaches to participation can be identified: a top-down and a bottom-up approach. The top-down approach refers to policies PLA in this Chapter: it applies averages to WTP combined with MC into P or MEC or both. The bottom up approach refers to projects PRO in this Chapter: it applies averages to WTP or MEC or both in CBA; it applies averages or majorities to preferences by people in MCA.

Table 4.24 summarizes the main features of participation to *new* institutions (e.g., agreements, policies) with or without relevant collective action problems and within top-down or a bottom-up approaches.

Table 4.24 Main characteristics of participation to new institutions towards sustainability

		Scale	Knowledge	Representativeness	Policies		Projects	
					Efficiency	Equity	CBA	MCA
No CAP	Bottom-up	Small	Check	Check	Small	Large	Large	Small
	Top-down	Large	Max	Max	Max	Check		
CAP	Bottom-up	Small	Check	Check	Small	Large	Small	Large
	Top-down	Large	Max	Max	Min	Check		
Paradigms					WS	SS	WS	SS

Abbreviations: CAP = collective action problems; SS = strong sustainability; WS = weak sustainability; CBA = Cost Benefit Analysis; MCA = Multi-Criteria Analysis

In particular, the maximum effectiveness (i.e., efficiency) in the top-down approach is achieved within weak sustainability based on averages, if markets are competitive: the equilibrium price is an acceptable measure of participation if markets are competitive (i.e., P is close to MC) and society are democratic (i.e., people can deliberately participate or not participate to the market). Moreover, the maximum effectiveness (i.e., equity) in the bottom-up approach is achieved within strong sustainability based on majorities, if a national referendum is implemented (Schläpfer, 2016). Finally, participation in CBA is related to markets (in terms of average values) to a greater extent than participation in MCA (in terms of majority values).

Note that policies are supported by public demand so they are consistent with the prevailing environmental ethics: the specific choice of policies (e.g. market-based vs. command-and-control) is based on the relative bargaining power of conflicting interests involved (e.g., public bureaucrats, workers, entrepreneurs) (Kirchgassner & Schneider, 2003). Moreover, whenever collective-action problems are relevant, values and decisions are biased, even if knowledge and representativeness are at maximum. Finally, agreements to solve a global problem might not be global: the specific choice of scales (i.e., local vs. regional) is based on the nested externalities from actions of decision units at different scales (Ostrom, 2012).

Thus, in many scenarios, *new* institutions can be unsuccessful in coping with environmental issues, if collective-action problems are relevant, where knowledge and representativeness must be checked. Note that environmental collective-action problems could be solved by relying on ethics of *existing* institutions such as religions, since any institution is based on ethical values and behavioral rules, where knowledge and representativeness can be assumed to hold, although effectiveness must be estimated statistically (Zagonari, 2022).

In particular, stakeholders involved in a sustainability decision on both policies PLA and projects PRO must be representative and informed, where sustainable business models within social corporate responsibility are in between PLA and PRO (Silvestre et al., 2022).

UN has stressed stakeholder participation in all subsequent definitions of sustainable development (e.g., UN, 1992, Agenda 21; UN, 2015, Agenda 2030). Thus, the involvement of stakeholders into sustainability science (i.e., environmental science applied in decision-making processes) has been an expanding trend in an increasing number of research areas (Mielke et al., 2016). In particular, many approaches have been adopted to integrate stakeholder participation: let us refer to *when*, *what*, *why*, and *how* labels to characterise these alternative approaches (Curseu and Schruiser, 2020).

As for *when*, stakeholders' engagement can be performed at different stages. If stakeholders' perspectives are included early in the decision-making process, they contribute to frame the problem. If stakeholders are involved late in the decision-making process, they contribute to evaluate or rank ready-made alternatives.

As for *what*, stakeholders' involvement can be substantive (i.e., participation is aimed at achieving better outcomes, and stakeholders bring in system knowledge through practical experience), normative (i.e., participation as an end in ad of itself, and stakeholders add orientation knowledge through their opinions) or instrumental (i.e., participation is aimed at securing particular interests, and stakeholders bring in system values and motivations).

As for *why*, stakeholders' engagement can be classified in four main typologies. Technocratic type, if the role of stakeholders is to provide issue-specific and objective information to be used by the experts' model or frame. Neo-liberal type, if stakeholders represent alternative or different interests. Functionalist type, if the role of stakeholders is to make experts more sensitive and concerned with social issues. Democratic type, if stakeholders represent all affected citizens and organisations.

As for *how*, stakeholders' involvement can be aimed at identifying the option that maximises positive economic, social and environmental impacts (i.e., minimum impacts based on bounded rationality), at finding a compromise solution (i.e., it is based on stakeholders' perspectives/attitudes and goals/concerns), and at identifying the option that maximises welfare or minimises costs (i.e., maximum outcomes based on perfect rationality).

Different combinations of alternative approaches are appropriate in different contexts. For example, if decisions are characterised by negative interdependence (i.e., some stakeholders' interests can be achieved only at the expense of other stakeholders' interests) and options are given (Ferretti et al., 2019), an instrumental and minimum-impact or maximum-outcome approaches seem to be suggested from *what* and *how*, respectively, by including stakeholders' perspectives/attitudes and goals/concerns (i.e., an intermediate involvement from *when*), to account for the negative interdependence: a balanced group of stakeholders should be invited, by making sure that the diversity of views and interests are actually expressed, to account for possible social conflicts in taking sustainability decisions (i.e., a democratic type from *why*).

Thus, stakeholders' sample representativeness and information gaps will affect sustainability decision-making (Scolobig and Lilliestam, 2016). These issues are crucial in contexts where decisions are based on a deontological approach (i.e., the choice of an action is based on its ethical principles) rather than on a consequentialist approach (i.e., the choice of an action is based on its expected achievements) (Linnerooth-Bayer et al., 2016).

To measure and cope with issues of stakeholders' representativeness and knowledge, by assuming symmetric and asymmetric responses for deontological consequentialist issues, respectively, I suggest to develop and apply conflict indicators to sustainable decision-making. Note that a Bayesian Belief Network approach quantifies the impacts on decisions of uncertainty about the group consensus from a diverse set of stakeholders, by assuming the prior distribution of responses (e.g., uniform distribution if no earlier information about how stakeholders might rate perceptions and concerns) (Laurila-Pant et al., 2019).

Many conflict indicators have been suggested in the literature (Fasth et al., 2018). Here, I suggest the following cardinal conflict index:

$$CCI = 1 - \left(\frac{\sum_{i=1}^{na} x_a}{x_t} - \frac{\sum_{i=1}^{nb} x_b}{x_t} \right)$$

with

$$x_t = \sum_{i=1}^{na} x_a + \sum_{i=1}^{nb} x_b$$

where subscripts a and b refer to responses above and below the average. In particular, index values in [0.75–1.25] suggest the presence of conflicts, index values in [0–0.75] suggest the absence of conflicts with scores above the average, and index values in [1.25–2] suggest the absence of conflicts with scores below the average. These values will highlight potential issues of stakeholders' representative samples and information gaps. Indeed, if questions in the questionnaire cover all the main features involved in a decision-making process; if all relevant stakeholders directly or indirectly involved in decision-making are invited to participate to a first meeting; if questions in the questionnaire represent all alternative perceptions and concerns about economic, social and environmental issues; then:

- If responses are equally distributed (i.e., presence of conflicts), the sample is representative for deontological issues (i.e., answers are personally arguable), although conflicts might arise due to information gaps on consequentialist issues (i.e., there is an objectively true answer): additional meetings with the initially invited stakeholders are suggested, to share different perceptions and concerns, and eventually to lead to a common perspective on consequentialist issues.
- If responses are unequally distributed (i.e., absence of conflicts) for deontological issues, the sample might not be representative: the involvement of additional stakeholders with respect to the initial sample is suggested.

In other words, a presence of conflicts should be empirically observed and not observed for questions theoretically characterised by a deontological and consequentialist approach, respectively.

The main (operational) strengths of the suggested methodology are:

1. My methodology can be applied to small samples. In other words, our methodology enables to save time and money.
2. My methodology enables to combine substantive rationality (i.e., combining different worldviews into a representative stakeholder objective) and instrumental rationality (i.e., choosing the option that best meets the representative stakeholder objective).
3. My methodology can be applied to any sustainable decision-making process. In other words, our methodology enables to reduce the impacts on decisions of stakeholders' representativeness and knowledge issues for irrelevant features.

Note that responses are assumed to be symmetric for deontological issues and non-symmetric for consequentialist issues, whereas no assumptions are made about unimodal or multimodal distribution of responses. By contrast, averaging alternative perceptions and representations by different stakeholders in a single model (e.g., VIKOR method on the distance from the utopia point or TOPSIS method on distances from utopia and nadir points) makes sense if responses can be assumed to be unimodal, whereas decisions based on majority rules (e.g., maximise consensus on mean or median options) about available options from representative and informed stakeholders should be preferred otherwise (Dowling et al., 2016).

As a summary of this remark, decisions on average could be inadequate (i.e., Min efficiency) if collective action problems are relevant so a per-capita inequality perspective (i.e., strong sustainability) must be adopted, although taking decisions in majority requires the sample of involved stakeholders to be representative and informed, for both policies PLA and projects PRO. Alternatively, if collective action problems are not relevant and decisions on average are adequate (i.e., Max efficiency), an average approach (i.e., weak sustainability) can be adopted, provided that markets are competitive and society are democratic, where people engaged in a referendum at national or regional level are representative by definition in a democratic society, although some people on some specific topics might not be properly informed.

Note that the choice of strong versus weak sustainability should be taken by stakeholders, whereas the adoption of multi-criteria analysis versus cost–benefit analysis should be chosen by experts. However, experts should clarify stakeholders to what extent decisions are affected by choosing strong versus weak sustainability and, consequently, multi-criteria analysis versus cost–benefit analysis.

4.5 Exercises

Exercises 1 to 3 share the following context: MPNB = P − 2 Q and MEC = Q, under the assumption that each consumer is willing to buy more than one unit of goods and services, provided that the price is smaller. Moreover, exercises 4 and 5 look for the relevant policies to have the amount of renewable and non-renewable resources at X = 0.5 and XT = 0.5, respectively. Finally, exercises 7 to 10 share the data specified in Table 4.25.

Table 4.25 Economic, social and environmental impacts of Project A and Project B

	Eco	Soc	Env	C_0	B_1	B_2	B_3	PDF	NPV	NPV
	f_{eco}	f_{soc}	f_{env}						Rich people	Poor people
Project A	4	2	1	1	5	2	1	N[7, 1]	3	4
Project B	1	3	5	2	3	3	5	N[9, 2]	7	2
Reliability	H	M	L							
	0.75	0.5	0.25							

Abbreviations: Eco = economic features, Soc = social features, Env = environmental features, C_0 = costs at time 0, B_1 = benefits at time 1, B_2 = benefits at time 2, B_3 = benefits at time 3, PDF = Probability Density Function, NPV = Net Present Value, N[7, 1] = Normal distribution with mean 7 and variance 1, N[9, 2] = Normal distribution with mean 9 and variance 2, H = high, M = medium, L = low. *Notes* Benefits, costs and NPV are in any monetary unit

Note that NPV are calculated at r = 0. Moreover, social benefits arise in period 2 and environmental benefits in period 3. Finally, the evaluation of economic, social and environmental impacts are characterised by high, medium and low reliability.

1. Calculate the optimal tax or optimal standard fine [tax* = P/3; fine* = P/3]
2. Calculate the optimal subsidy in the short-run and long-run [sub_{SR}* = P/3; sub_{LR}* = (1/3)[(P + $\sqrt{3}$ $\sqrt{(6\,FC - P^2)}$]
3. Calculate the equilibrium price of permit [P_p = (1/3) c_Y (3 Y_0 − 1)]
4. Calculate the optimal a, b, W and P without interaction (i.e., a monopolistic firm) such that the stock of a renewable resource in equilibrium is X* = (1/2) Max X = 0.5 for a low and high concern about it as depicted by β = 1 and β = 2, respectively; calculate the optimal W and P with interaction (i.e., competitive firms) such that the stock of a renewable resource in equilibrium is X* (1/2) Max X = 0.5 for a low and high concern about it as depicted by β = 1 and β = 2, respectively [Monopolistic firm: if β = 1, a* = 1.1, if β = 2, a* = 1.357; if β = 1, b* = 0.828, if β = 2, b* = 0.472; if β = 1, W* = 0.666, if β = 2, W* = 1.333; if β = 1, no P*, if β = 2, P* = 0.375; for competitive firms, if β = 1, W* = 0.25, if β = 2, W* = 0.5; if β = 1, no P*, if β = 2, P* = 1]
5. Calculate the optimal the optimal P0 or PT such that the stock of a non-renewable resource at final time T is XT* = (1/2) Max X = 0.5 for a low and high concern about it as depicted by β = 1 and β = 2, respectively [if β = 1, P0* = 0.454, if β = 2, P0* = 0.476; if β = 1, PT* = 1.372, if β = 2, PT* = 1.559]
6. If 25 Billion CM is the global CO2 emissions in 1990 to be shared by developed and developing countries, based on the OECD (DC) and non-OECD (LDC) emissions in 2020 at 35 and 65%, respectively, with the bargaining power at γ = 0.5, calculate the pollution rights in terms of Nash bargaining, Kalai-Smorodinsky, Rawls, Sovereignty, Responsibility, Solidarity [Nash short-run is DC = 17.90, LDC = 7.10; Nash long-run is DC = 20.5, LDC = 4.5; Kalai-Smorodinsky is DC = 22.14, LDC = 2.86; Rawls is DC = 12.5, LDC = 12.5; Sovereignty is DC = 16.25, LDC = 8.75; Responsibility is DC = 8.75, LDC = 16.25; Solidarity is DC = 2.86, LDC = 22.14. Note that the Harsanyi

Table 4.26 CBA and MCA ranking of Project A and Project B in simple contexts

	NPV	BCR	IRR	MAUT	TOPSIS	VIKOR			ELECTRE		PROMETHEE
	(r = 0)					S	R	Q	c(i,j)	d(i,j)	Φ(i)
Project A	7	**6.94**	**4.40**	**0.81**	*0.230*	0.33	0.16	0	**0.66**	0.5	**0.5**
Project B	**9**	4.48	1.50	0.46	*0.230*	**0.66**	**0.66**	1	0.33	0.5	0.125

Abbreviations: NPV = Net Present Value, BCR = Benefit Cost Ratio, IRR = Internal Rate of Return, MAUT = Multi-Attribute Utility Theory, TOPSIS = The Technique for Order of Preference by Similarity to Ideal Solution, VIKOR = The Multi-Criteria Optimisation and Compromise Solutions (Serbian), ELECTRE = The ELimination Et Choix Traduisant La REalite, PROMETHEE = The Preference Ranking Organisation METHod for Enrichment Evaluation. *Notes* Discount rate at r = 0; S, R, Q, c(i,j), d(i,j) and Φ(i) are detailed in text. Bold text identifies the chosen project (in rows) according to the specified methodology (in columns).

Table 4.27 CBA and MCA ranking of Project A and Project B in complicated contexts

	EU	EV	FS	IM	SWF	IW
	$U(x) = \sqrt{x}$			Gini	ε = 0.5	U(x) = Log[x]
Project A	**2.63**	**7**	**4.25**	**0.07**	13.92	**10.00**
Project B	1.94	**9**	3.50	0.27	**16.49**	8.66

Abbreviations: EU = Expected Utility, EV = Expected Value, FS = Fuzzy Set, IM = Inequality Measure, SWF = Social Welfare Function, IW = Inequality Weights. *Notes* weco = 2/3, wsoc = 1/6, wenv = 1/6. Bold text identifies the chosen project (in rows) according to the specified methodology (in columns).

equilibrium if $SWF = \{[(Y/tyn)^{an}]^{1-ep} + [(Y/tys)^{as}]^{1-ep}\}^{1/(1-ep)}$ with ep = 0.5 is YN = 8.93, YS = 16.07]

7. Choose among projects A and B in terms on NPV, BCR, IRR
8. Choose among projects A and B in terms on MAUT, TOPSIS, VIKOR, ELECTRE and PROMETHEE
9. Choose among projects A and B in terms on EU, EV, FS
10. Choose among projects A and B in terms on IM, IW, SWF.

Results of exercise 7 and exercise 8 are summarised in Table 4.26.

Results of exercise 9 and exercise 10 are summarised in Table 4.27.

References

Abbott, J. K., & Wilen, J. E. (2011). Dissecting the tragedy: A spatial model of behavior in the commons. *Journal of Environmental Economics and Management, 62*, 386–401.

Abrell, J., et al. (2018). How robust is the uniform emissions pricing rule to social equity concerns? *Journal of Environmental Economics and Management, 92*, 783–914.

Adler, M. D. (2016). Benefit–cost analysis and distributional weights: An overview. *Review of Environmental Economics and Policy, 10*, 264–285.

Ahlheim, M. (1998). Contingent valuation and the budget constraint. *Ecological Economics, 27*, 205–211.

Aldy, J. E., et al. (2017). Comparing emissions mitigation efforts across countries. *Climate Policy, 17*, 501–515.

Almeida-Lenero, L., et al. (2017). Not the same for everyone: Community views of Mexico's payment for environmental services programmes. *Environmental Conservation, 44*, 201–211.

Althouse, J., et al. (2020). Ecological macroeconomics in the open economy: Sustainability, unequal exchange and policy coordination in a center-periphery model. *Ecological Economics, 172*. Art. No. 106628.

Anderson, K., et al. (2020). A factor of two: How the mitigation plans of 'climate progressive' nations fall far short of Paris-compliant pathways. *Climate Policy, 20*, 1290–1304.

Antelo, M., & Loureiro, M. L. (2009). Asymmetric information, signaling and environmental taxes in oligopoly. *Ecological Economics, 68*, 1430–1440.

Anthoff, D., et al. (2009). Equity weighting and the marginal damage costs of climate change. *Ecological Economics, 68*, 836–849.

Arguedas, C., & Van Soest, D. P. (2009). On reducing the windfall profits in environmental subsidy programs. *Journal of Environmental Economics and Management, 58*, 192–205.

Aronsson, T. (1999). On cost benefit rules for green taxes. *Environmental & Resource Economics, 13*, 31–43.

Aronsson, T., & Blomquist, S. (2003). On environmental taxation under uncertain environmental damage. *Environmental & Resource Economics, 24*, 183–196.

Aswani, S., et al. (2018). Marine resource management and conservation in the Anthropocene. *Environmental Conservation, 45*, 192–202.

Bai, C., et al. (2020). Will income inequality influence the abatement effect of renewable energy technological innovation on carbon dioxide emissions? *Journal of Environmental Management, 264*. Art. No. 110482.

Barcena-Ruiz, J. C., & Campo, M. L. (2017). Taxes versus standards under cross-ownership. *Resource and Energy Economics, 50*, 36–50.

Bauer, N., et al. (2020). Quantification of an efficiency–sovereignty trade-off in climate policy. *Nature, 588*, 261–266.

Baumgartner, S., et al. (2017). Income inequality and willingness to pay for environmental public goods. *Journal of Environmental Economics and Management, 85*, 35–611.

Baumol, W. J., Oates, W. E. (1988). The theory of environmental policy, Cambridge: Cambridge University Press.

Bayinder-Upmann, T. (2004). On the double dividend under imperfect competition. *Environmental & Resource Economics, 28*, 169–194.

Beckerman, W., & Pasek, J. (1997). Plural values and environmental valuation. *Environmental Values, 6*, 65–86.

Belfiori, M. E. (2018). Climate change and intergenerational equity: Revisiting the uniform taxation principle on carbon energy inputs. *Energy Policy, 121*, 292–299.

BenDor, T., et al. (2009). Ecological and economic sustainability in fishery management: A multi-agent model for understanding competition and cooperation. *Ecological Economics, 68*, 1061–1073.

Bene, C., & Doyen, L. (2000). Storage and viability of a fishery with resource and market dephased seasonalities. *Environmental & Resource Economics, 15*, 1–26.

Benites-Lazaro, L. L., & Mello-Théry, N. A. (2019). Empowering communities? Local stakeholders' participation in the clean development mechanism in Latin America. *World Development, 114*, 254–266.

Berger, K., et al. (1992). The oil market and international agreements on CO2 emissions. *Resources and Energy, 14*, 315–336.

Bernard, A., et al. (2008). A two-level dynamic game of carbon emission trading between Russia, China, and Annex B countries. *Journal of Econimic Dynamics and Control, 32*, 1830–1856.

Blais-McPherson, M., & Rudiak-Gould, P. (2017). Strengthening inter-disciplinary and inter-ideological collaboration on REDD: A cultural theory approach. *Global Environmental Change, 42*, 13–23.
Blamey, R. (1998). Contingent valuation and the activation of environmental norms. *Ecological Economics, 24*, 47–72.
Bohringer, C., et al. (2012). The role of border carbon adjustment in unilateral climate policy: Overview of an Energy Modeling Forum study (EMF 29). *Energy Economics, 34*, S97–S110.
Bolis, I., et al. (2017). Are we making decisions in a sustainable way? A comprehensive literature review about rationalities for sustainable development. *Journal of Cleaner Production, 145*, 310–322.
Bonilla-Alicea, R. J., & Fu, K. (2019). Systematic map of the social impact assessment field. *Sustainability, 11*. Art. No. 4106.
Boom, J. T., & Dijkstra, B. R. (2009). Permit trading and credit trading: A comparison of cap-based and rate-based emissions trading under perfect and imperfect competition. *Environmental & Resource Economics, 44*, 107–136.
Boran, I. (2017). two concepts of wrongful harm: A conceptual map for the warsaw international mechanism for loss and damage. *Ethics, Policy and Environment, 20*, 195–207.
Borner, J., et al. (2016). Will up-scaled forest conservation incentives in the Peruvian Amazon produce cost-effective and equitable outcomes? *Environmental Conservation, 43*, 407–416.
Boyca, J. K., et al. (2016). Measuring environmental inequality. *Ecological Economics, 124*, 114–123.
Brecard, D. (2011). Environmental tax in a Green Market. *Environmental & Resource Economics, 49*, 387–403.
Briggs, R. J. (2011). Prices vs. quantities in a dynamic problem: Externalities from resource extraction. *Resource and Energy Economics, 33*, 843–854.
Brink, E., & Wamsler, C. (2017). Collaborative governance for climate change adaptation: Mapping citizen–municipality interactions. *Environmental Policy and Governance, 28*, 82–97.
Buchholz, W., et al. (2016). Equity as a prerequisite for stability of cooperation on global public good provision. *Environmental & Resource Economics, 65*, 61–781.
Budolfson, M., et al. (2017). The comparative importance for optimal climate policy of discounting, inequalities and catastrophes. *Climate Change, 145*, 481–494.
Cameron, T. A. (1988). A new paradigm for valuing non-market goods using referendum data: Maximum likelihood estimation by censored logistic regression. *Journal of Environmental Economics and Management, 15*, 355–367.
Campbell, H. F., & Brown, R. P. C. (1983). *Benefit-cost analysis*. Cambridge: CUP.
Caparros, A. (2009). Delayed carbon sequestration and rising carbon prices. *Climate Change, 96*, 421–441.
Cappellan-Perez, I., et al. (2015). More growth? On unfeasible options to overcome critical energy constraints and climate change. *Sustainability Science, 10*, 397–411.
Carlsson, F. (2000). Environmental taxation and strategic commitment in duopoly models. *Environmental & Resource Economics, 15*, 243–256.
Carlsson, F., et al. (2010). Dealing with ignored attributes in choice experiments on valuation of Sweden's environmental quality objectives. *Environmental & Resource Economics, 47*, 65–89.
Carson, R. T., et al. (2001). Contingent valuation: Controversies and evidence. *Environmental & Resource Economics, 19*, 173–210.
Cavaliere, A. (2000). Overcompliance and voluntary agreements. *Environmental & Resource Economics, 17*, 195–202.
Chee, Y. E. (2004). An ecological perspective on the valuation of ecosystem services. *Biological Conservation, 120*, 549–565.
Chen, C., et al. (2013). An inexact robust optimization method for supporting carbon dioxide emissions management in regional electric-power systems. *Energy Economics, 40*, 441–456.
Chen, Y., et al. (2020). Quantifying economics impacts of climate change under nine future emission scenarios within CMIP6. *Science of the Total Environment, 703*. Art. No. 134950.

Chesney, M., & Taschini, L. (2012). The endogenous price dynamics of emission allowances and an application to CO 2 option pricing. *Applied Mathematical Finance, 19*, 447–475.

Chu, H., & Lai, C. C. (2014). Abatement R&D, market imperfections, and environmental policy in an endogenous growth model. *Journal of Economic Dynamics & Control, 41*, 20–37.

Coase, R. (1960). The problem of social cost. *Journal of Law and Economics, 3*, 1–44.

Common, M., et al. (1997). Do existence values for cost benefit analysis exist? *Environmental & Resource Economics, 9*, 225–238.

Costa, D., et al. (2019). A systematic review of life cycle sustainability assessment: Current state, methodological challenges, and implementation issues. *Science of the Total Environment, 686*, 774–787.

Curseu P. L., Schruiser, S. G. L. (2020). Participation and goal achievement of multiparty collaborative systems dealing with complex problems: a natural experiment, Sustainability 12: art. no. 987.

D'Amato, A., & Dijkstra, B. R. (2015). Technology choice and environmental regulation under asymmetric information. *Resource and Energy Economics, 41*, 224–247.

Dachary-Bernard, J., & Rambonilaza, T. (2012). Choice experiment, multiple programmes contingent valuation and landscape preferences: How can we support the land use decision making process? *Land Use Policy, 29*, 846–854.

Daigneault, A., et al. (2017). Equitably slicing the pie: Water policy and allocation. *Ecological Economics, 131*, 449–459.

Dawson, N. M., et al. (2018). Barriers to equity in REDD+: Deficiencies in national interpretation processes constrain adaptation to context. *Environmental Science & Policy, 88*, 1–9.

De Cian, E., et al. (2016). Alleviating inequality in climate policy costs: An integrated perspective on mitigation, damage and adaptation. *Environmental Research Letters, 11*. Art. No. 074015.

Diaz-Balteiro, L., & Romero, C. (2008). Valuation of environmental goods: A shadow value perspective. *Ecological Economics, 64*, 517–520.

Dong, F., et al. (2018). How can China allocate CO2 reduction targets at the provincial level considering both equity and efficiency? Evidence from its Copenhagen accord pledge. *Resources, Conservation & Recycling, 130*, 31–43.

Dooley, K., & Gupta, A. (2017). Governing by expertise: The contested politics of (accounting for) land-based mitigation in a new climate agreement. *International Environmental Agreements: Politics, Law and Economics, 17*, 483–500.

Dovie, D. B. K. (2019). Case for equity between Paris Climate agreement's Co-benefits and adaptation. *Science of the Total Environment, 656*, 732–739.

Dowling, A. W., et al. (2016). A framework for multi-stakeholder decision making and conflict resolution. *Computers & Chemical Engineering, 90*, 136–150.

Durand-Lasserve, O., et al. (2010). Uncertain long-run emissions targets, CO2 price and global energy transition: A general equilibrium approach. *Energy Policy, 38*, 5108–5122.

Duro, J. A. (2016). Intercountry inequality on greenhouse gas emissions and world levels: An integrated analysis through general distributive sustainability indexes. *Ecological Indicators, 66*, 173–179.

Ebert, U. (1998). Relative standards: A positive and normative analysis. *Journal of Economics, 67*, 17–38.

Ebert, U., & Von Dem Hagen, O. (1998). Pigouvian taxes under imperfect competition if consumption depends on emissions. *Environmental & Resource Economics, 12*, 507–513.

Elevitch, C. R., & Johnson, C. R. (2020). A procedure for ranking parameter importance for estimation in predictive mechanistic models. *Ecological Modelling, 4191*. Art. No. 108948.

Etxano, I., & Villalba Eguiluz, U. (2021). Twenty-five years of social multi-criteria evaluation (SMCE) in the search for sustainability: Analysis of case studies. *Ecological Economics, 188*. Art. No. 107131.

Faccioli, M., et al. (2016). Do we care about sustainability? An analysis of time sensitivity of social preferences under environmental time-persistent effects. *Journal of Environmental Management, 177*, 356–364.

Fang, C., & Ma, T. (2020). Stylized agent-based modeling on linking emission trading systems and its implications for China's practice. *Energy Economics, 92*. Art. No. 104916.

Farnsworth, K. D., et al. (2015). The complexity of biodiversity: A biological perspective on economic valuation. *Ecological Economics, 120*, 350–354.

Fauzi, R. T., et al. (2019). Exploring the current challenges and opportunities of life cycle sustainability assessment. *Sustainability, 11*. Art. No. 636.

Ferguson, I. (2017). Stochastic economic analyses of even-aged timber growing, New Zealand Journal of Forestry Science 47: Art. No. 9.

Ferretti, V., et al. (2019). Studying the generation of alternatives in public policy making processes. *European Journal of Operational Research, 273*, 353–363.

Fischer, C. (2005). Project-based mechanisms for emission reductions: Balancing trade-offs with baselines. *Energy Policy, 33*, 1807–1823.

Fitter, A. H. (2013). Are ecosystem services replaceable by technology? *Environmental & Resource Economics, 55*, 513–524.

Fontes, J., Tarne, P., Traverso, M., Bernstein, P., et al. (2018). Product social impact assessment. *International Journal of Life Cycle Assessment, 23*, 547–555.

França, W. T., et al. (2021). Integrating life cycle assessment and life cycle cost: A review of environmental-economic studies. *International Journal of Life Cycle Assessment, 26*, 244–274.

Fasth, T. et al. (2018). Measuring conflicts using cardinal ranking: an application to decision analytic conflict evaluations, Advances in Operations Research DOI 10.11555/2018/8290434.

Fullerton, D., & Muehlegger, E. (2019). Who bears the economic burdens of environmental regulations? *Review of Environmental Economics and Policy, 13*, 62–82.

Gayer, T., & Viscusi, W. (2016). Determining the proper scope of climate change policy benefits in U.S. Regulatory analyses: Domestic versus global approaches. *Review of Environmental Economics and Policy, 10*, 245–263.

Gewirtzman, J., et al. (2018). Financing loss and damage: Reviewing options under the Warsaw International Mechanism. *Climate Policy, 18*, 1076–1086.

Giergiczny, M., et al. (2012). Including cost income ratio into utility function as a way of dealing with 'exploding' implicit prices in mixed logit models. *Journal of Forest Economics, 18*, 370–380.

Gren, I. M. (2008). Adaptation and mitigation strategies for controlling stochastic water pollution: An application to the Baltic Sea. *Ecological Economics, 66*, 337–347.

Gren, I. M., et al. (2012). Stochastic carbon sinks for combating carbon dioxide emissions in the EU. *Energy Economics, 34*, 1523–1531.

Groeneveld, R. A. (2020). Welfare economics and wicked problems in coastal and marine governance. *Marine Policy, 117*. Art. No. 193945.

Gupta, J., et al. (2020). Re-imagining the driver–pressure–state–impact–response framework from an equity and inclusive development perspective. *Sustainability Science, 15*, 503–520.

Haas, J. C., et al. (2019). How fair can incentive-based conservation get? The interdependence of distributional and contextual equity in Vietnams' payments for Forest Environmental Services Program. *Ecological Economics, 160*, 205–214.

Hagem, C., et al. (2006). Market power with interdependent demand: Sale of emission permits and natural gas from Russia. *Environmental & Resource Economics, 34*, 211–227.

Haiata-Falah, C. (2016). Uncertainty and speculators in an auction for emissions permits. *Journal of Regulatory Economics, 49*, 315–343.

Hajkowicz, S. (2007). Can we put a price tag on nature? Rethinking approaches to environmental valuation. *Australasian Journal of Environmental Management, 14*, 22–26.

Hanemann, W. M. (1984). Welfare evaluations in contingent evaluation experiments with discrete responses. *Americal Journal of Agricultural Economics, 66*, 332–346.

Harris, P. G. (2017). China's Paris pledge on climate change: Inadequate and irresponsible. *Journal of Environmental Studies and Science, 7*, 102–107.

Hart, R. (2004). Growth, environment and innovation—A model with production vintages and environmentally oriented research. *Journal of Environmental Economics and Management, 48*, 1078–1098.

Haurie, A., & Viguier, L. (2003). A stochastic dynamic game of carbon emissions trading. *Environmental Modeling and Assessment, 8*, 239–248.
Healy, N., & Barry, J. (2017). Politicizing energy justice and energy system transitions: Fossil fuel divestment and a "just transition." *Energy Policy, 108*, 451–459.
Hedenus, F., & Azar, C. (2009). Bioenergy plantations or long-term carbon sinks? A model based analysis. *Biomass and Bioenergy, 33*, 1693–1702.
Heijnen, P., & Kooreman, P. (2006). Modelling strategic responses to car and fuel taxation. *Journal of Transport Economics and Policy, 40*, 203–223.
Holz, C., et al. (2018). Fairly sharing 1.5: National fair shares of a 1.5 °C-compliant global mitigation effort. *International Environmental Agreements: Politics, Law and Economics, 18*, 117–134.
Hoogmartens, R., et al. (2014). Bridging the gap between LCA, LCC and CBA as sustainability assessment tools. *Environmental Impact Assessment Review, 48*, 27–33.
Huang, C., et al. (2018). Interstate pollution spillover and setting environmental standards. *Journal of Cleaner Production, 170*, 1544–1553.
Huettner, M., et al. (2009). A comparison of baseline methodologies for 'Reducing Emissions from Deforestation and Degradation'. *Carbon Balance and Management, 4*. Art. No. 4.
Huggel, C., et al. (2016). Reconciling justice and attribution research to advance climate policy. *Nature Clinical Practice Endocrinology & Metabolism, 6*, 901–908.
Ibanez-Lopez, A.S. & Moratilla-Soria, B.Y. (2017). An assessment of Spain's new alternative energy support framework and its long-term impact on wind power development and system costs through behavioral dynamic simulation, Energy 138 (C): 629–646.
Iofrida, N., et al. (2018). Can social research paradigms justify the diversity of approaches to social life cycle assessment? *International Journal of Life Cycle Assessment, 23*, 464–480.
Isyaku, U., et al. (2017). Framing justice in REDD+ governance: Centring transparency, equity and legitimacy in readiness implementation in West Africa. *Environmental Conservation, 44*, 212–220.
Jabbri, M., et al. (2020). Differentiating countries based on the sustainable development proximities using the SDG indicators. *Environment, Development and Sustainability, 22*, 6405–6423.
Johansson, P. O. (1992). Altruism in cost-benefit analysis. *Environmental & Resource Economics, 2*, 605–613.
Jorgensen, A., et al. (2013). Analysis of the link between a definition of sustainability and the life cycle methodologies. *International Journal of Life Cycle Assessment, 18*, 1440–1449.
Kahneman, D., & Knetsch, J. L. (1992). Valuing public goods: The purchase of moral satisfaction. *Journal of Environmental Economics and Management, 22*, 57–70.
Kahraman, C., et al. (2017). Fuzzy economic analysis methods for environmental economics. *Intelligent Systems Reference Library, 113*, 315–346.
Karydas, C., & Zhang, L. (2019). Green tax reform, endogenous innovation and the growth dividend. *Journal of Environmental Economics and Management, 97*, 158–181.
Kirchgassner, G., & Schneider, F. (2003). On the political economy of environmental policy. *Public Choice, 115*, 369–396.
Kirchhoff, S. (2000). Green business and blue angels. *Environmental & Resource Economics, 15*, 403–420.
Klenert, D., et al. (2018). Environmental taxation, inequality and Engel's law: The double dividend of redistribution. *Environmental & Resource Economics, 71*, 605–624.
Klinsky, S., et al. (2017). Operationalizing equity and supporting ambition: Identifying a more robust approach to 'respective capabilities.' *Climate and Development, 9*, 287–297.
Kneese, A. V. (1986). Environmental economics. *Science of the Total Environment, 56*, 155–169.
Koch, N., et al. (2017). Permits vs. offsets under investment uncertainty. *Resource and Energy Economics, 49*, 33–471.
Konishi, Y. (2011). Efficiency properties of binary ecolabeling. *Resource and Energy Economics, 33*, 798–819.
Kopittke, P. M., et al. (2019). Soil and the intensification of agriculture for global food security. *Environment International, 132*. Art. No. 105078.

Krupnik, A., & McLaughlin, D. (2012). Valuing the impacts of climate change on terrestrial ecosystem services. *Climate Change Economics, 3*. Art. No. 4.

Kuosmanen, T., & Kortelainen, M. (2007). Valuing environmental factors in cost-benefit analysis using data envelopment analysis. *Ecological Economics, 62*, 56–65.

Lahn, B. (2018). In the light of equity and science: Scientific expertise and climate justice after Paris. *International Environmental Agreements: Politics, Law and Economics, 18*, 29–43.

Lange, A., & Schwirplies, C. (2017). (Un) fair delegation: Exploring the strategic use of equity rules in international climate negotiations. *Environmental & Resource Economics, 67*, 505–533.

Larkin, A., et al. (2018). What if negative emission technologies fail at scale? Implications of the Paris agreement for big emitting nations. *Climate Policy, 18*, 690–714.

Laurila-Pant, M., et al. (2019). Incorporating stakeholders' values into environmental decision support: A Bayesian belief network approach. *Science of the Total Environment, 697*, 134026.

Lee, T. C. (2011). Endogenous market structures in non-cooperative international emissions trading. *Mitigation and Adaptation Strategies for Global Change, 16*, 663–675.

Lefebvre, M., et al. (2020). Adoption of environment-friendly agricultural practices with background risk: Experimental evidence. *Environmental & Resource Economics, 76*, 405–428.

Lindfors, A. (2021). Assessing sustainability with multi-criteria methods: A methodologically focused literature review. *Environmental and Sustainability Indicators, 12*. Art. No. 100149.

Linnerooth-Bayer, J. et al. (2016). Expert engagement in participatory processes: translating stakeholder discourses into policy options, Natural Hazards 81: 69–88.

Liu, L., et al. (2017). An equity-based framework for defining national responsibilities in global climate change mitigation. *Climate and Development, 9*, 152–153.

Lyssenko, N., & Martínez-Espiñeira, R. (2012). 'Been there done that': Disentangling option value effects from user heterogeneity when valuing natural resources with a use component. *Environmental Management, 50*, 819–836.

MacKenzie, I. A., & Ohndorf, M. (2012). Optimal monitoring of credit-based emissions trading under asymmetric information. *Journal of Regulatory Economics, 42*, 180–203.

Macombe, C., et al. (2018). Extended community of peers and robustness of social LCA. *International Journal of Life Cycle Assessment, 23*, 492–506.

Madariaga, B., & McConnell, K. E. (1987). Exploring existence value. *Water Resources Research, 23*, 936–942.

Maloney, M. T., & Yandle, B. (1984). Estimation of the cost of air pollution control regulation. *Journal of Environmental Economics and Management, 11*, 244–263.

Markkanen, S., & Anger-Kraavi, A. (2019). Social impacts of climate change mitigation policies and their implications for inequality. *Climate Policy, 19*, 827–844.

Martinet, V., & Blanchard, F. (2009). Fishery externalities and biodiversity: Trade-offs between the viability of shrimp trawling and the conservation of Frigatebirds in French Guiana. *Ecological Economics, 68*, 2960–2968.

Masri, S., et al. (2020). Social and spatial distribution of soil lead concentrations in the City of Santa Ana, California: Implications for health inequities. *Science of the Total Environment, 74315*. Art. No. 140764.

Matsumura, T., & Yamagishi, A. (2017). Long-run welfare effect of energy conservation regulation. *Economic Letters, 154*, 64–68.

McCabe, A., & Halog, A. (2018). Exploring the potential of participatory systems thinking techniques in progressing SLCA. *International Journal of Life Cycle Assessment, 23*, 739–750.

Medin, H., et al. (2001). The assumption of equal marginal utility of income: How much does it matter? *Ecological Economics, 36*, 397–411.

Mejean, A., et al. (2020). Catastrophic climate change, population ethics and intergenerational equity. *Climate Change, 163*, 873–890.

Meunier, G. (2011). Emission permit trading between imperfectly competitive product markets. *Environmental & Resource Economics, 50*, 347–364.

Mi, Z., et al. (2019). Assessment of equity principles for international climate policy based on an integrated assessment model. *Natural Hazards, 95*, 309–323.

Miah, J. H., et al. (2017). A hybridised framework combining integrated methods for environmental life cycle assessment and life cycle costing. *Journal of Cleaner Production, 168*, 846–866.
Mielke, J. et al. (2016). Stakeholder involvement in sustainability science – a critical review, Energy Research & Social Science 17: 71–81.
Montgomery, W. D. (1972). Markets in licenses and efficient pollution control programs, Journal of Economic Theory 5 (3): 395–418.
Nahuelhual, L., et al. (2018). Patterns of ecosystem services supply across farm properties: Implications for ecosystem services-based policy incentives. *Science of the Total Environment, 634*, 941–950.
Nannerup, N. (1998). Strategic environmental policy under incomplete information. *Environmental & Resource Economics, 11*, 61–78.
Newburn, D., & Berck, P. (2011). Exurban development. *Journal of Environmental Economics and Management, 62*, 323–336.
Okereke, C., & Coventry, P. (2016). Climate justice and the international regime: Before, during, and after Paris. *Wiley Interdisciplinary Reviews: Climate Change, 7*, 834–851.
Onat, N. C., et al. (2017). Systems thinking for life cycle sustainability assessment: A review of recent developments, applications, and future perspectives. *Sustainability, 9*. Art. No. 706.
Orlov, A., & Grethe, H. (2012). Carbon taxation and market structure: A CGE analysis for Russia. *Energy Policy, 51*, 696–707.
Ostrom, E. (2012). Nested externalities and polycentric institutions: Must we wait for global solutions to climate change before taking actions at other scales? *Economic Theory, 49*, 353–369.
Pauw, W. P., et al. (2020). Conditional nationally determined contributions in the Paris Agreement: Foothold for equity or Achilles heel? *Climate Policy, 20*, 468–484.
Paylor, A. (2017). The social–economic impact of shale gas extraction: A global perspective. *Third World Quarterly, 38*, 340–355.
Pérez-Cirera, V., & Lovett, J. C. (2006). Power distribution, the external environment and common property forest governance: A local user groups model. *Ecological Economics, 59*, 341–352.
Petti, L., et al. (2018). Systematic literature review in social life cycle assessment. *International Journal of Life Cycle Assessment, 23*, 422–431.
Piga, C. (2003). Pigouvian taxation in tourism. *Environmental & Resource Economics, 26*, 343–359.
Pigou, A. C. (1920). The economics of welfare, London: Macmillan.
Pizzol, M., et al. (2015). Monetary valuation in life cycle assessment: A review. *Journal of Cleaner Production, 86*, 170–179.
Plottu, E., & Plottu, B. (2007). The concept of total economic value of environment: A reconsideration within a hierarchical rationality. *Ecological Economics, 61*, 52–61.
Pollok, L., et al. (2021). Social life cycle assessments: A review on past development, advances and methodological challenges. *Sustainability, 13*. Art. No. 10286.
Prato, T. (1999). Multiple attribute decision analysis for ecosystem management. *Ecological Economics, 30*, 207–222.
Price, C. (2000). Valuation of unpriced products. Contingent valuation, cost-benefit analysis and participatory democracy. *Land Use Policy, 17*, 187–196.
Ramos Huarachi, D. A., et al. (2020). Past and future of social life cycle assessment: Historical evolution and research trends. *Journal of Cleaner Production, 264*. Art. No. 121506.
Rao, N. D., et al. (2017). Improving poverty and inequality modelling in climate research. *Nature—Climate Change, 7*, 857–862.
Rege, M. (2000). Strategic policy and environmental quality: Helping the domestic industry to provide credible information. *Environmental & Resource Economics, 15*, 279–296.
Reichenbach, J., & Requate, T. (2012). Subsidies for renewable energies in the presence of learning effects and market power. *Resource and Energy Economics, 34*, 236–254.
Rentschler, J., et al. (2018). On imperfect competition and market distortions: The causes of corporate under-investment in energy and material efficiency. *International Economics and Economic Policy, 15*, 159–183.
Riddel, M. (2011). Uncertainty and measurement error in welfare models for risk changes. *Journal of Environmental Economics and Management, 61*, 341–354.

Ritchie, J., & Reay, D. S. (2017). Delivering the two degree global climate change target using a flexible ratchet framework. *Climate Policy, 17*, 1031–1045.
Roberts, E., & Pelling, M. (2020). Loss and damage: An opportunity for transformation? *Climate Policy, 20*, 758–771.
Roberts, T. J., et al. (2017). How will we pay for loss and damage? *Ethics, Policy & Environment, 20*, 208–226.
Rogers, M. F., et al. (1997). The precautionary principle for environmental management: A defensive-expenditure application. *Journal of Environmental Management, 51*, 343–360.
Roseland, M., & Spiliotopoulou, M. (2016). Converging urban agendas: Toward healthy and sustainable communities. *Social Sciences, 5*. Art. No. 28.
Rozenberg, J., et al. (2020). Instrument choice and stranded assets in the transition to clean capital. *Journal of Environmental Economics and Management, 100*. Art. No. 102183.
Rubin, J. D. (1996). A model of intertemporal emission trading, banking, and borrowing. *Journal of Environmental Economics and Management, 31*, 269–286.
Salliou, N., et al. (2017). A participatory Bayesian Belief Network approach to explore ambiguity among stakeholders about socio-ecological systems. *Environmental Modelling and Software, 96*, 199–209.
Sandsmark, M., & Vennemo, H. (2007). A portfolio approach to climate investments: CAPM and endogenous risk. *Environmental & Resource Economics, 37*, 681–695.
Sartzetakis, E. S., & Tsigars, P. D. (2009). Uncertainty and the double dividend hypothesis. *Environment and Development Economics, 14*, 565–585.
Scarborough, H. (2011). Intergenerational equity and the social discount rate. *Australian Journal of Agricultural and Resource Economics, 55*, 145–158.
Schenker, O. (2013). Exchanging goods and damages: The role of trade on the distribution of climate change costs. *Environmental & Resource Economics, 54*, 261–282.
Schennach, S. M. (2000). The economics of pollution permit banking in the context of Title IV of the 1990 Clean Air Act Amendments. *Journal of Environmental Economics and Management, 40*, 189–210.
Scheufele, G., & Bennett, J. (2012). Response strategies and learning in discrete choice experiments. *Environmental & Resource Economics, 52*, 435–453.
Schläpfer, F., et al. (2004). Validation of stated preferences for public goods: A comparison of contingent valuation survey response and voting behaviour. *Ecological Economics, 51*, 1–16.
Schläpfer, F., & Brauer, I. (2007). Theoretical incentive properties of contingent valuation questions: do they matter in the field? *Ecological Economics, 62*, 451–460.
Schläpfer, F., & Getzner, M. (2020). Beyond current guidelines: A proposal for bridging behavioural economics to the design and analysis of stated preferences surves. *Ecological Economics, 176*. Art. No. 106720.
Schläpfer, F., & Schmitt, M. (2007). Anchors, endorsements, and preferences: A field experiment. *Resource and Energy Economics, 29*, 229–243.
Schläpfer, F. (2016). Democratic valuation (DV): Using majority voting principles to value public services. *Ecological Economics, 122*, 36–42.
Schoonbeek, L., & De Vries, F. P. (2009). Environmental taxes and industry monopolization. *Journal of Regulatory Economics, 36*, 94–106.
Scolobig, A., Lilliestam, J. (2016). Comparing approaches for the integration of stakeholder perspectives in environmental decision making, Resources 5: art. no. 37.
Scrieciu, S. T., et al. (2013). Pushing the boundaries of climate economics: Critical issues to consider in climate policy analysis. *Ecological Economics, 85*, 155–165.
Selden, T. M., & Terrones, M. E. (1993). Environmental legislation and enforcement: A voting model under asymmetric information. *Journal of Environmental Economics and Management, 24*, 212–228.
Shao, J., et al. (2019). Adoption of electric vehicles. *Journal of Transport Economics and Policy, 53*, 175–198.
Sharma, R., & Gupta, K. (2020). Life cycle modelling for environmental management: A review of trends and linkages. *Environmental Monitoring and Assessment, 192*. Art. No. 51.

Shen, B., et al. (2021). Green technology adoption in textiles and apparel supply chains with environmental taxes. *International Journal of Production Research, 59*, 4157–4174.

Shrivastava, M. K., & Bhaduri, S. (2019). Market-based mechanism and 'climate justice': Reframing the debate for a way forward. *International Environmental Agreements: Politics, Law and Economics, 19*, 497–513.

Silvestre, W. J., et al. (2022). Strategic sustainability integration: Merging management tools to support business model decisions. *Business Strategy and the Environment, 31*, 2052–2067.

Singh, R., & Weninger, Q. (2009). Bioeconomies of scope and the discard problem in multiple-species fisheries. *Journal of Environmental Economics and Management, 58*, 72–92.

Singh, R., et al. (2006). Fisheries management with stock growth uncertainty and costly capital adjustment. *Journal of Environmental Economics and Management, 52*, 582–599.

Soretz, S. (2007). Efficient dynamic pollution taxation in an uncertain environment. *Environmental & Resource Economics, 36*, 57–84.

Stavins, R. N. (1996). Correlated uncertainty and policy instrument choice. *Journal of Environmental Economics and Management, 30*, 218–232.

Stoeckl, N., et al. (2018). The crowding out of complex social goods. *Ecological Economics, 144*, 65–72.

Stranlund, J. K., et al. (2019). Tying enforcement to prices in emissions markets: An experimental evaluation. *Journal of Environmental Economics and Management, 98*. Art. No. 102246.

Stuhlmacher, M. et al. (2019). Cap-and-Trade and Emissions Clustering: A Spatial-Temporal Analysis of the European Union Emissions Trading Scheme, Journal of Environmental Management 249: 109352.

Subramanian, K., et al. (2018). Relevance and feasibility of the existing social LCA methods and case studies from a decision-making perspective. *Journal of Cleaner Production, 171*, 690–703.

Suhariyanto, T. T., et al. (2017). Multi-life cycle assessment for sustainable products: A systematic review. *Journal of Cleaner Production, 165*, 677–696.

Sunnevag, K. J. (2003). Auction design for the allocation of emission permits in the presence of market power. *Environmental & Resource Economics, 26*, 385–400.

Sun, H. et al. (2019). Trade Openness and Carbon Emissions: Evidence from Belt and Road Countries, Sustainability 11 (9): Art. No. 2682.

Sureau, S., et al. (2018). Social life-cycle assessment frameworks: A review of criteria and indicators proposed to assess social and socioeconomic impacts. *International Journal of Life Cycle Assessment, 23*, 904–920.

Talbot, B. (2018). Collective action problems and conflicting obligations. *Philosophical Studies, 175*, 2239–2261.

Taschini, L. (2010). Environmental economics and modeling marketable permits. *Asia-Pacific Financial Markets, 17*, 325–343.

Tilton, J. E. (2016). Global climate policy and the polluter pays principle: A different perspective. *Resources Policy, 50*, 117–1181.

Tokede, O., & Traverso, M. (2020). Implementing the guidelines for social life cycle assessment: past, present, and future. *International Journal of Life Cycle Assessment, 25*, 1910–1929.

Toniolo, S., et al. (2019). International standards with a life cycle perspective: Which dimension of sustainability is addressed? *International Journal of Life Cycle Assessment, 24*, 1765–1777.

Traverso, M., et al. (2018). Towards social life cycle assessment: A quantitative product social impact assessment. *International Journal of Life Cycle Assessment, 23*, 597–606.

Tsur, Y., & Zemel, A. (2008). Regulating environmental threats. *Environmental & Resource Economics, 39*, 297–310.

Ueckerdt, F., et al. (2019). The economically optimal warming limit of the planet. *Earth System Dynamics, 10*, 741–763.

UN (United Nations) (1992). Agenda 21: The United Nations programme of action from Rio. https://www.un.org/esa/dsd/agenda21.

UN (United Nations) (2015). Transforming our world: the 2030 Agenda for Sustainable Development. https://sdgs.un.org/2030agenda.

Van Zyl, H., et al. (2018). The cost of achieving South Africa's 'fair share' of global climate change mitigation. *Climate Policy, 18*, 1327–1339.

Van der Wilde, C. P., & Newel, J. P. (2021). Ecosystem services and life cycle assessment: A bibliometric review. *Resources, Conservation and Recycling, 169*. Art. No. 105461.

Vatn, A. (2004). Environmental valuation and rationality. *Land Economics, 80*, 1–18.

Venkatachalam, L. (2004). The contingent valuation method: A review. *Environmental Impact Assessment Review, 24*, 89–124.

Verchere, A. (2017). The middle-class collapse and the environment. *Ecological Economics, 131*, 510–523.

Vetter, H. (2009). Product variety, scale economies, and environmental taxes, B.E. *Journal of Theoretical Economics, 9*. Article No. 18.

Walsh, R. G., et al. (1990). Estimating the public benefits of protecting forest quality. *Journal of Environmental Management, 30*, 175–189.

Wang, G., et al. (2017). Impacts of climate change on water resources in the Yellow River basin and identification of global adaptation strategies. *Mitigation and Adaptation Strategies for Global Change, 22*, 67–83.

Weidema, B. P. (2018). The social footprint—A practical approach to comprehensive and consistent social LCA. *International Journal of Life Cycle Assessment, 23*, 700–709.

Whitehead, J. C. (2016). Plausible responsiveness to scope in contingent valuation. *Ecological Economics, 128*, 17–22.

Winkler, H. (2020). Putting equity into practice in the global stocktake under the Paris Agreement. *Climate Policy, 20*, 124–132.

Wirl, F. (2014). Taxes versus permits as incentive for the intertemporal supply of a clean technology by a monopoly. *Resource and Energy Economics, 36*, 248–269.

Wong, G. Y., et al. (2019). Narratives in REDD+ benefit sharing: Examining evidence within and beyond the forest sector. *Climate Policy, 19*, 1038–1051.

Wulf, C. et al. (2019). Review of sustainability assessment approaches based on life cycles. *Sustainability, 11*. Art. No. 5717.

Xu, L., et al. (2016). Carbon emission permit price volatility reduction through financial options. *Energy Economics, 53*, 248–260.

Yang, Q., et al. (2018). Damage to buildings and structures due to recent devastating wind hazards in East Asia. *Natural Hazards, 92*, 1321–1353.

Zagonari, F. (1998). International pollution problems: Unilateral initiatives by environmental groups in one country. *Journal of Environmental Economics and Management, 3*, 315–326.

Zagonari, F. (2016). Choosing among weight-estimation methods for multi-criterion analysis: A case study for the design of multi-purpose offshore platforms. *Applied Soft Computing, 39*, 1–10.

Zagonari, F. (2019). Learning and dynamic choices under uncertainty: From weighted regret and rejoice to expected utility. *Managerial and Decision Economics, 40*, 292–308.

Zagonari, F. (2020). Environmental sustainability is not worth pursuing unless it is achieved for ethical reasons. *Nature—Palgrave Communications, 6*. Art. No. 108.

Zagonari, F. (2021). Decommissioning vs. reusing offshore gas platforms within ethical decision-making for sustainable development: Theoretical framework with application to the Adriatic Sea. *Ocean and Coastal Management, 199*. Art. No. 105409.

Zagonari, F. (2022). Pope Francis vs. Patriarch Bartholomew to achieve global environmental sustainability: Theoretical insights supported by empirical results. *Development and Change* (under review).

Zagonari, F., & Rossi, C. (2014). A Negotiation Support System for disputes between Iraq and Turkey over the Tigris-Euphrates basin. Journal of Hydrology, 514: 65–84.

Zhao, H. et al. (2019). Inequality of household consumption and air pollution-related deaths in China. *Nature Communications, 10*. Art. No. 4337.

Zimm, C., & Nakicenovic, N. (2020). What are the implications of the Paris Agreement for inequality? *Climate Policy, 20*, 458–467.

Chapter 5
Discussion

Section 4.1.1 in Chap. 4 showed that if one or more assumptions behind the Economic General Equilibrium Model do not hold, then efficiency (i.e., maximisation of total welfare) is not achieved, *even* after implementing environmental policy measures. Some of these measures achieve do *not* efficiency within market-based policies, such as taxes under asymmetric information and imperfect competition or permits under uncertainty; subsidies within market-based policies achieve efficiency under imperfect competition at different optimal prices and quantities than those that would be advocated without policies; other policy measures, such as standards under asymmetric information or other command-and-control policies, do *not* achieve efficiency. This requires some measure of the inequality for the sustainability burden. In terms of acronyms, asymmetric information (ASY) or imperfect competition (IMP) or uncertainty (UNC) without equity (EQU) (in terms of inequality) represent a mistake (MIS1 hereafter) (i.e., ASY or IMP or UNC with EQU is correct).

In Sect. 4.1.1 in Chap. 4, I showed that if environmental issues in terms of pollution production *Y* and resource use *X* involve a decision interaction (i.e., if pollution produced or resources used by one decision-maker affects welfare or options of pollution production and resource use for another decision-maker), then *X* or *Y* might not be socially optimal at equilibrium *and* the shadow prices of *X* and *Y* might differ from the equilibrium market price; that is, sometimes positive (i.e., larger than 0) optimal *X* or *Y* equilibria do not exist, and sometimes they exist at positive (i.e., larger than 0) shadow prices that differ from the equilibrium market prices. This requires an agreement between decision-makers to attain the socially optimal equilibrium. In terms of acronyms, interaction (REL) without equity (EQU) (in terms of agreement) represent a mistake (MIS2 hereafter) (i.e., REL with EQU is correct).

Section 4.1.2 in Chap. 4 showed that if the environmental issues, both in terms of pollution production *Y* and resource use *X*, involve stocks (STO), then the shadow prices of *X* and *Y* might differ from the equilibrium market price. In that situation, Sect. 4.2.2 in Chap. 4 showed that monetary assessments (i.e., utility-based approaches UA, dose responses DR, replacement costs RC) will be inadequate. This requires the development of a dynamic model to assess the impacts of *X* and *Y* over

F. Zagonari, *Environmental Ethics, Sustainability and Decisions*,
https://doi.org/10.1007/978-3-031-21182-9_5

time in investment projects. In terms of acronyms, utility-based approaches (UA) or dose responses (DR) or replacement costs (RC) applied to environmental stocks (STO) or decision interactions (REL) without a dynamic model (DYN) represent a mistake (MIS3 hereafter) within cost–benefit analysis (CBA). That is, opportunity cost (OC) or preventive cost (PC) a dynamic model (DYN) would be correct if sensitivity analyses are performed.

In the same situation, Sect. 4.2.2 in Chap. 4 showed that accounting for stocks with subjective methods (SM) (i.e., using the analytical hierarchy process AHP, the revised Simos procedure RSP, linear regression LR, or factor analysis FA) will be inadequate. This requires the development of a dynamic model to assess the impacts of *X* and *Y* over time in investment projects. In terms of acronyms, subjective methods (SM) applied to environmental stocks (STO) without a dynamic model (DYN) represent a mistake (MIS4 hereafter) within multi-criteria analysis (MCA). That is, subjective methods (SM) with a dynamic model (DYN) would be correct if sensitivity analyses are performed.

Similarly, I highlighted some concerns about policy measures (i.e., standards are unsuitable for efficiency, and taxes, subsidies, and permits are unsuitable for equity) and about investment projects in easy contexts: in particular, temporal and geographical information systems, fuzzy sets, a social accounting matrix, and inequality weights are unsuitable in cost–benefit analysis; time discounting, spatial discounting, expected utility, a computable general equilibrium model, and social welfare functions are unsuitable in multi-criteria analysis. In the problematic contexts within cost–benefit analysis, I suggested time discounting where time is relevant, spatial discounting where space is relevant, expected utility where uncertainty is relevant, a computable general equilibrium model where linkages are relevant, and a social welfare function where inequalities are relevant. Similarly, I suggested a temporal information system where time is relevant, use of a geographical information system where space is relevant, fuzzy sets where uncertainty is relevant, a social accounting matrix where linkages are relevant, and inequality weights where inequality is relevant.

Note that relative weights based on the subjective methods (SM) used in multi-criteria analysis (MCA) are likely to change to a smaller extent in the long-run than monetary assessments in the utility-based approaches (UA) or dose response (DR) or replacement cost (RC) methods used in cost–benefit analysis (CBA), and the three-dimension simplex lets us compare the estimated values of the relative weights with the whole range of weight values. Moreover, I will not distinguish equity expressed in terms of inequality from equity expressed in terms of agreement, although this can lead to a size underestimation of MIS2. Finally, a decision interaction is not relevant in multi-criteria analysis, since the focus is on minimising the standardised impacts rather than maximising total or average welfare, as in cost–benefit analysis.

The purpose of this chapter is twofold. First, to evaluate the four types of possible mistake and all previous concerns in the literature by determining the frequencies of these mistakes and concerns and whether these frequencies are increasing or decreasing over time, both overall and in the most important scientific journals that focus on environmental sustainability. To do so, I constructed a representative dataset of empirical articles and performed descriptive and estimative statistical analyses in Sect. 5.1. Second, to suggest some solutions to those mistakes by referring to the methodologies discussed in Chap. 4. To do so, I will discuss my own 25 empirical articles in terms of the environmental issues they tackled and the analytical methodologies they applied (Sect. 5.2).

In particular, I will use the insights described in Chap. 4 to classify the empirical articles in terms of the environmental issues they tackled, the objectives they pursued (e.g., efficiency EFF and equity EQU), and the policies that were implemented within the policy measures (PLA) or the methodologies that I used within investment projects (PRO) to achieve the pursued objectives.

For *environmental issues*, I will apply the classifications for pollution (POL) versus resources (RES), for stocks (STO) versus flows (FLO), for a decision interaction (REL) versus independence (IND), and for air (AIR) versus land (LAN) versus water (WAT). I will stress the risk context (RIS), and will refer to policies to achieve efficiency for renewable resources (REN) (i.e., nature conservation GRO) and for non-renewable resources (NON) (i.e., technological support TEC).

The resulting environmental issues are classified as in Table 5.1.

Note that pollution versus resources is based on the second and third laws of thermodynamics: any transformation process produces some undesired outcomes such as pollution (second law) and no transformation process is completely reversible to obtain the total amount of resources originally used so some quantity of new resources are needed for new transformation processes (third law). Moreover, decision interaction versus independence is based on elementary results in game theory. Finally, both resources and pollution imply distribution issues. For example, these issues arise whenever ecological systems can absorb a maximum amount of pollution, and this total must be distributed among polluters, or whenever ecological systems provide a maximum amount of resources that must be allocated among the users.

For the *policies* that are implemented within policy measures (PLA), I will characterise each article in terms of a policy measure to achieve efficiency for both pollution and resources (i.e., taxes TAX, standards STA, subsidies SUB, permits PER, exploitation rights RIG, protected areas GRO, technological support TEC) and policy measures to achieve equity for both individuals and countries (i.e., a Nash bargaining solution NAS, a Kalai–Smorodinksi solution KAL, a Rawls solution RAW, Harshanyi solution HAR, capability CAP, sovereignty SOV, responsibility DUT, Kyoto Protocol KYP, Paris Agreement PAA, a debt-for-nature swap DNS, Reducing Emissions from Deforestation RED).

Note that interventions in policy measures are implemented and then evaluated, whereas interventions in investment projects are evaluated and then implemented.

For the *methodologies* that are used within investment projects (PRO), I will characterise each article in terms of methodologies applied to cost–benefit analysis (CBA) (i.e., investment projects PRO to achieve efficiency EFF) in problematic contexts and in terms of monetary assessment in easy contexts. For easy contexts, dose responses (DR), replacement costs (RC), opportunity costs OC), and preventive costs (PC) within production-based approaches (PA), but hedonic pricing (HP), travel costs (TC), contingent valuation (CV), and choice experiments (CE) within

Table 5.1 Classification of environmental issues

Environmental issue	POL/RES	STO/FLO	WHERE	Context	Policy measure
Sea level rise	RES	STO	WAT	RIS	
Flooding	RES	FLO	WAT	RIS	
Invasive aquatic species	POL	STO	WAT	RIS	
Sea, river or lake pollution	POL	STO	WAT		
Climate change	POL	STO	AIR		
Wind, wave, tidal energy	POL	STO	AIR		TEC
River management	RES	FLO	WAT		
Hydroelectric energy	RES	FLO	WAT		TEC
Lake and groundwater management	RES	STO	WAT		
Coastal and land erosion	RES	STO	LAN	RIS	
RED	POL	STO	AIR		GRO
Biomass and geothermal energy	POL	STO	LAN		TEC
Wastewater reuse and desalination	POL	FLO	WAT		TEC
Waste recycling	POL	STO	LAN		TEC
Forest fires	RES	STO	REN	RIS	
Solid waste management	POL	STO	LAN		
Invasive terrestrial species	POL	STO	LAN	RIS	
Acid rain	POL	STO	LAN	RIS	
Land and wetland use	RES	STO	LAN		
Copper reuse	RES	STO	NON		TEC
Aquaculture	RES	STO	WAT		GRO
Coral reef restoration	POL	STO	WAT		GRO
Magnetic fields	POL	STO	AIR	RIS	
Ecosystem management	RES	STO	REN		
Policies in Table 4.1 Sect. 4.1.1.1	POL	FLO	AIR		TAX, STA, SUB, PER
Policies in Table 4.3 Sect. 4.1.1.2	RES	STO	REN		TAX, STA, SUB, RIG, GRO

(continued)

Table 5.1 (continued)

Environmental issue	POL/RES	STO/FLO	WHERE	Context	Policy measure
Policies in Table 4.3 Sect. 4.1.1.2	RES	STO	NON		TAX, STA, SUB, RIG, TEC

POL = pollution, RES = resources, STO = stocks, FLO = flows, REL = a decision interaction, IND = a decision independence, AIR = air, LAN = land, WAT = water, RIS = a risky context, REN = renewable resources, GRO = protected areas, NON = non-renewable resources, TEC = technological support, RED = Reduce Emissions from Deforestation, TAX = taxes, STA = standards, SUB = subsidies, PER = permits, RIG = exploitations rights

the utility-based approaches (UA). For problematic contexts, time discounting (TD), spatial discounting (SD), expected utility (EU), the capital assessment pricing model (CAPM), a computable general equilibrium model CGEM), and social welfare functions (SWF). For investment projects (PRO) to achieve equity (EQU), I will consider the methodologies applied to multi-criteria analysis (MCA) in problematic contexts and the assessment of relative weights in easy contexts. For easy contexts, I will consider the analytical hierarchy process (AHP), the revised Simos procedure (RSP), linear regression (LR), and factor analysis (FA). For problematic contexts, I will consider use of a temporal information system (TIS), use of a geographical information system (GIS), expected values (EV), fuzzy sets (FS), a social accounting matrix (SAM), and inequality weights (IW).

Note that I will also distinguish each article in terms of asymmetric information ASY, uncertainty UNC, and imperfect competition IMP, whereas methodologies other than those I have specified will be identified with a context label (i.e., a dynamic model DYN, a spatial model SPA, uncertainty UNC, linkages LIN, inequalities INE). Moreover, the use of water in production activities includes groundwater recharge, noise is considered an aspect of pollution flows for air, and waste recycling includes energy production. Finally, I will also distinguish each article based on its scale (i.e., international INT, national NAT, regional REG, and local LOC).

Consequently, I have classified the empirical articles as follows:

- *efficiency* (EFF), if the article refers to policy measures (PLA) to achieve efficiency
- *equity* (EQU), if the article refers to a policy measure (PLA) to achieve equity
- *a policy measure* (PLA), if the article includes an attempt to achieve efficiency or equity
- *cost–benefit analysis* (CBA), if the article applies one or more cost–benefit analysis methodologies to monetary assessment in an easy or problematic context
- *multi-criteria analysis* (MCA), if the article applies one or more multi-criteria analysis methodologies to weight estimation in an easy or problematic context
- *an investment project* (PRO) if the article uses cost–benefit analysis or multi-criteria analysis
- *an assessment* (ASS) if the article aims to estimate the status quo that will be used as a reference value in implementation of a policy measure (PLA) or investment project (PRO)

In other words, once the environmental issue is identified, an article can estimate the starting point for the assessment (ASS), can tackle the environmental issue with a policy measure (PLA) based on efficiency (EFF) or equity (EQU), or can tackle the environmental issue with an investment project (PRO) based on cost–benefit analysis (CBA) or multi-criteria analysis (MCA).

Note that I assumed, unless specified otherwise by the authors of an article, that analysis of policy measures occurs at a national level. Moreover, I disregarded articles that focused only on population, metrics, or participation. Finally, I assumed that, unless specified otherwise by the authors of an article, analysis of investment projects occurs at regional level.

Based on these characteristics of the articles, I defined three levels of analysis:

- "High level" refers to pollution, resources, flows, stocks, policy measures, and investment projects (i.e., POL, RES, FLO, STO, PLA, PRO)
- "Middle level" is based on efforts to achieve efficiency or equity using cost–benefit analysis or multi-criteria analysis (i.e., EFF, EQU, CBA, MCA)
- "Low level" refers to the use of taxes, subsidies, permits, standards, exploitation rights, protected areas, and technological support (i.e., TAX, SUB, PER, STA, RIG, GRO, TEC) to support policy measures (PLA) designed to achieve efficiency (EFF); and it refers to the use of the Kyoto Protocol, the Paris Agreement, a debt-for-nature swap, or Reducing Emissions from Deforestation (i.e., KYP, PAA, DNS, RED) for policy measures (PLA) designed to achieve equity (EQU). This "low level" includes production-based approaches (PA) and utility-based approaches (UA) for cost–benefit analysis (CBA) within investment projects (PRO); and it includes subjective methods (SM) for multi-criteria analysis (MCA) within investment projects (PRO).

Note that, independently from the insights of Chap. 4, I also recorded the journal that published the article and the year of publication, together with the country or countries in which the case study occurred or the studies were performed. I did not specify the country if the article referred to new technologies, since new technologies can be implemented in any country.

5.1 Literature Problems

In this Section, I will evaluate the possible mistakes and concerns in the literature on environmental sustainability by determining the frequency of a given type of fault and specifying whether the frequency is increasing or decreasing over time, in specific regions, in specific journals, and in discussions of specific environmental issues. To do so, I will first obtain a representative sample of empirical articles; I will then perform descriptive and estimative statistical analyses of their characteristics.

Chapter 4 highlighted the following *concerns* (CON):

- Time discounting (TD) is unsuitable in multi-criteria analysis (MCA), since impacts at different times are incommensurable (i.e., impacts are measured in percentages with respect to different references statuses), whereas a temporal information system (TIS) (i.e., benefits and costs in different times are complementary) is inappropriate in cost–benefit analysis (CBA) because the goal is to maximise the (time-discounted) total welfare.
- Spatial discounting (SD) is unsuitable in multi-criteria analysis (MCA) because impacts in different regions are incommensurable (i.e., impacts are measured in percentages with respect to different references statuses), whereas the use of a geographical information system (GIS) is inappropriate (i.e., benefits and costs in different regions are complementary) in cost–benefit analysis (CBA) because the goal is to maximise the (spatially discounted) total welfare.
- The expected values (EV) method is inappropriate in cost–benefit analysis (CBA), unless risk neutrality is assumed, whereas fuzzy sets (FS) are inadequate in cost–benefit analysis (CBA) because they do not maximise the expected total welfare. The capital asset pricing model (CAPM) is inappropriate in multi-criteria analysis (MCA), since many impacts are not evaluated in monetary terms, whereas the expected utility (EU) model is unnecessary in multi-criteria analysis (MCA), since the goal is not maximisation of the expected total welfare.
- Input–output models (IOM) are unsuitable in cost–benefit analysis (CBA) because they assume fixed prices, whereas a computable general equilibrium model (CGEM) is unnecessary in multi-criteria analysis (MCA) because the goal is not maximisation of the total welfare.
- Inequality weights (IW) are unsuitable in cost–benefit analysis (CBA) because they do not maximise the total welfare; social welfare functions (SWF) are superfluous for multi-criteria analysis (MCA) because the goal is not maximisation of the total welfare.
- If a decision interaction (REL) is not relevant, market-based policies (i.e., taxes TAX, subsidies SUB, permits PER, technological support for non-renewable resources TEC) are suitable for efficiency (EFF) under weak sustainability (WS), whereas command-and-control policies (i.e., standards STA, protected areas for renewable resources GRO, exploitation rights for resources RIG) are suitable for achieving equity (EQU) under strong sustainability (SS).
- If a decision interaction (REL) is relevant, market-based agreements (i.e., Nash bargaining NAS, Harshanyi HAR, the Paris Agreement PAA) are suitable for equity (EQU) under weak sustainability (WS), whereas command-and-control agreements (i.e., Kalai–Smorodinski KAL, Rawls RAW, Kyoto Protocol KYP) are suitable for equity (EQU) under strong sustainability (SS).

Note that I have disregarded articles based on scenarios because scenarios are inadequate both in cost–benefit analysis (i.e., they don't maximise the expected total welfare) and in multi-criteria analysis (i.e., they don't minimise the weighted sum of the impacts). Moreover, the willingness to pay (WTP) method is useful in cost–benefit analysis to account for the risk of species extinction (i.e., resources, stocks, renewable resources risk), since risk is a flow variable (i.e., it refers to a period of time, although it is often a long period). Similarly, willingness to pay (WTP) method is useful in cost–benefit analysis for participation, since participation is a flow variable (i.e., it refers to a period of time, and often a short period). Finally, some methodologies are suitable in both cost–benefit analysis and multi-criteria analysis (e.g., inequality measures).

Chapter 4 also highlighted the following *mistakes* (MIS):

- asymmetric information (ASY), uncertainty (UNC), or imperfect competition (IMP) without equity (EQU) (in terms of inequality)
- a decision interaction (REL) without equity (EQU) (in terms of agreement)
- environmental stocks (STO) or a decision interaction (REL) with utility-based approaches (UA) or dose response (DR) and replacement cost (RC) studies without a dynamic model (DYN) within cost–benefit analysis (CBA)
- environmental stocks (STO) with subjective methods (SM) without a dynamic model (DYN) within multi-criteria analysis (MCA)

Thus, I will statistically analyse the significance of mistakes and concerns over time and among journals, with the mistakes and concerns *logically* specified as follows:

Mistakes (about objectives):

1. (ASY or UNC or IMP) and no EQU
2. REL and no EQU
3. (STO or REL) and (UA or DR or RC) and no DYN in CBA
4. STO and SM and no DYN in MCA

Concerns (about policies or projects):

1. STA, GRO, RIG and EFF
2. TAX, SUB, PER, TEC and EQU
3. CBA and TIS
4. CBA and GIS
5. CBA and FS
6. CBA and IOM
7. CBA and IW
8. MCA and TD
9. MCA and SD
10. MCA and EU
11. MCA and CGEM
12. MCA and SWF
13. PAA and EFF
14. KYP and EFF

To obtain a *representative sample* of empirical articles from peer-reviewed journals, I identified 835 English articles (i.e., excluding reviews and books) in Scopus by searching for the following words in the abstract, title, or keywords:

- Always "data" to identify empirical articles
- Middle-level specificity (i.e., efficiency *or* equity *or* cost–benefit analysis *or* multi-criteria analysis)
- Low-level specificity (i.e., each single policy *and* context for policy measures; each single methodology *or* context for investment projects)

Note that the evaluation of efficiency of policy measures is useless without context (i.e., asymmetric information, uncertainty, imperfect competition) because all policies are equivalent in terms of efficiency. In other words, if policy measures differ, some contexts might be relevant, but I cannot argue against the hypotheses assumed by the authors. Consequently, I will disregard all articles focused on a partial evaluation of efficiency, since estimation of the marginal impact of a policy evaluates its reliability rather than its feasibility (Zagonari, 2019a, b, c), while the non-linear properties of the Earth system suggest that linear responses (e.g., national policies targeted only at environmental degradation and resource use within national territories) are inadequate.

Moreover, the evaluation of investment projects without participation is useless, since participation is essential for sustainability. Indeed, participation in policy measures is assessed by the equilibrium prices. Consequently, I will disregard articles without stakeholder involvement, under the assumption that estimation of monetary assessments for cost–benefit analysis and of relative weights for multi-criteria analysis do a good job of representing how stakeholders can participate in environmental decisions, at least if environmental stocks and a decision interaction are not crucial.

Finally, I did not distinguish ex-ante investment projects from ex-post investment projects, since the difference depends on the available dataset. However, policy measures are often evaluated after implementation, although taxes can be earlier evaluated by a computable general equilibrium model; and investment projects are often evaluated before implementation, although protected areas for renewable resources may be later evaluated by ecological models.

Thus, contexts in investment projects (i.e., time, space, uncertainty, linkages, inequalities) play the same role as contexts in policy measures (i.e., asymmetric information, uncertainty, imperfect competition). In other words, I will exclude papers about investment projects simply based on decision rules if the contexts are not specified or if participation is not evaluated; I will also exclude papers on policy measures based simply on taxes, standards, subsidies, permits, exploitation rights, protected areas for renewable resources, and technological support for non-renewable resources if the contexts are not specified.

Note that the purpose of my sampling procedure was not to create an exhaustive set of articles (i.e., I do not expect to have included all relevant articles), but rather to support a comprehensive classification of a wide range of articles. That is, I expect that articles I missed and future articles can be classified within my framework. In addition, a sample of 835 empirical articles seems to be a reasonable size, since searching using "environmental sustainability" and "data" in Scopus identified 2759 English articles published before 2021, mainly in the areas of economics, econometrics and finance (244 articles) and environmental science (1503 articles). In

Table 5.2 Classification of empirical articles on environmental sustainability

Environmental science	1503
Social sciences	709
Business, management and accounting	529
Energy	521
Engineering	457
Agricultural and biological sciences	418
Economics, econometrics, and finance	244
Earth and planetary sciences	112
Arts and humanities	57
Multidisciplinary	96
Total	4646

Notes articles from Scopus published before 2021

other words, my classification covered around 50% of the articles in environmental science, economics, econometrics, and finance (i.e., 835/1747) and around 20% of all articles (i.e., 835/4646).

Table 5.2 classifies the empirical articles from Scopus according to the subjects supplied by Scopus.

Some methodological remarks are noteworthy here:

- I considered the whole period for articles focused on environmental issues from 1987 to 2020.
- I included papers focused on sectoral decisions to cope with general environmental problems (e.g., electric cars, generation of energy from dairy wastes, airport traffic).
- I included cost–benefit analysis articles that focused on either the demand side or the supply side without an overall decision based on benefits and costs.
- I included articles aimed at coping with an existing environmental issue (e.g., green technologies such as wind, wave, and tidal energy to reduce greenhouse-gas emissions) by disregarding articles aimed at designing solutions with potential environmental impacts (e.g., urban planning, agriculture planning).
- I excluded articles focused on individual companies, since that topic is closer to life-cycle assessment, which is in its theoretical and empirical infancy for research on both efficiency and equity.
- I disregarded articles on health problems, even if they were due to environmental issues (e.g., safe drinking water), since their goal was not to solve environmental problems.
- I excluded articles on finance, since their goal was not to solve environmental problems, although finance could have environmental implications for both (international) efficiency and equity.

- I excluded articles based on a survey because this is an embryonic methodology to estimate participation within cost–benefit analysis. Similarly, I excluded articles based on environmental impact assessment because this is an embryonic methodology to estimate impacts within multi-criteria analysis.
- I excluded articles on trade, since their goal was not to solve environmental issues, although trade can have huge environmental impacts and can exploit differences between trading partners in their environmental status and legislation.
- I excluded articles based on benefit transfers or meta-analysis, since their goal was not to provide innovative decision methodologies.

For the *statistical analyses*, the relatively large sample of articles with various categories of mistakes suggested the need to perform a panel probit analysis with year as a continuous variable and dummy variables for the most popular journals. In contrast, the relatively small sample of articles related to concerns suggested the need to perform arithmetical analysis.

For the *descriptive* statistical analysis, I will analyse the relative dynamics and importance of environmental issues and decision methodologies at different levels:

- At the highest level (i.e., policy measures PLA, investment projects PRO, pollution POL, resources RES), I will express the relative importance as a percentage of the total number of articles.
- At the middle level (i.e., efficiency EFF, equity EQU, cost–benefit analysis CBA, multi-criteria analysis MCA), I will express both the relative importance as a percentage of the total number of articles and the relative dynamics over time in terms of linear regression slopes, together with pollution POL and resource RES relationships.
- At the lowest level (i.e., taxes TAX, standards STA, subsidies SUB, permits PER, exploitation rights RIG, protected areas GRO for policy measures PLA; the analytical hierarchy process AHP and linear regression LR for multi-criteria analysis MCA within investment projects PRO; contingent valuation CV and choice experiments CE for cost–benefit analysis CBA within investment projects PRO), I will show the relative dynamics over time in terms of linear regression slopes.

Note that I will not apply a descriptive statistical analysis to journals (e.g., the 15 most popular) and countries (e.g., the 15 most popular), since identifying which journal is focused on which topic in which country is irrelevant for my second purpose (i.e., to suggest solutions to mistakes in the literature in Sect. 5.2). Moreover, I will disregard relationships between pollution POL or resources RES and policy measures PLA at the lowest level, since the three main market-based policies to achieve efficiency (EFF) (i.e., taxes TAX, subsidies SUB, and permits PER) are mostly linked to pollution (POL), whereas the three main command-and-control policies to achieve efficiency (EFF) (i.e., standards STA, protected areas GRO, and exploitation rights RIG) are mostly linked to resources (RES). Finally, I will not analyse the relative importance of decision methodologies at the lowest level, since the analytical hierarchy process (AHP) and linear regression (LR) on one side and contingent valuation (CV) and choice experiments (CE) on the other side are the most

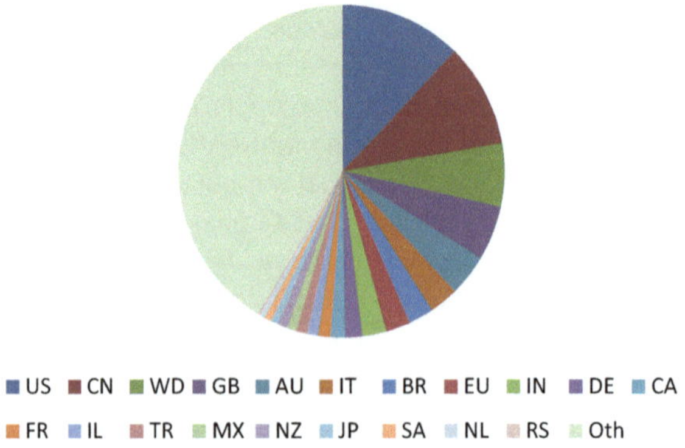

Fig. 5.1 The 20 most popular countries in the selected case studies. Abbreviations: US = USA, CN = China, WD = world, GB = Great Britain, AU = Australia, IT = Italy, BR = Brazil, EU = European Union, IN = India, DE = Germany, CA = Canada, FR = France, IL = Israel, TR = Turkey, MX = Mexico, NZ = New Zealand, JP = Japan, SA = Saudi Arabia, NL = Netherland, RS = Russia, Oth = other countries. Number of case studies are as follows: US = 102, CN = 84, WD = 52, GB = 44, AU = 34, IT = 24, BR = 21, EU = 21, IN = 20, DE = 14, CA = 12, FR = 10, IL = 9, TR = 9, MX = 8, NZ = 8, JP = 6, SA = 6, NL = 4, RS = 2, Oth = 345

popular methodologies for multi-criteria analysis (MCA) and cost–benefit analysis (CBA), respectively.

The dataset is available from the author on reasonable request.

Figures 5.1, 5.2, 5.3, 5.4, 5.5, 5.6, 5.7, 5.8, 5.9, 5.10, 5.11 provide a graphical analysis of the sampled empirical articles.

Figure 5.1 shows that the 5 and the 20 most popular countries accounted for 37% and 57% of case studies in my sample of articles, respectively. Note that the "Other" category also includes case studies that occurred in more than one country.

Figure 5.2 shows that the 10 and 15 most popular countries accounted for 48% and 54% of the case studies in my sample of the articles. Thus, most case studies have been performed in relatively few countries.

Figure 5.3 shows that stock issues (STO) have been analysed to a greater extent than flow issues (FLO) (i.e., in percentages of articles, $24 + 11 + 24 + 22 > 5 + 4 + 6 + 6$), with the number of articles on policy measures (PLA) slightly larger than the number of articles on investment projects (PRO) (i.e., in percentages, $5 + 24 + 6 + 24 > 4 + 11 + 6 + 22$).

Figure 5.4 shows that climate change (i.e., identified by AIR, POL, STO) and biodiversity losses (i.e., identified REN, RES, STO) are the most-analysed environmental issues (i.e., in percentages of articles, 29 and 35, respectively). This confirms insights obtained in Chap. 2. However, the water resource and pollution (for both flows and stocks) (i.e., in percentages, WAT, POL, FLO; WAT, POL, STO; WAT, RES, FLO; WAT, RES, STO at 4, 5, 13 and 3, respectively), air pollution (i.e., in percentages, AIR, PLO, FLO at 4), and land resource and pollution (i.e., in percentages, LAN, POL, STO; LAN, RES, STO at 4 and 3, respectively) were also common environmental topics.

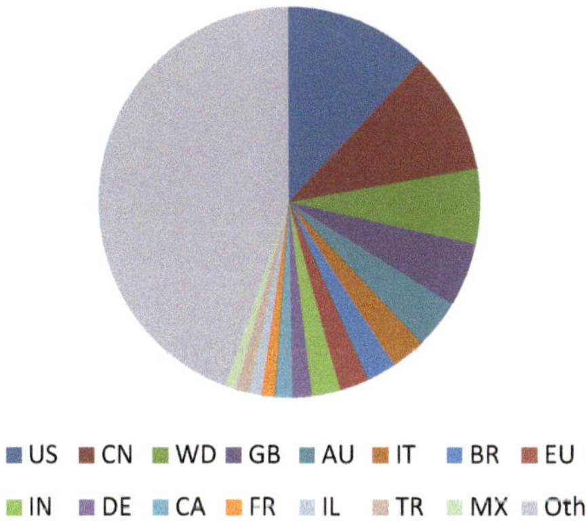

Fig. 5.2 The 15 most popular countries in the selected case studies. Abbreviations: US = USA, CN = China, WD = world, GB = Great Britain, AU = Australia, IT = Italy, BR = Brazil, EU = European Union, IN = India, DE = Germany, CA = Canada, FR = France, IL = Israel, TR = Turkey, MX = Mexico, Oth = other countries. Number of case studies are as follows: US = 102, CN = 84, WD = 52, GB = 44, AU = 34, IT = 24, BR = 21, EU = 21, IN = 20, DE = 14, CA = 12, FR = 10, IL = 9, TR = 9, MX = 8, Oth = 371

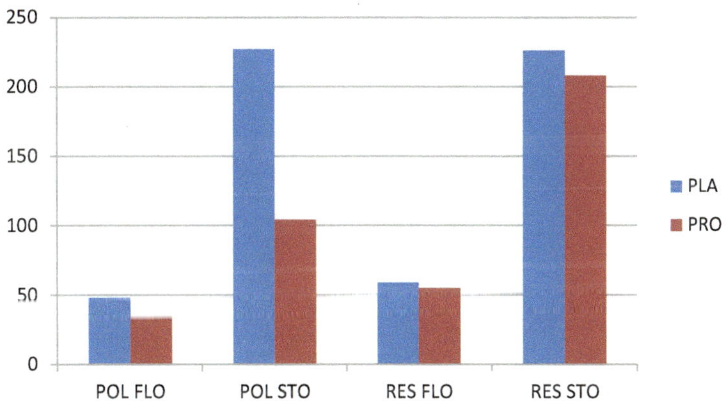

Fig. 5.3 No. articles on PLA and PRO versus POL FLO, POL STO, RES FLO, RES STO. Abbreviations: PLA = policy measures, PRO = investment projects, POL = pollution, FLO = flows, STO = stocks, RES = resources. *Notes* From left to right bars, percentages of articles are 5, 4, 24, 11, 6, 6, 24, and 22, respectively

Figure 5.5 shows that climate change (i.e., identified by AIR, POL, STO with 86% of articles) was the most popular environmental topic analysed in terms of decision interactions (REL) (i.e., in percentages, 86), although some articles focused on trans-boundary water resources (WAT, RES) (flows FLO to a greater extent than stocks STO, with percentages at 8 and 2, respectively) and there were few articles on trans-boundary renewable resources (i.e., REN, RES, STO at 3% of articles).

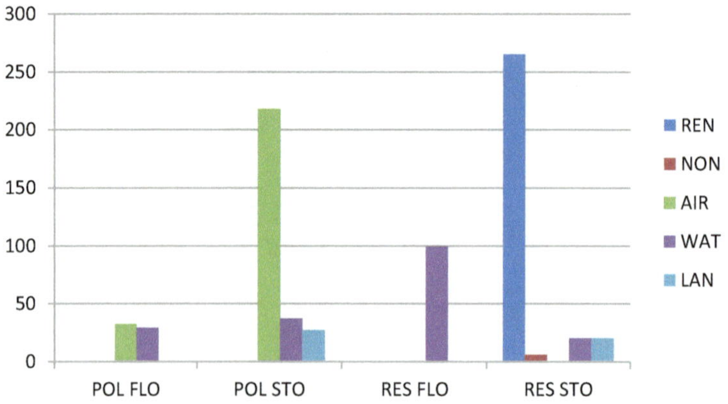

Fig. 5.4 No. articles on REN, NON, AIR, WAT, LAN versus POL FLO, POL STO, RES FLO, RES STO. Abbreviations: REN = renewable resources, NON = non-renewable resources, AIR = air, WAT = water, LAN = land, POL = pollution, FLO = flows, STO = stocks, RES = resources. *Notes* From left to right bars, positive percentages of articles are 4, 4, 29, 5, 4, 13, 35, 1, 3, and 3, respectively

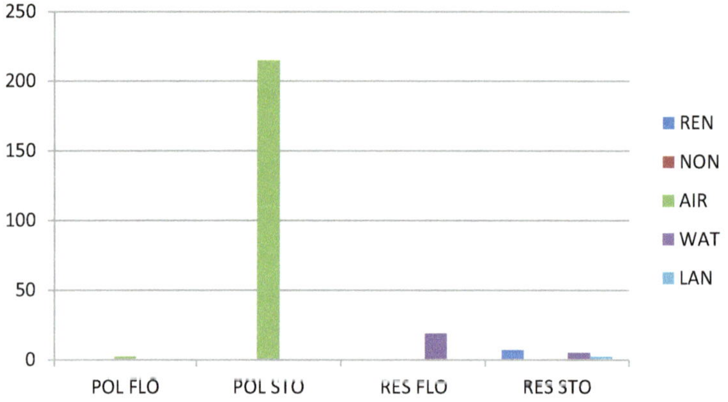

Fig. 5.5 No. articles on REL for REN, NON, AIR, WAT, LAN versus POL FLO, POL STO, RES FLO, RES STO. Abbreviations: REL = decision interactions, REN = renewable resources, NON = non-renewable resources, AIR = air, WAT = water, LAN = land, POL = pollution, FLO = flows, STO = stocks, RES = resources. *Notes* From left to right bars, positive percentages of articles are 1, 86, 8, 3, 2, and 1, respectively

Figure 5.6 shows that efficiency (i.e., in percentages of articles, 4 + 3 + 17 + 8 + 4 + 3 + 20 + 18) was always analysed to a greater extent than equity (i.e., in percentages, 1 + 1 + 8 + 3 + 2 + 2 + 3 + 3) for all four main environmental topics, both for policies (PLA) (i.e., efficiency EFF at 4 + 17 + 4 + 20 > equity EQU at 1 + 8 + 2 + 3) and projects (PRO) (i.e., cost–benefit analysis CBA at 3 + 8 + 3 + 18 > multi-criteria analysis MCA at 1 + 3 + 2 + 3), although the overall difference was

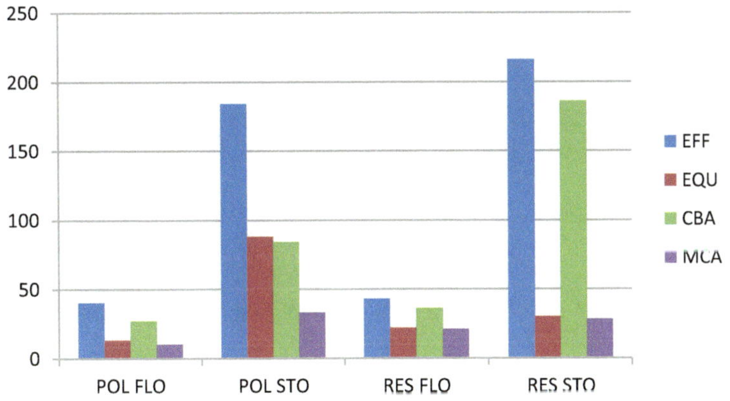

Fig. 5.6 No. articles on EFF, EQU, CBA, MCA versus POL FLO, POL STO, RES FLO, RES STO. Abbreviations: EFF = efficiency, EQU = equity, CBA = cost–benefit analysis, MCA = multi-criteria analysis, POL = pollution, FLO = flows, STO = stocks, RES = resources. *Notes* From left to right bars, positive percentages of articles are 4, 1, 3, 1, 17, 8, 8, 3, 4, 2, 3, 2, 20, 3, 18, and 3, respectively

larger for resource stocks (i.e., in percentages, 20 + 18 > 3 + 3) than for pollution stocks (i.e., in percentages, 17 + 8 > 8 + 3).

Figure 5.7 shows that the focus on equity (EQU) has been increasing over time (both for pollution and for resources, with linear regression slopes at 0.86 and 0.82, respectively), although efficiency (EFF) was the most popular approach in many years (to a greater extent in recent years). Note that articles based on all assumptions behind the Economic General Equilibrium Model were excluded from the sample. As a result, an increasing attention on efforts to achieve efficiency (both for pollution POL and for resources RES) means an increasing number of articles that specify assumptions other than those behind the model.

Figure 5.8 shows that cost–benefit analysis (CBA) has been applied since the beginning of the study period and has focused on resources to a greater extent than on pollution (i.e., linear regression slopes at 0.28 and 0.51, respectively), whereas multi-criteria analysis (MCA) is a more recent methodology and has focused on resources and pollution to similar extents (i.e., linear regression slopes at 0.16 and 0.19, respectively).

Figure 5.9 shows that, within the market-based policies, taxes (TAX) and subsidies (SUB) have been continuously analysed at roughly similar *high* levels (i.e., linear regression slopes at 0.23 and 0.27, respectively), whereas the focus on permits has decreased since about 2010. Within the command-and-control policies, standards (STA) for pollution and protected areas (GRO) for renewable resources have been continuously analysed at roughly similar *medium* levels (i.e., linear regression slopes at 0.13 and 0.09, respectively), whereas the focus on exploitation rights (RIG) has been smaller but relatively constant (i.e., the linear regression slope at 0.03).

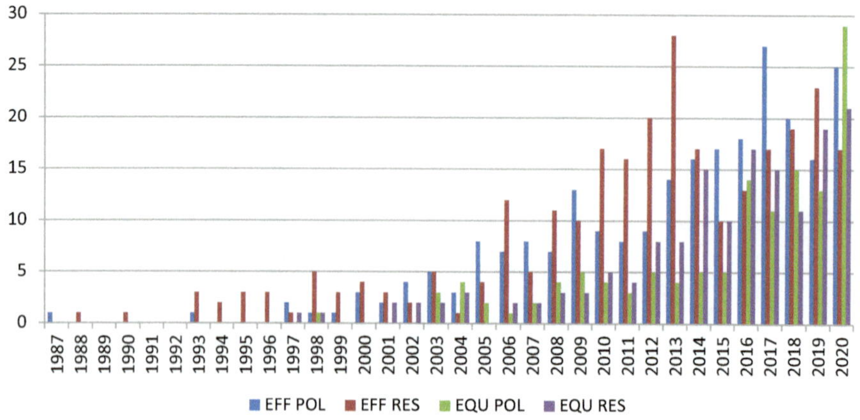

Fig. 5.7 No. articles over time on EFF versus EQU for POL and RES. Abbreviations: EFF = efficiency, POL = pollution, RES = resources, EQU = equity. *Notes* From left to right time series, linear regressions are as follows: y = 0.8018x − 7.4768; y = 0.7144x − 4.6114; y = 0.8603x − 14.485; and y = 0.8281x − 12.146, respectively, where y = No. of articles and x = year

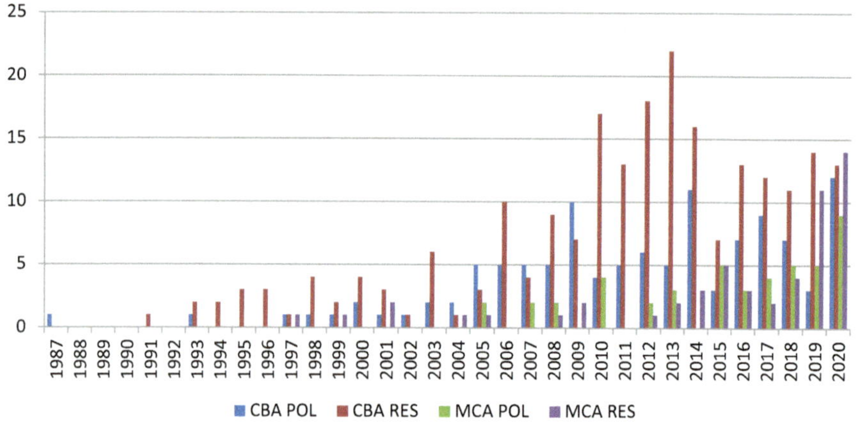

Fig. 5.8 No. articles over time on CBA versus MCA for POL and RES. Abbreviations: CBA = cost–benefit analysis, POL = pollution, RES = resources, MCA = multi-criteria analysis. *Notes* From left to right time series, linear regressions are as follows: y = 0.2834x − 1.5775; y = 0.5173x − 2.5241; y = 0.1644x − 1.5241; and y = 0.1962x − 1.8449, respectively, where y = No. of articles and x = year

Figure 5.10 shows that international agreements on both pollution and resources have been mostly analysed immediately after their introduction, then the focus on them decreased few years later.

Figure 5.11 shows that choice experiments (CE) for cost–benefit analysis (CBA) and the analytical hierarchy process (AHP) for multi-criteria analysis (MCA) have been more popular in recent years (i.e., linear regression slopes at 0.29 and 0.51,

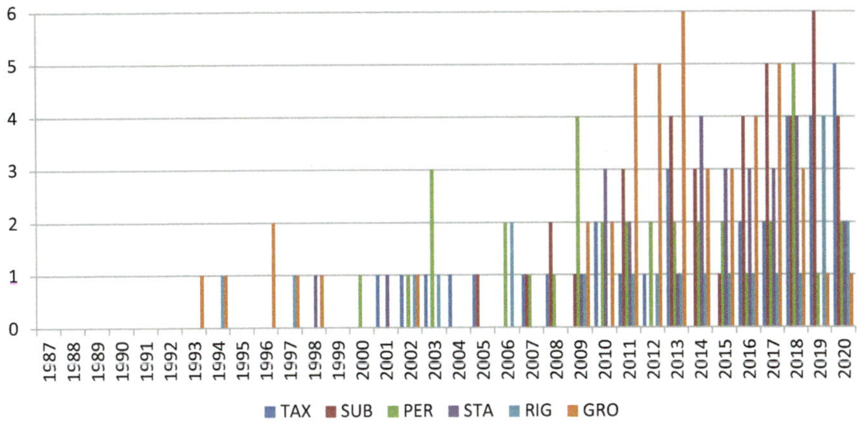

Fig. 5.9 No. articles over time on policies to achieve efficiency (TAX, SUB, PER, STA, RIG, GRO). Abbreviations: TAX = taxes, SUB = subsidies, PER = permits, STA = standards, RIG = exploitation rights, GRO = protected areas. *Notes* From left to right time series, linear regressions are as follows: y = 0.238x − 3.5577; y = 0.2744x − 4.4708; y = 0.0785x + 0.2509; y = 0.1391x − 0.901; y = 0.0332x + 0.5035; and y = 0.0913x + 0.5518, respectively, where y = No. of articles and x = year

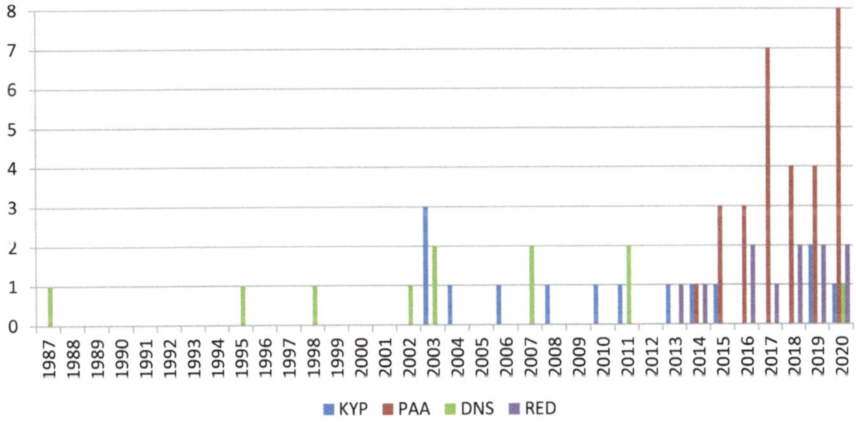

Fig. 5.10 No. articles over time on policies to achieve equity (KYP, PAA, DNS, RED). Abbreviations: KYP = the Kyoto Protocol (1997), PAA = the Paris Agreement (2015), DNS = debt-for-nature swaps (1987), RED = reducing emissions from deforestation (2015)

respectively). Contingent valuation (CV) for monetary assessment within CBA was mostly applied between 2005 and 2015, whereas linear regression (LR) to assess relative weights within MCA has mostly been applied since 2013.

In summary, Figs. 5.1, 5.2, 5.3, 5.4, 5.5, 5.6, 5.7, 5.8, 5.9, 5.10, 5.11 (descriptive statistics) highlighted the *expected* relative numbers and dynamics of articles that

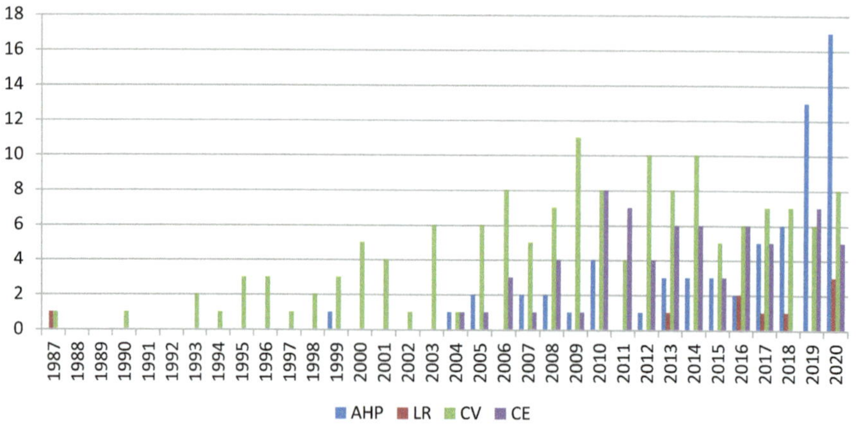

Fig. 5.11 No. articles on CV and CE for CBA and on AHP and LR for MCA. Abbreviations: AHP = analytical hierarchy process, LR = linear regression, CV = contingent valuation, CE = choice experiments, CBA = cost–benefit analysis, MCA = multi-criteria analysis. *Notes* From left to right time series, linear regressions are as follows: y = 0.5109x − 8.9675; y = 0.0267x + 0.8094; y = 0.2459x + 0.2548; and y = 0.2989x − 3.408, respectively, where y = No. of articles and x = year

focused on different environmental issues based on alternative goals (i.e., efficiency vs. equity) and alternative methodologies (i.e., cost–benefit analysis vs. multi-criteria analysis). In other words, the sample of empirical articles seems to be a representative sample of the published articles on environmental sustainability.

The results reveal the following numbers of concerns:

1. standards and efficiency = 0 (STA and EFF = 0)
2. taxes, subsidies, permits, and equity = 0 (TAX, SUB, PER & EQU = 0)
3. cost–benefit analysis and a temporal information system = 0 (CBA & TIS = 0)
4. cost–benefit analysis and a geographical information system = 5 (CBA & GIS = 5)
5. cost–benefit analysis and fuzzy sets = 1 (CBA & FS = 1)
6. cost–benefit analysis and input–output model = 10 (CBA & IOM = 10)
7. cost–benefit analysis and inequality weights = 1 (CBA & IW = 1)
8. multi-criteria analysis and time discounting = 0 (MCA & TD = 0)
9. multi-criteria analysis and spatial discounting = 0 (MCA & SD = 0)
10. multi-criteria analysis and expected utility = 2 (MCA & EU = 2)
11. multi-criteria analysis and computable general equilibrium model = 0 (MCA & CGEM = 0)
12. multi-criteria analysis and a social welfare function = 0 (MCA & SWF = 0)
13. the Kyoto Protocol and efficiency = 6 (KYP & EFF = 6)
14. the Paris Agreement and efficiency = 7 (PAA & EFF = 7)

Thus, there seem to be few concerns in the sample, apart from the use of a geographical information systems (GIS) and input–output models (IOM) within cost–benefit analysis (CBA).

Note that the number of articles on the Paris Agreement in policies designed to achieve equity (i.e., 23 out of 30) is greater than the number of articles on the Kyoto Protocol in policies designed to achieve equity (i.e., 8 out of 14). This suggests that weak sustainability has prevailed over strong sustainability in the scientific debate about international agreements.

Figure 5.12 highlights that the 10 and 15 most popular journals accounted for 41% and 50% of the articles in my sample, respectively.

Figure 5.13 highlights that the 5 and the 20 most popular journals accounted for 27% and 56% of the articles in my sample, respectively. Thus, the most 20 popular journals do not cover a significantly larger percentage of the sampled articles than the 15 most popular journals, with quite few articles in the last five of the 20 most popular journals. That is, *Ecological Indicators*, *Marine Policy*, *Environmental Modelling and Software*, *Waste Management*, and *Ocean and Coastal Management* appear fewer than 10 times each. Thus, in my estimative statistical analysis, I will use 15 dummy variables (one each for the most 15 popular journals) and will combine all other journals into a single dummy variable to be used as a reference group.

Tables 5.3, 5.4, 5.5, 5.6 present robust probit regressions for the panel data on mistakes 1 to 4, and can be used to estimate the overall significance of these mistakes and the relative probability of observing each mistake in each of the most popular 15 journals. Note that negative coefficients for a journal mean a smaller frequency of the specified type of mistake with respect to other journals. Moreover, the values of confidence interval represents the minimum and maximum values of coefficients which include its mean at 95%: these values can be translated into probability of

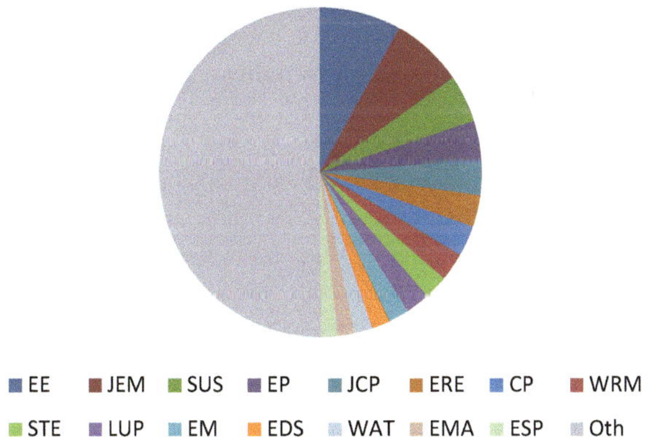

Fig. 5.12 The 15 most popular journals in the selected articles. Abbreviations: EE = *Ecological Economics*, JEM = *Journal of Environmental Management*, SUS = *Sustainability*, EP = *Environmental Policy*, JCP = *Journal of Cleaner Production*, ERE = *Environmental and Resource Economics*, CP = Climate Policy, WRM = *Water Resources Mana*gement, STE = *Science of the Total Environment*, LUP = *Land Use Policy*, EM = *Environmental Management*, EDS = *Environment, Development and Sustainability*, WAT = *Water*, EMA = *Environmental Modelling and Assessment*, ESP = *Environmental Science and Policy,* Oth = all journals not included in the top 15

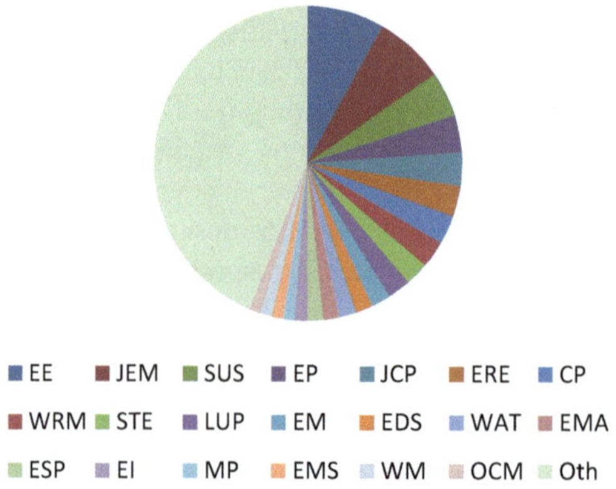

Fig. 5.13 The 20 most popular journals in the selected articles. Abbreviations: EE = *Ecological Economics*, JEM = *Journal of Environmental Management*, SUS = *Sustainability*, EP = *Environmental Policy*, JCP = *Journal of Cleaner Production*, ERE = *Environmental and Resource Economics*, CP = Climate Policy, WRM = Water Resources Management, STE = *Science of the Total Environment*, LUP = *Land Use Policy*, EM = Environmental Management, EDS = *Environment, Development and Sustainability*, WAT = *Water*, EMA = *Environmental Modelling and Assessment*, ESP = *Environmental Science and Policy*, EI = *Ecological Indicators*, MP = *Marine Policy*, EMS = *Environmental Modelling and Software*, WM = *Waste Management*, OCM = *Ocean and Coastal Management*, Oth = all journals not included in the top 20

observing a mistake by using the probit probability function. Finally, positive coefficients for a journal mean a larger frequency of the specified type of mistake with respect to other journals.

Table 5.3 shows that mistake 1 (MIS1) was explained overall by the variables used (Prob > χ^2 is smaller than 0.001) and it was highly significant ($p < 0.001$ for CONS) at around 20% (i.e., the regression constant CONS = −1.09 represents a value of 0.2 in the probit probability function). The value of rho (i.e., the proportion of the total variance contributed by the panel-level variance component) smaller than 0.001 (i.e., the panel-level variance component is unimportant and the panel estimator is not different from the pooled estimator) suggests that the use of dummy variables for the 15 most popular journals made the pooled estimation (i.e., a time series of cross-sections where observations in each cross-section do not refer to the same journal) similar to the panel estimation (i.e., a time series of cross-sections where the same journals are observed at multiple points in time). In particular, journals where mistake 1 (MIS1) was observed with a higher and significant probability were *Sustainability*, *Environmental Policy*, *Environmental and Resource Economics*, *Water Resources Management*, *Science of the Total Environment*, and *Water*, whereas journals where mistake 1 (MIS1) was observed with a lower but significant probability were *Ecological Economics*, *Climate Policy*, *Land Use Policy*, and *Environmental Management*, although the four latter journals were correct in 3, 3, 1, and 1 cases, respectively. Note that I have excluded *Environment, Development and Sustainability* because their dummy variables perfectly predicted MIS1 = 0 (i.e., articles in *Environment,*

Development and Sustainability never mention assumptions other than those behind the Economic General Equilibrium Model EGEM such as asymmetric information ASY, uncertainty UNC and imperfect competition IMP so ASY = UNC = IMP = 0).

Table 5.4 shows that mistake 2 (MIS2) is explained overall by the variables used (Prob > χ^2 is smaller than 0.001) and it is highly significant ($p < 0.001$ for CONS) at around 20% (i.e., the regression constant CONS = −1.02 amounts to 0.2 in the probit probability function), where using dummy variables for the 15 most popular journals made the pooled estimation similar to the panel estimation (i.e., rho = 0). In particular, journals where mistake 2 (MIS2) was observed with a higher and significant probability were *Ecological Economics, Sustainability, Environmental Policy, Journal of Cleaner Production, Environmental and Resource Economics,* and *Environment, Development and Sustainability*, whereas journals where mistake 2 (MIS2) was observed with a lower and significant probability were *Water Resources Management, Science of the Total Environment, Land Use Policy*, and *Environmental*

Table 5.3 Average frequency of mistake 1 in the 15 most popular journals used as dummy variables

MIS1	Coef	Robust std. err	z	P > |z|	[95% conf. interval]	
EED	−0.6077245	0.0314445	−19.33	0.000	−0.6693547	−0.5460944
JEMD	−0.0137119	0.0314445	−0.44	0.663	−0.075342	0.0479183
SUSD	0.076996	0.0314445	2.45	0.014	0.0153659	0.1386261
EPD	0.0870876	0.0314445	2.77	0.006	0.0254575	0.1487176
JCPD	0.0295114	0.0314445	0.94	0.348	−0.0321186	0.0911415
ERED	0.0770079	0.0314445	2.45	0.014	0.0153779	0.138638
CPD	−0.307974	0.0314445	−9.79	0.000	−0.369604	−0.246344
WRMD	0.3160541	0.0314445	10.05	0.000	0.254424	0.3776841
STED	0.7610582	0.0314445	24.20	0.000	0.6994282	0.8226882
LUPD	−0.5712582	0.0314444	−18.17	0.000	−0.6328882	−0.5096282
EMD	−0.4675906	0.0314444	−14.87	0.000	−0.5292206	−0.4059606
WATD	0.2554786	0.0314445	8.12	0.000	0.1938486	0.3171086
EMAD	0.0136636	0.0314444	−0.43	0.664	−0.0752936	0.0479664
ESPD	0.0295386	0.0314444	0.94	0.348	−0.0320914	0.0911686
CONS	−1.09713	0.0314444	−34.89	0.000	−1.15876	−1.0355

Number of observations = 835, probability > χ^2 smaller than 0.001, rho smaller than 0.001. Abbreviations for dummy (d) variables: EE = *Ecological Economics*, JEM = *Journal of Environmental Management*, SUS = *Sustainability*, EP = *Environmental Policy*, JCP = *Journal of Cleaner Production*, ERE = *Environmental and Resource Economics*, CP = Climate Policy, WRM = Water Resources Management, STE = *Science of the Total Environment*, LUP = *Land Use Policy*, EM = Environmental Management, EDS = *Environment, Development and Sustainability*, WAT = *Water*, EMA = *Environmental Modelling and Assessment*, ESP = *Environmental Science and Policy*. Other abbreviations: CONS = probit regression constant, rho = the proportion of the total variance contributed by the journal variance component. *Notes* MIS1 (on PLA & EFF) in Boolean logic can be depicted as follows: (ASY or UNC or IMP) & no EQU

Monitoring and Assessment, although the latter four journals were correct in 6, 1, 1, and 3 cases, respectively. Note that I excluded *Environmental Management* and *Water* because their dummy variables perfectly predicted MIS2 = 0 (i.e., articles in *Environmental Management* and *Water* never focus on decision interactions so REL = 0).

Table 5.5 shows that mistake 3 (MIS3) was explained overall by the variables used (Prob > χ^2 is smaller than 0.001) and it was highly significant ($p < 0.001$ for CONS) at around 30% (i.e., the regression constant CONS = −0.69 amounts to 0.3 in the probit probability function), where using dummy variables for the 15 most popular journals made the pooled estimation similar to the panel estimation (i.e., rho = 0). In particular, journals in which mistake 3 (MIS3) was observed with a higher and significant probability were *Ecological Economics, Journal of Environmental Management, Environmental Policy, Environmental and Resource Economics, Land Use Policy, Environmental Management*, and *Environmental Science and Policy*, whereas journals in which mistake 3 (MIS3) was observed with a lower and significant probability were *Sustainability, Journal of Cleaner Production, Water Resources*

Table 5.4 Average frequency of mistake 2 in the 15 most popular journals used as dummy variables

MIS2	Coef	Robust std. err	z	P > |z|	[95% conf. interval]	
EED	0.1486721	0.0583339	2.55	0.011	0.0343398	0.2630044
JEMD	−0.0885304	0.0583339	−1.52	0.129	−0.2028627	0.0258019
SUSD	0.367789	0.0583339	6.30	0.000	0.2534567	0.4821213
EPD	1.511018	0.0583339	25.90	0.000	1.396685	1.62535
JCPD	0.4562924	0.0583339	7.82	0.000	0.3419601	0.5706247
ERED	0.519839	0.0583339	8.91	0.000	0.4055067	0.6341713
CPD	0.0277833	0.0583339	0.48	0.634	−0.0865489	0.1421156
WRMD	−0.3374962	0.0583339	−5.79	0.000	−0.4518284	−0.2231639
STED	−0.2298783	0.0583339	−3.94	0.000	−0.3442106	−0.115546
LUPD	−0.64615	0.0583339	−11.08	0.000	−0.7604822	−0.5318177
EDSD	0.5915139	0.0583339	10.14	0.000	0.4771816	0.7058462
EMAD	−0.4788447	0.0583339	−8.21	0.000	−0.593177	−0.3645124
ESPD	−0.0453293	0.0583339	−0.78	0.437	−0.1596616	0.069003
CONS	−1.022241	0.0583339	−17.52	0.000	−1.136574	−0.9079089

Number of observations = 835, probability > χ^2 smaller than 0.001, rho smaller than 0.001. Abbreviations for dummy (d) variables: EE = *Ecological Economics*, JEM = *Journal of Environmental Management*, SUS = *Sustainability*, EP = *Environmental Policy*, JCP = *Journal of Cleaner Production*, ERE = *Environmental and Resource Economics*, CP = Climate Policy, WRM = Water Resources Management, STE = *Science of the Total Environment*, LUP = *Land Use Policy*, EM = Environmental Management, EDS = *Environment, Development and Sustainability*, WAT = *Water*, EMA = *Environmental Modelling and Assessment*, ESP = *Environmental Science and Policy*. Other abbreviations: CONS = probit regression constant, rho = the proportion of the total variance contributed by the journal variance component. *Notes* MIS2 (on PLA & EQU) in Boolean logic can be depicted as follows: REL & no EQU

Management, *Water*, and *Environmental Monitoring and Assessment*. In other words, the former journals have improperly used cost–benefit analysis (CBA) less often, whereas the latter journals have improperly used cost–benefit analysis (CBA) more often, although the latter five journals were correct in 0, 2, 0, 0, and 0 cases, respectively. Note that I have excluded *Climate Policy* because its dummy variable perfectly predicted mistake MIS3 = 0 (i.e., articles in *Climate Policy* never used utility-based approaches UA or dose response DR or replacement cost RC studies within a dynamic DYN cost–benefit analysis CBA so UA = 0 or DR = 0 or RC = 0 and DYN = 0).

Table 5.6 shows that mistake 4 (MIS4) was explained overall by the variables used (Prob > χ^2 is smaller than 0.001) and was highly significant ($p < 0.001$ for CONS) at around 10% (i.e., the regression constant CONS = −1.56 amounts to 0.1 in the probit probability function), where using dummy variables for the 15 most popular journals made the pooled estimation similar to the panel estimation (i.e., rho = 0). In particular, journals where mistake 4 (MIS4) was observed with a higher and significant probability were *Environmental Policy*, *Journal of Cleaner*

Table 5.5 Average frequency of mistake 3 in the 15 most popular journals used as dummy variables

MIS3	Coef	Robust std. err	z	P > |z|	[95% conf. interval]	
EED	0.4711462	0.0641586	7.34	0.000	0.3453977	0.5968947
JEMD	0.6524063	0.0641585	10.17	0.000	0.5266579	0.7781547
SUSD	−0.57267	0.0641587	−8.93	0.000	−0.6984188	−0.4469213
EPD	0.1152441	0.0641586	1.80	0.072	−0.0105045	0.2409927
JCPD	−0.5474319	0.0641588	−8.53	0.000	−0.6731809	−0.421683
ERED	0.5003697	0.0641587	7.80	0.000	0.374621	0.6261184
WRMD	−0.6652928	0.0641588	−10.37	0.000	−0.7910418	−0.5395439
STED	0.0608056	0.0641588	0.95	0.343	−0.0649434	0.1865545
LUPD	0.7541469	0.0641588	11.75	0.000	0.628398	0.8798958
EMD	0.6206609	0.0641588	9.67	0.000	0.4949118	0.7464099
EDSD	0.0715371	0.0641589	1.11	0.265	−0.054212	0.1972862
WATD	−0.4163141	0.0641589	−6.49	0.000	−0.5420631	−0.290565
EMAD	−0.8066579	0.0641588	−12.57	0.000	−0.9324067	−0.680909
ESPD	0.128518	0.0641589	2.00	0.045	0.0027689	0.2542671
CONS	−0.694251	0.0641585	−10.82	0.000	−0.8199994	−0.5685027

Number of observations = 835, probability > χ^2 smaller than 0.001, rho smaller than 0.001. Abbreviations for dummy (d) variables: EE = *Ecological Economics*, JEM = *Journal of Environmental Management*, SUS = *Sustainability*, EP = *Environmental Policy*, JCP = *Journal of Cleaner Production*, ERE = *Environmental and Resource Economics*, CP = Climate Policy, WRM = Water Resources Management, STE = *Science of the Total Environment*, LUP = *Land Use Policy*, EM = Environmental Management, EDS = *Environment, Development and Sustainability*, WAT = *Water*, EMA = *Environmental Modelling and Assessment*, ESP = *Environmental Science and Policy*. Other abbreviations: CONS = probit regression constant, rho = the proportion of the total variance contributed by the journal variance component. *Notes* MIS3 (on PRO & EFF) in Boolean logic can be depicted as follows: (STO or REL) & (UA or DR or RC) & no DYN

Production, *Climate Policy*, *Water Resources Management*, and *Environmental Monitoring and Assessment*, whereas journals where mistake 4 was observed with a lower but significant probability were *Ecological Economics*, *Journal of Environmental Management*, and *Sustainability*. In other words, the former journals improperly used multi-criteria analysis (MCA) less often, whereas the latter journals improperly used multi-criteria analysis (MCA) more often, although the latter journals were correct in 0, 1, and 0 cases, respectively. Note that I have excluded *Environmental and Resource Economics*, *Environmental Management*, and *Environmental Science and Policy* because their dummy variables perfectly predicted MIS4 = 0 (i.e., articles in these journals never used subjective methods SM within a dynamic DYN multi-criteria analysis MCA so SM = 0 and DYN = 0).

In summary, Tables 5.3, 5.4, 5.5, 5.6 (i.e., static estimative statistical analyses) show that all four mistakes were significant overall, although to a greater or smaller extent in different journals. Moreover, all journals had at least one significant positive or negative impact on mistakes (i.e., each journal in the top 15 journals differed from the reference average). Finally, all journals had at least one increased frequency

Table 5.6 Average frequency of mistake 4 in the 15 most popular journals used as dummy variables

MIS4	Coef	Robust std. err	z	P > |z|	[95% conf. interval]	
EED	−0.614335	0.0723711	−8.49	0.000	−0.7561797	−0.4724902
JEMD	−0.2703333	0.0723711	−3.74	0.000	−0.4121781	−0.1284885
SUSD	−0.3855278	0.0723711	−5.33	0.000	−0.5273726	−0.243683
EPD	0.676431	0.0723711	9.35	0.000	0.5345862	0.8182758
JCPD	0.3217113	0.0723711	4.45	0.000	0.1798665	0.4635561
CPD	0.1585072	0.0723711	2.19	0.029	0.0166625	0.300352
WRMD	0.4392397	0.0723711	6.07	0.000	0.2973949	0.5810845
STED	−0.0562757	0.0723711	−0.78	0.437	−0.1981205	0.0855691
LUPD	−0.1048102	0.0723711	−1.45	0.148	−0.246655	0.0370346
EDSD	0.0624938	0.0723711	0.86	0.388	−0.079351	0.2043385
WATD	0.0624938	0.0723711	0.86	0.388	−0.079351	0.2043385
EMAD	0.4528064	0.0723711	6.26	0.000	0.3109616	0.5946512
CONS	−1.563578	0.0723711	−21.60	0.000	−1.705422	−1.421733

Number of observations = 835, probability > χ^2 smaller than 0.001, rho smaller than 0.001. Abbreviations for dummy (d) variables: EE = *Ecological Economics*, JEM = *Journal of Environmental Management*, SUS = *Sustainability*, EP = *Environmental Policy*, JCP = *Journal of Cleaner Production*, ERE = *Environmental and Resource Economics*, CP = Climate Policy, WRM = Water Resources Management, STE = *Science of the Total Environment*, LUP = *Land Use Policy*, EM = Environmental Management, EDS = *Environment, Development and Sustainability*, WAT = *Water*, EMA = *Environmental Modelling and Assessment*, ESP = *Environmental Science and Policy*. Other abbreviations: CONS = probit regression constant, rho = the proportion of the total variance contributed by the journal variance component. *Notes* in MIS4 (on PRO & EQU) in Boolean logic can be depicted as follows: STO & SM & no DYN of mistake (i.e., each of the 15 most popular journals was incorrect more than the reference average for at least for one mistake).

Tables 5.7, 5.8, 5.9, 5.10 test whether these mistakes were increasing or decreasing overall and in which journal. Note that negative and positive coefficients for a journal mean a decreasing and increasing (respectively) frequency of the specified type of mistake with respect to other journals.

Table 5.7 shows that mistake 1 (MIS1) was explained overall by the variables used (Prob > χ^2 is smaller than 0.001), it was *not* significant net of its trend ($p = 0.890$ for CONS) at around 10% (i.e., the regression constant CONS = -1.42 amounts to 0.1 in the probit probability function) and it was constant (a positive coefficient for year but $p = 0.975$), where using dummy variables for the 15 most popular journals made the pooled estimation similar to the panel estimation (i.e., rho = 0). In particular, journals where mistake 1 (MIS1) was observed with an increasing and significant probability were *Sustainability, Environmental Policy, Environmental and Resource Economics, Water Resources Management, Science of the Total Environment,* and *Water* (the same journals in Table 5.3), whereas journals where mistake 1 (MIS1) was observed with a decreasing and significant probability were *Ecological Economics, Climate Policy, Land Use Policy,* and *Environmental Management* (the same journals in Table 5.5). Note that I excluded *Environment, Development and Sustainability* because its dummy variable perfectly predicted MIS1 = 0.

Table 5.8 shows that mistake 2 (MIS2) was explained overall by the variables used (Prob > χ^2 is smaller than 0.001), it was *not* significant net of its trend ($p = 0.428$ for CONS) at around 0% (i.e., the regression constant CONS = -18.96 amounts to 0 in the probit probability function) and it was constant (a positive coefficient for year but $p = 0.454$), where using dummy variables for the 15 most popular journals made the pooled estimation similar to the panel estimation (i.e., rho = 0). In particular, journals where mistake 2 (MIS2) was observed with an increasing and significant probability were *Ecological Economics, Sustainability, Environmental Policy, Journal of Cleaner Production, Environmental and Resource Economics,* and *Environment, Development and Sustainability* (the same journals in Table 5.4), whereas journals where mistake 2 (MIS2) was observed with a decreasing and significant probability were *Water Resources Management, Science of the Total Environment, Land Use Policy,* and *Environmental Monitoring and Assessment* (the same journals in Table 5.6). Note that I excluded *Environmental Management* and *Water* because their dummy variables perfectly predicted MIS2 = 0.

Table 5.9 shows that mistake 3 (MIS3) was explained overall by the variables used (Prob > χ^2 is smaller than 0.001), it was highly significant net of its trend ($p < 0.001$ for CONS) at around 100% (i.e., the regression constant CONS = 79.40 amounts to 1 in the probit probability function) but it was decreasing (a negative coefficient for year with $p < 0.001$), where using dummy variables for the 15 most popular journals made the pooled estimation similar to the panel estimation (i.e., rho = 0). In particular, journals where mistake 3 (MIS3) was observed with an increasing and significant probability were *Ecological Economics, Journal of Environmental Management, Environmental Science and Policy, Environmental and Resource Economics, Science of the Total Environment, Land Use Policy, Environmental Management,*

Table 5.7 Frequency changes of mistake 1 in the 15 most popular journals over time

MIS1	Coef	Robust std. err	z	P > |z|	[95% conf. interval]	
YEA	0.0001617	0.0051061	0.03	0.975	−0.009846	0.0101695
EEY	−0.0003024	0.0000183	−16.48	0.000	−0.0003384	−0.0002664
JEMY	−6.10e−06	0.0000166	−0.37	0.712	−0.0000386	0.0000263
SUSY	0.000038	0.0000206	1.84	0.065	−2.38e−06	0.0000784
EPY	0.0000436	0.0000157	2.77	0.006	0.0000128	0.0000744
JCPY	0.0000145	0.0000209	0.70	0.485	−0.0000263	0.0000554
EREY	0.0000385	0.0000157	2.46	0.014	7.85e−06	0.0000692
CPY	−0.0001535	0.0000186	−8.24	0.000	−0.00019	−0.000117
WRMY	0.0001571	0.000016	9.83	0.000	0.0001258	0.0001884
STEY	0.0003778	0.0000169	22.29	0.000	0.0003446	0.000411
LUPY	−0.0002836	0.0000164	−17.31	0.000	−0.0003157	−0.0002515
EMY	−0.0002324	0.0000167	−13.93	0.000	−0.0002651	−0.0001997
WATY	0.0001263	0.00002	6.32	0.000	0.0000871	0.0001655
EMAY	−6.21e−06	0.0000171	−0.36	0.716	−0.0000396	0.0000272
ESPY	0.0000152	0.0000159	0.96	0.338	−0.0000159	0.0000463
CONS	−1.423058	10.27703	−0.14	0.890	−21.56566	18.71955

Number of observations = 835, probability > χ^2 smaller than 0.001, rho smaller than 0.001. Abbreviations for yearly (y) variables: EE = *Ecological Economics*, JEM = *Journal of Environmental Management*, SUS = *Sustainability*, EP = *Environmental Policy*, JCP = *Journal of Cleaner Production*, ERE = *Environmental and Resource Economics*, CP = Climate Policy, WRM = Water Resources Management, STE = *Science of the Total Environment*, LUP = *Land Use Policy*, EM = Environmental Management, EDS = *Environment, Development and Sustainability*, WAT = *Water*, EMA = *Environmental Modelling and Assessment*, ESP = *Environmental Science and Policy.* Other abbreviations: yea = temporal trend, CONS = probit regression constant, rho = the proportion of the total variance contributed by the journal variance component. *Notes* MIS1 (on PLA & EFF) in Boolean logic can be depicted as follows: (ASY or UNC or IMP) & no EQU

and *Environment, Development and Sustainability*, whereas journals where mistake 3 (MIS3) was observed with a decreasing and significant probability were *Sustainability*, *Journal of Cleaner Production*, *Water Resources Management*, *Water*, and *Environmental Monitoring and Assessment* (the same journals in Table 5.5). Note that I excluded *Climate Policy* because its dummy variable perfectly predicted MIS3 = 0.

Table 5.10 shows that mistake 4 (MIS4) was explained overall by the variables used (Prob > χ^2 is smaller than 0.001), it was highly significant net of its trend (p = 0.04 for CONS) at around 0% (i.e., the regression constant CONS = −96.27 amounts to 0 in the probit probability function) but it was increasing (a positive coefficient of year with $p = 0.043$), where using dummy variables for the 15 most popular journals made the pooled estimation similar to the panel estimation (i.e., rho = 0). In particular, journals where mistake 4 (MIS4) was observed with an increasing and significant probability were *Environmental Policy*, *Water Resources Management*,

Table 5.8 Frequency changes of mistake 2 in the 15 most popular journals over time

MIS2	Coef	Robust std. err	z	P > |z|	[95% conf. interval]	
Yea	0.0089119	0.0118976	0.75	0.454	−0.0144071	0.0322308
EEY	0.0000911	0.0000231	3.93	0.000	0.0000457	0.0001364
JEMY	−0.0000324	0.0000231	−1.40	0.161	−0.0000776	0.0000128
SUSY	0.0001627	0.0000493	3.30	0.001	0.0000661	0.0002593
EPY	0.0007555	0.0000251	30.15	0.000	0.0007064	0.0008047
JCPY	0.0002059	0.0000503	4.09	0.000	0.0001073	0.0003045
EREY	0.0002625	0.0000255	10.31	0.000	0.0002126	0.0003124
CPY	−3.16e−06	0.000046	−0.07	0.945	−0.0000933	0.000087
WRMY	−0.0001694	0.0000305	−5.55	0.000	−0.0002292	−0.0001096
STEY	−0.00012	0.0000339	−3.54	0.000	−0.0001866	−0.0000535
LUPY	−0.0003277	0.0000361	−9.08	0.000	−0.0003984	−0.000257
EDSY	0.0002777	0.0000442	6.28	0.000	0.000191	0.0003643
EMAY	−0.0002255	0.0000229	−9.86	0.000	−0.0002703	−0.0001807
ESPY	−0.0000169	0.0000243	−0.70	0.487	−0.0000646	0.0000308
CONS	−18.96644	23.92046	−0.79	0.428	−65.84969	27.9168

Number of observations = 835, probability > χ^2 smaller than 0.001, rho smaller than 0.001. Abbreviations for yearly (y) variables: EE = *Ecological Economics*, JEM = *Journal of Environmental Management*, SUS = *Sustainability*, EP = *Environmental Policy*, JCP = *Journal of Cleaner Production*, ERE = *Environmental and Resource Economics*, CP = Climate Policy, WRM = Water Resources Management, STE = *Science of the Total Environment*, LUP = *Land Use Policy*, EM = Environmental Management, EDS = *Environment, Development and Sustainability*, WAT = *Water*, EMA = *Environmental Modelling and Assessment*, ESP = *Environmental Science and Policy.* Other abbreviations: yea = temporal trend, CONS = probit regression constant, rho = the proportion of the total variance contributed by the journal variance component. *Notes* MIS2 (on PLA & EQU) in Boolean logic can be depicted as follows: REL & no EQU

and *Environmental Monitoring and Assessment*, whereas journals where mistake 4 (MIS4) was observed with a decreasing and significant probability were *Ecological Economics*, *Journal of Environmental Management*, *Sustainability*, and *Land Use Policy*. Note that I have excluded *Environmental and Resource Economics*, *Environmental Management*, and *Environmental Science and Policy* because their dummy variables perfectly predict MIS4 = 0.

In summary, Tables 5.3, 5.4, 5.5, 5.6, 5.7, 5.8, 5.9, 5.10 (i.e., static and dynamic estimative statistical analyses) show that mistake 1 (i.e., the lack of a focus on equity if asymmetric information or uncertainty or imperfect competition are relevant) was medium overall (i.e., it was significant in 20% of the cited references) and constant (i.e., it did not increase significantly, with $p > 0.90$); mistake 2 (i.e., the lack of focus on equity if interaction is relevant) was medium overall (i.e., it was significant in 20% of the cited references) and constant (i.e., it did not increase significantly, with $p > 0.40$); mistake 3 (i.e., the improper use of utility-based approaches or some production-based approaches to monetary assessment within cost–benefit analysis)

Table 5.9 Frequency changes of mistake 3 in the 15 most popular journals over time

MIS3	Coef	Robust std. err	z	P > |z|	[95% conf. interval]	
Yea	−0.0397935	0.0086973	−4.58	0.000	−0.0568398	−0.0227472
EEY	0.0001652	0.0000325	5.08	0.000	0.0001014	0.000229
JEMY	0.0002742	0.0000296	9.26	0.000	0.0002161	0.0003322
SUSY	−0.0001847	0.0000322	−5.75	0.000	−0.0002478	−0.0001217
EPY	0.0000358	0.0000278	1.29	0.197	−0.0000186	0.0000903
JCPY	−0.0001653	0.0000328	−5.04	0.000	−0.0002296	−0.0001011
EREY	0.0002449	0.0000267	9.17	0.000	0.0001926	0.0002972
WRMY	−0.00032	0.0000265	−12.05	0.000	−0.000372	−0.000268
STEY	0.0000698	0.000027	2.58	0.010	0.0000168	0.0001228
LUPY	0.0004101	0.0000269	15.27	0.000	0.0003575	0.0004628
EMY	0.0002655	0.0000294	9.02	0.000	0.0002078	0.0003232
EDSY	0.0001135	0.0000299	3.79	0.000	0.0000548	0.0001722
WATY	−0.0001122	0.0000316	−3.55	0.000	−0.0001741	−0.0000503
EMAY	-0.0004857	0.0000358	−13.55	0.000	−0.0005559	−0.0004154
ESPY	0.0000627	0.0000268	2.34	0.019	0.0000101	0.0001153
CONS	79.40479	17.51266	4.53	0.000	45.08061	113.729

Number of observations = 835, probability > χ^2 smaller than 0.001, rho smaller than 0.001. Abbreviations for yearly (y) variables: EE = *Ecological Economics*, JEM = *Journal of Environmental Management*, SUS = *Sustainability*, EP = *Environmental Policy*, JCP = *Journal of Cleaner Production*, ERE = *Environmental and Resource Economics*, CP = Climate Policy, WRM = Water Resources Management, STE = *Science of the Total Environment*, LUP = *Land Use Policy*, EM = Environmental Management, EDS = *Environment, Development and Sustainability*, WAT = *Water*, EMA = *Environmental Modelling and Assessment*, ESP = *Environmental Science and Policy*. Other abbreviations: yea = temporal trend, CONS = probit regression constant, rho = the proportion of the total variance contributed by the journal variance component. *Notes* MIS3 (on PRO & EFF) in Boolean logic can be depicted as follows: (STO or REL) & (UA or DR or RC) & no DYN

was large overall (i.e., it was significant in 30% of the cited references) but decreased significantly ($p < 0.10$); and mistake 4 (i.e., the improper use of subjective methods SM for assessment of relative weights within multi-criteria analysis MCA) was small overall (i.e., it was significant in 10% of the cited references) but increased significantly (i.e., $p < 0.10$). However, apart from *Environmental Policy* (significant and constant) and *Science of the Total Environment* and *Environment, Development and Sustainability* (non-significant and increasing) for mistake 3, and apart from *Journal of Cleaner Production* and *Climate Policy* (significant and constant) and *Land Use Policy* (non-significant and decreasing) for mistake 4, journals where these mistakes were relatively frequent also showed an increasing frequency, whereas in journals where the mistakes were relatively infrequent, they were also decreasing. In other words, the frequency of mistakes depended on the journal (i.e., panel data estimates were better than pooled estimates in the 15 most popular journals) and there was *no* convergence of journals towards minimizing these four types of mistake (i.e.,

Table 5.10 Frequency changes of mistake 4 in the 15 most popular journals over time

MIS4	Coef	Robust std. err	z	P > |z|	[95% conf. interval]	
Yea	0.047014	0.0232516	2.02	0.043	0.0014417	0.0925863
EEY	−0.0001924	0.00004	−4.82	0.000	−0.0002707	−0.0001141
JEMY	−0.0000784	0.000027	−2.91	0.004	−0.0001312	−0.0000256
SUSY	−0.0002781	0.0000631	−4.41	0.000	−0.0004017	−0.0001544
EPY	0.000394	0.0000301	13.10	0.000	0.000335	0.0004529
JCPY	0.0000708	0.0000642	1.10	0.270	−0.0000551	0.0001966
CPY	3.46e−06	0.0000579	0.06	0.952	−0.00011	0.000117
WRMY	0.0002174	0.0000314	6.93	0.000	0.0001559	0.0002789
STEY	−0.0000594	0.0000419	−1.42	0.156	−0.0001415	0.0000227
LUPY	−0.0000783	0.0000418	−1.88	0.061	−0.0001602	3.54e−06
EDSY	−0.0000512	0.0000627	−0.82	0.415	−0.0001741	0.0000718
WATY	−0.0000549	0.0000623	−0.88	0.378	−0.000177	0.0000672
EMAY	0.0003089	0.0000349	8.86	0.000	0.0002405	0.0003772
CONS	−96.27218	46.81449	−2.06	0.040	−188.0269	−4.517467

Number of observations = 835, probability > χ^2 smaller than 0.001, rho smaller than 0.001. Abbreviations for yearly (y) variables: EE = *Ecological Economics*, JEM = *Journal of Environmental Management*, SUS = *Sustainability*, EP = *Environmental Policy*, JCP = *Journal of Cleaner Production*, ERE = *Environmental and Resource Economics*, CP = Climate Policy, WRM = Water Resources Management, STE = *Science of the Total Environment*, LUP = *Land Use Policy*, EM = Environmental Management, EDS = *Environment, Development and Sustainability*, WAT = *Water*, EMA = *Environmental Modelling and Assessment*, ESP = *Environmental Science and Policy*. Other abbreviations: yea = temporal trend, CONS = probit regression constant, rho = the proportion of the total variance contributed by the journal variance component. *Notes* in MIS4 (on PRO & EQU) in Boolean logic can be depicted as follows: STO & SM & no DYN

researchers who made mistake 1 to mistake 4 were more likely to submit to and publish their papers in the same journals).

Table 5.11 summarizes the significant coefficients (i.e., $p < 0.10$) for each of the 15 popular journals for each mistake. Note that there were no journals with only negative coefficients. In particular, since *Ecological Economics* and *Environmental and Resource Economics* represent the two main fields in the literature, I calculated the distance of each journal from these two journals instead of the distances of each journal from each other journal. Moreover, since the coefficients come from a probit estimation, I used the sum of their absolute values instead of the square root of the sum of quadratic distances. Finally, since both policy measures and investment projects are important in analysing sustainability, I attached the same relative weight (i.e., 1) to each mistake instead of emphasising policy measures or investment projects.

Note that in calculating distances between journals, all mistakes have the same importance (i.e., standardisation is not needed *and* I used the same weight for each mistake), since significant coefficients refer to the *absolute* probability of making

Table 5.11 Distances of journals from *EE* and *ERE* with DEA incorrectness

Journal	MIS1	MIS2	MIS3	MIS4	EE distance	ERE distance	DEA incorrectness
EE	−0.607	0.148	0.471	−0.614	0.00	**1.70**	0.938
ERE	0.077	0.519	0.5	0	**1.70**	0.00	1
CP	−0.307	0	0	0.158	1.69	1.56	0.776
LUP	−0.571	−0.646	0.754	0	1.73	2.07	1
EM	−0.467	0	0.62	0	1.05	1.18	1
WRM	0.316	−0.337	−0.665	0.439	3.60	2.70	0.858
STE	0.716	−0.229	0	0	2.79	1.89	0.917
EMA	0	−0.478	−0.806	0.452	3.58	2.83	0.866
SUS	0.076	0.367	−0.572	−0.385	2.17	1.61	0.575
JCP	0	0.456	−0.547	0.321	2.87	1.51	0.788
WAT	0.255	0	−0.416	0	2.51	1.61	0.671
JEM	0	0	0	−0.27	1.57	1.37	0.734
EP	0.87	1.511	0.115	0.676	4.49	2.85	1
ESP	0	0	0.128	0	1.71	0.97	0.802
EDS	0	0.591	0	0	2.14	0.65	0.774

Abbreviations: EE = *Ecological Economics*, ERE = *Environmental and Resource Economics*, DEA = Data Envelopment Analysis, MIS1 = mistake 1, MIS2 = mistake 2, MIS3 = mistake 3, MIS4 = mistake 4, JEM = *Journal of Environmental Management*, SUS = *Sustainability*, EP = *Environmental Policy*, JCP = *Journal of Cleaner Production*, CP = *Climate Policy*, WRM = *Water Resources Management*, STE = *Science of the Total Environment*, LUP = *Land Use Policy*, EM = *Environmental Management*, EDS = *Environment, Development and Sustainability*, WAT = *Water*, EMA = *Environmental Modelling and Assessment*, ESP = *Environmental Science and Policy*. *Notes* distances are calculated as the unweighted sum of the differences between the significant estimated coefficients

a mistake. In contrast, in comparing the frequency of incorrectness of the journals, I applied Data Envelopment Analysis DEA, in which mistakes are properly standardised because a *relative* ranking is obtained.

Let us assume that *Ecological Economics* and *Environmental and Resource Economics* have a significantly different focus. Indeed, *Ecological Economics* is not concerned with the assumptions behind the Economic General Equilibrium Model (i.e., it is more focused on equity), whereas *Environmental and Resource Economics* did not apply multi-criteria analysis (i.e., it is more focused on efficiency), although *Ecological Economics* and *Environmental and Resource Economics* had similar frequencies of making mistakes 2 and 3. Let us use 1.70 (i.e., the mutual distance between *Ecological Economics* and *Environmental and Resource Economics*) as the threshold value to state whether a journal is similar or dissimilar to *Ecological Economics* or *Environmental and Resource Economics*, within a range of 10% (i.e., a distance $\geq$ 1.53 means they are different). We can then refer to the smallest distance to *Ecological Economics* or *Environmental and Resource Economics* for each journal to determine which journal it is closer to.

In summary, three groups of journals are identified as *significantly* different (i.e., distances are based on significant coefficients): ecological economics (i.e., *Ecological Economics, Environmental Management, Land Use Policy*); environmental economics (i.e., *Environmental and Resource Economics, Climate Policy, Journal of Environmental Management, Environmental Science and Policy, Environment, Development and Sustainability, Journal of Cleaner Production*); and other sustainability journals (i.e., *Water Resources Management, Science of the Total Environment, Environmental Monitoring and Assessment, Sustainability, Water, Environmental Policy*).

Note that *some* other sustainability journals scored better (i.e., they had a larger distance from the incorrectness frontier identified by DEA) than *Ecological Economics* and *Environmental and Resource Economics*. Moreover, both *Ecological Economics* and *Environmental and Resource Economics* were similar in terms of their incorrectness score, with *Environmental and Resource Economics* at 100% and *Ecological Economics* at 94%. Finally, *many* journals close to the two journals scored worse (i.e., they had a larger distance from the incorrectness frontier identified by DEA) than other sustainability journals.

Figures 5.14 and 5.15 visualize the distances of the 15 most popular journals from *Ecological Economics* and *Environmental and Resource Economics*, respectively, in terms of the four mistakes.

Thus, the distances from *Ecological Economics* were mainly accounted for by mistakes 3 and 4, whereas the distances from *Environmental and Resource Economics* were mainly accounted for by mistakes 2 and 3.

Tables 5.12, 5.13, 5.14, 5.15 test whether mistakes 1 to 4 were more or less frequent for some environmental issues, by identifying them as a combination of pollution, resources, flows, stocks, renewable resources, air, water, land as suggested in Sect. 5.1.

Table 5.12 shows that mistake 1 (MIS1) was explained overall by the variables used (Prob > χ^2 is smaller than 0.001). Note that the residual variance (rho < 0.001) suggests that the relevance of this mistake type across environmental issues does *not* depend on journals (i.e., the pooled estimation was similar to the panel estimation). Moreover, the non-significant constant highlights that many articles still assume the assumptions behind the Economic General Equilibrium Model. Finally, the significant result for land ($p = 0.009$ for LAN) at around 60% (i.e., the regression coefficient LAN = 0.70 represents a value of 0.60 in the probit probability function) was attached to research on solid waste management.

Table 5.13 shows that mistake 2 (MIS2) was explained overall by the variables used (Prob > χ^2 is smaller than 0.001). Note that the residual variance (rho = 0.139) suggests that the relevance of this mistake type across environmental issues depends on journals (i.e., the panel estimation would be better than the pooled estimation). Moreover, the significantly ($p < 0.001$ for AIR) increased frequency of mistake 2 for air at around 95% (i.e., the regression coefficient AIR = 4.15 represents a value of 0.95 in the probit probability function) was due to articles on climate change.

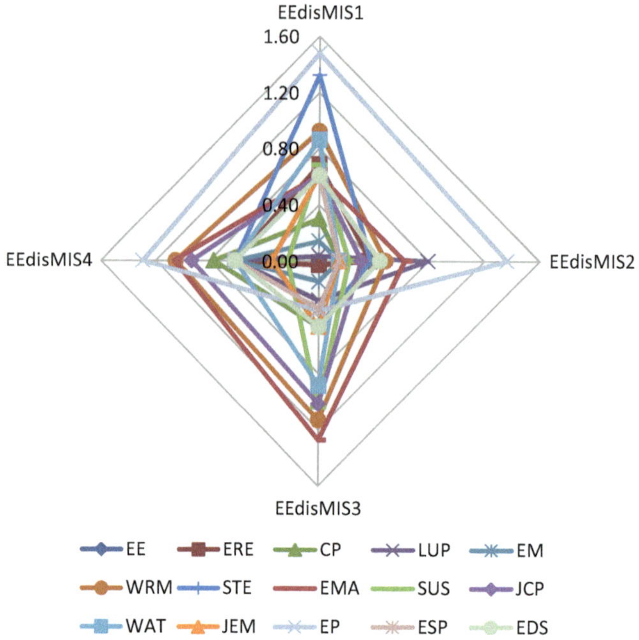

Fig. 5.14 Distances from *Ecological Economics* in terms of the four mistakes. Abbreviations: EEdis = distance from Ecological Economics, MIS1 = mistake 1, MIS2 = mistake 2, MIS3 = mistake 3, MIS4 = mistake 4, EE = Ecological Economics, JEM = Journal of Environmental Management, SUS = Sustainability, EP = Environmental Policy, JCP = Journal of Cleaner Production, ERE = Environmental and Resource Economics, CP = Climate Policy, WRM = Water Resources Management, STE = Science of the Total Environment, LUP = Land Use Policy, EM = Environmental Management, EDS = Environment, Development and Sustainability, WAT = Water, EMA = Environmental Modelling and Assessment, ESP = Environmental Science and Policy

Finally, the significant result for water ($p = 0.023$ for WAT) at around 90% (i.e., the regression coefficient WAT = 2.45 represents a value of 0.90 in the probit probability function) was attached to research on trans-boundary aquifers.

Table 5.14 shows that mistake 3 (MIS3) was explained overall by the variables used (Prob > χ^2 is smaller than 0.001). Note that the residual variance (rho = 0.148) suggests that the relevance of this mistake type across environmental issues depends on journals (i.e., the panel estimation would be better than the pooled estimation). Next, the significantly ($p < 0.001$ for REN) increased frequency of mistake 3 for renewable resources at around 80% (i.e., the regression coefficient REN = 1.52 represents a value of 0.80 in the probit probability function) highlighted that cost–benefit analysis has been improperly applied to biodiversity loss.

Table 5.15 shows that mistake 4 (MIS4) was explained overall by the variables used (Prob > χ^2 is smaller than 0.001). Note that the residual variance (rho = 0.049) suggests that the relevance of this mistake type across environmental issues does *not* depend on journals (i.e., the pooled estimation was similar to the panel estimation).

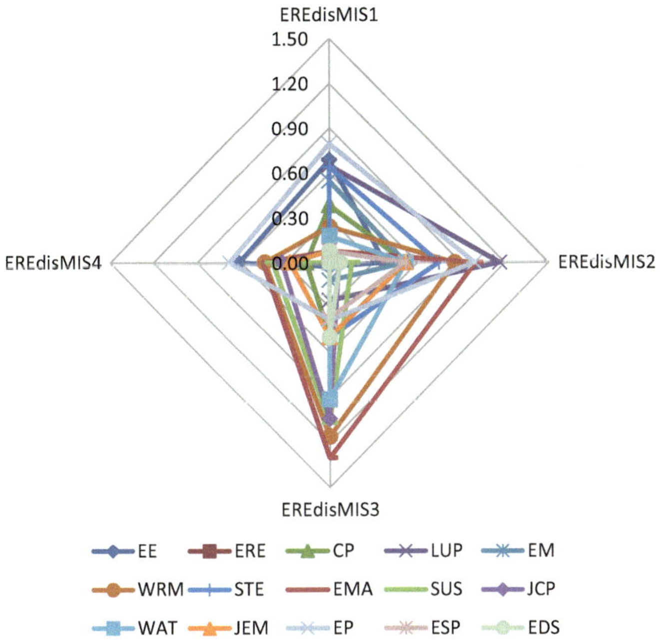

Fig. 5.15 Distances from *Environmental and Resource Economics* in terms of the four mistakes. Abbreviations: EREdis = distance from Environmental and Resource Economics, MIS1 = mistake 1, MIS2 = mistake 2, MIS3 = mistake 3, MIS4 = mistake 4, EE = Ecological Economics, JEM = Journal of Environmental Management, SUS = Sustainability, EP = Environmental Policy, JCP = Journal of Cleaner Production, ERE = Environmental and Resource Economics, CP = Climate Policy, WRM = Water Resources Management, STE = Science of the Total Environment, LUP = Land Use Policy, EM = Environmental Management, EDS = Environment, Development and Sustainability, WAT = Water, EMA = Environmental Modelling and Assessment, ESP = Environmental Science and Policy

Table 5.12 Average frequency of mistake 1 for environmental topics

MIS1	Coef	Robust std. err	z	P > |z|	[95% conf. interval]	
RES	−0.6585714	0.4765336	−1.38	0.167	−1.59256	0.2754172
POL	−0.5281146	0.5456598	−0.97	0.333	−1.597588	0.5413589
FLO	−0.1474779	0.1568929	−0.94	0.347	−0.4549823	0.1600265
REN	−0.2492266	0.2915701	−0.85	0.393	−0.8206934	0.3222402
AIR	−0.0072969	0.2987832	−0.02	0.981	−0.5929012	0.5783074
WAT	0.5488777	0.3955193	1.39	0.165	−0.2263259	1.324081
LAN	0.7055145	0.2710416	2.60	0.009	0.1742827	1.236746
CONS	−0.6011911	0.5732887	−1.05	0.294	−1.724816	0.5224341

Number of observations = 835, probability > χ^2 smaller than 0.001, rho smaller than 0.001. Abbreviations: RES = resources, POL = pollution, FLO = flows, REN = renewable resources, AIR = air, WAT = water, LAN = land, CONS = probit regression constant, rho = the proportion of the total variance contributed by the journal variance component

Table 5.13 Average frequency of mistake 2 for environmental topics

MIS2	Coef	Robust std. err	z	P > |z|	[95% conf. interval]	
RES	−0.3732022	0.6303701	−0.59	0.554	−1.608705	0.8623004
POL	−0.5306937	0.7175249	−0.74	0.460	−1.937017	0.8756293
FLO	−1.358731	0.2506493	−5.42	0.000	−1.849995	−0.8674675
REN	1.289596	1.249263	1.03	0.302	−1.158914	3.738107
AIR	4.152028	1.072116	3.87	0.000	2.050719	6.253337
WAT	2.459428	1.078188	2.28	0.023	0.3462182	4.572638
LAN	−0.6525945	0.4225036	−1.54	0.122	−1.480686	0.1754973
CONS	−3.370332	0.8861168	−3.80	0.000	−5.107089	−1.633575

Number of observations = 835, probability > χ^2 smaller than 0.001, rho = 0.139. Abbreviations: RES = resources, POL = pollution, FLO = flows, REN = renewable resources, AIR = air, WAT = water, LAN = land, CONS = probit regression constant, rho = the proportion of the total variance contributed by the journal variance component

Table 5.14 Average frequency of mistake 3 for environmental topics

MIS3	Coef	Robust std. err	z	P > |z|	[95% conf. interval]	
RES	0.3864359	0.3226569	1.20	0.231	−0.2459599	1.018832
POL	0.8861477	0.3975654	2.23	0.026	0.1069339	1.665361
FLO	−1.906151	0.121803	−15.65	0.000	−2.144881	−1.667422
REN	1.525182	0.2233861	6.83	0.000	1.087353	1.963011
AIR	0.0591601	0.1994016	0.30	0.767	−0.3316599	0.44998
WAT	0.5174212	0.3826683	1.35	0.176	−0.2325949	1.267437
LAN	0.0726683	0.2483702	0.29	0.770	−0.4141284	0.5594649
CONS	−1.898871	0.429384	−4.42	0.000	−2.740449	−1.057294

Number of observations = 835, probability > χ^2 smaller than 0.001, rho = 0.148. Abbreviations: RES = resources, POL = pollution, FLO = flows, REN = renewable resources, AIR = air, WAT = water, LAN = land, CONS = probit regression constant, rho = the proportion of the total variance contributed by the journal variance component

Moreover, the significantly ($p = 0.038$ for WAT) increased frequency of mistake 4 for water at around 55% (i.e., the regression coefficient WAT = 0.27 represents a value of 0.55 in the probit probability function) was due to articles on groundwater management. Finally, the significant result for land ($p = 0.076$ for LAN) at around 55% (i.e., the regression coefficient LAN = 0.26 represents a value of 0.55 in the probit probability function) was attached to research on waste management (i.e., multi-criteria analysis seems to have been improperly applied to both groundwater and waste management).

Thus, mistake 1 (measured by the number of articles or the frequency of articles that contained this type of mistake) was slightly underestimated (non-significant CONS at 0.60 with $p = 0.294$) and to a similar extent by all journals (rho = 0):

Table 5.15 Average frequency of mistake 4 for environmental topics

MIS4	Coef	Robust std. err	z	P > |z|	[95% conf. interval]	
RES	0.6461203	0.5279042	1.22	0.221	−0.3885528	1.680793
POL	0.8876572	0.589084	1.51	0.132	−0.2669261	2.042241
FLO	−1.410838	0.242511	−5.82	0.000	−1.886151	−0.935525
REN	0.1755069	0.1736636	1.01	0.312	−0.1648675	0.5158814
AIR	0.1287034	0.1118226	1.15	0.250	−0.0904649	0.3478717
WAT	0.2768412	0.1336854	2.07	0.038	0.0148226	0.5388599
LAN	0.2635129	0.1485062	1.77	0.076	−0.0275539	0.5545797
CONS	−2.392259	0.6740158	−3.55	0.000	−3.713306	−1.071213

Number of observations = 835, probability > χ^2 smaller than 0.001, rho = 0.049. Abbreviations: RES = resources, POL = pollution, FLO = flows, REN = renewable resources, AIR = air, WAT = water, LAN = land, CONS = probit regression constant, rho = the proportion of the total variance contributed by the journal variance component

renewable resources (only stocks) and air (mainly stocks) were (not significantly) less relevant for mistake 1, since the assumptions behind the Economic General Equilibrium Model are plausibly irrelevant for these environmental issues.

Mistake 2 was important overall (significant CONS at −3.37 with $p < 0.001$) and to a different extent by different journals (rho = 0.139): renewable resources (mainly independent) and land (only independent) were (not significantly) less relevant for mistake 2, since decision interactions are plausibly irrelevant for these environmental issues.

Mistake 3 was important overall (significant CONS at −1.89 with $p < 0.001$) and to a different extent for different journals (rho = 0.148): air (mainly stocks) and land (only stocks) were (not significantly) less relevant for mistake 3, since cost–benefit analysis is less likely applied to climate change and waste management.

Mistake 4 was important overall (significant CONS at −2.39 with $p < 0.001$) and to a different extent by different journals (rho = 0.049): air (mainly stocks) and renewable resources (only stocks) were (not significantly) less relevant for mistake 4, since multi-criteria analysis is less likely applied to climate change and biodiversity loss.

In summary, there were significant differences in the average frequency of mistakes across environmental issues.

Two methodological remarks are worthy here:

- I did not distinguish stocks from a decision interaction in mistake 3 because the application of utility-based approaches or dose response or replacement cost studies are mistakes in both cases and because few of the sampled articles focused on a decision interaction without stocks.
- I did not use stocks as an independent variable for estimates of environmental topics to make all four estimates comparable, since the dependent variable mistake

4 is crucially based on stocks (i.e., mistake 4 = stocks and multi-criteria analysis and no dynamic model and subjective methods).

To summarize this Section, we can say that all four mistakes were statistically significant, although to a different extent for different journals. In particular, in terms of mistakes, journals could be separated into journals around ecological economics (i.e., *Ecological Economics, Environmental Management, Land Use Policy*), around environmental economics (i.e., *Environmental and Resource Economics, Climate Policy, Journal of Environmental Management, Environmental Science and Policy, Environment, Development and Sustainability, Journal of Cleaner Production*), and other sustainability journals (i.e., *Water Resources Management, Science of the Total Environment, Environmental Monitoring and Assessment, Sustainability, Water, Environmental Policy*), although *Ecological Economics* and *Environmental and Resource Economics* had similar incorrectness scores. Moreover, mistake 1 was infrequent (i.e., the number of articles with mistake 1 was small) and its frequency increased slowly, although this could be due to the lack of specification of unmet assumptions. Mistake 2 was infrequent (i.e., the number of articles with mistake 2 was small) and its frequency increased slowly, although this could be due to the lack of specification of relevant interactions. Mistake 3 was frequent (i.e., the number of articles with mistake 3 was large) and its frequency decreased quickly. Mistake 4 was infrequent (i.e., the number of articles with mistake 4 was small) and its frequency increased quickly. Finally, all four mistakes were statistically significant, although to a different extent for different environmental issues. In particular, mistake 1 was significant for solid waste management, mistake 2 for climate change and groundwater management, mistake 3 for biodiversity loss, and mistake 4 for solid waste management and groundwater management.

5.2 Suggested Solutions

Section 5.1 highlighted the relevance and dynamics of mistakes in the sustainability science literature. The purpose of this Section is to suggest some *logical* solutions to these specific mistakes and to provide *methodological* suggestions for sustainability research in more general settings (i.e., where those mistakes are uncommon). I will also refer to my 25 empirical articles to support both the logical solutions and the methodological suggestions. Note that most concerns turn out to be practically non-significant in the sustainability science literature, and the few practically significant concerns have a straightforward solution, such as the proper application of spatial discounting instead of using a geographical information system and the use of a computable general equilibrium model instead of using an input–output model within cost–benefit analysis.

For the two alternative approaches to sustainability that I identified in Chap. 3 (i.e., weak sustainability vs. strong sustainability) based on the environmental ethics

I discussed in Chap. 2 (i.e., maximisation of total or average welfare for weak sustainability vs. minimisation of individual resource inequality for strong sustainability), we can define solutions to the specific mistakes identified in Chap. 4 that were highlighted as being statistically significant in the sustainability science literature in Sect. 5.1. These logical solutions can be summarised as follows:

- Mistake 1 on policy measures and efficiency: (asymmetric information or uncertainty or imperfect competition) and no equity
- Solution 1: if asymmetric information or uncertainty or imperfect competition, then equity in terms of welfare and of resources for weak and strong sustainability, respectively
- Mistake 2 on policy measures and efficiency: a decision interaction and no equity
- Solution 2: if a decision interaction, then equity in terms of welfare and of resources for weak and strong sustainability, respectively
- Mistake 3 on investment projects and efficiency: (stocks or a decision interaction) and utility-based approaches and cost–benefit analysis
- Solution 3: if stocks & cost–benefit analysis, then production-based approaches (i.e., opportunity cost or preventive cost) and a dynamic model; if a decision interaction and cost–benefit analysis, then production-based approaches (i.e., opportunity cost or preventive cost) and dynamic game models with sensitivity analyses for monetary assessment; and if stocks and a decision interaction and cost–benefit analysis, then production-based approaches (i.e., opportunity cost or preventive cost) and dynamic game models with sensitivity analyses for monetary assessment
- Mistake 4 on investment projects and equity: stocks and subjective methods and multi-criteria analysis
- Solution 4: if stocks and multi-criteria analysis, then subjective methods and dynamic models with sensitivity analyses for relative weights

Note that Solution 3 is not applicable to strong sustainability, whereas Solution 4 is not applicable to weak sustainability.

Table 5.16 summarises the same logical solutions by using acronyms.

In summary, in the most complicated contexts (i.e., asymmetric information, uncertainty, imperfect competition, stocks, or a decision interaction), strong sustainability seems to be more appropriate. Indeed, the preferences and technologies of future generations are uncertain, but strong sustainability is based on the precautionary principle. Asymmetric information, uncertainty, and imperfect competition are likely in real settings, but strong sustainability is not focused on efficiency. Relative weights in multi-criteria analysis *probably* change to a smaller extent than monetary preferences in cost–benefit analysis, and decision interactions are irrelevant for strong sustainability. Sensitivity analyses for relative weights in multi-criteria analysis can be based on all possible values by using the three-dimension simplex (i.e., relative weights sum up to 1), whereas sensitivity analyses for monetary assessments in cost–benefit analysis could miss some values that are *possibly* relevant in the long-run. Note that a decision-support system enables a sensitivity analysis performed by stakeholders for relative weights as well as for uncertain parameter values.

Table 5.16 Logical solutions to the four mistakes

Mistakes	Goals	Problems	Solutions
MIS1	PLA & EFF	(ASY or UNC or IMP) & no EQU	if ASY or UNC or IMP then EQU in terms of welfare and of resource for WS and SS, respectively
MIS2	PLA & EFF	REL & no EQU	if REL then EQU in terms of welfare and of resource for WS and SS, respectively
MIS2	PRO & EFF	(STO or REL) & UA & & DR & RC & CBA	if STO & CBA then PA (i.e., OC or PC) & dynamic models with sensitivity analyses on monetary assessments; if REL & CBA then PA (i.e., OC or PC) & dynamic game models with sensitivity analyses on monetary assessments; if STO & REL & CBA then PA (i.e., OC or PC) & dynamic game models with sensitivity analyses on monetary assessments
MIS4	PRO & EQU	STO & SM & MCA	if STO & MCA then SM & dynamic models with sensitivity analyses on relative weights

Abbreviations: MIS1 = mistake 1, MIS2 = mistake 2, MIS3 = mistake 3, MIS4 = mistake 4, PLA = policy measures, EFF = efficiency, ASY = asymmetric information, UNC = uncertainty, IMP = imperfect competition, EQU = equity, REL = decisional interaction, STO = stocks, UA = utility approaches, DR = dose response, RC = replacement cost, OC = opportunity cost, PC = preventive cost, CBA = cost–benefit analysis, PRO = investment projects, SM = subjective methods, MCA = multi-criteria analysis; WS = weak sustainability; SS = strong sustainability

By referring to the two alternative approaches to sustainability (i.e., weak sustainability or efficiency vs. strong sustainability or equity) and the two main characterisations of environmental issues (i.e., flows vs. stocks), the methodological suggestions for policy measures (i.e., a top-down approach) in more general settings are as follows:

- For environmental flow problems, market-based policies should be suggested within weak sustainability (to maximise welfare of the current generation); command-and-control policies should be suggested within strong sustainability (to minimise intra-generational inequality).
- For environmental stock problems, command-and-control policies should be suggested within strong sustainability (to minimise inter-generational inequality); market-based policies within weak sustainability are inadequate to maximise the welfare of current and future generations.
- For environmental decision interaction problems, command-and-control agreements (e.g., the Kyoto Protocol based on the Kalai–Smorodinsky equilibrium)

should be suggested within strong sustainability to maximise intra-generational equity in term of pollution or resources; in contrast, market-based agreements, such as the Paris Agreement based on a Nash bargaining equilibrium, should be suggested within weak sustainability to maximise intra-generational equity in terms of welfare.

The methodological suggestions for investment projects (i.e. a bottom-up approach) in more general settings are as follows:

- For an environmental issue in terms of flows, multi-criteria analysis should be adopted with the estimation of relative weights by involving stakeholders using methods such as the analytical hierarchy process or linear regression. Alternatively, if cost–benefit analysis is applied, stakeholder participation should be estimated by relying on utility approaches, with contingent valuation and choice experiments preferred to the hedonic pricing and travel cost methods.
- For an environmental issue in terms of stocks, a dynamic model within multi-criteria analysis should be combined with sensitivity analyses for decisions with respect to relative weights to be estimated by involving stakeholders. Indeed, stakeholder relative weights could be biased due to a lack of information and changes over time thanks to additional information so that decisions should be robust with respect to potential changes in relative weights. Alternatively, if cost–benefit analysis is applied, then sensitivity analyses for decisions should be performed within a dynamic model for a relevant range of monetary assessments, with the assessments obtained by applying production approaches such as the opportunity cost and preventive cost approaches; however, stakeholder monetary assessments could be biased due to a lack of information and could change over time thanks to additional information.
- For an environmental issue in terms of a decision interaction, a game model within multi-criteria analysis should be combined with sensitivity analyses for decisions with respect to relative weights, with the weights estimated by involving stakeholders. Alternatively, if cost–benefit analysis is applied, then sensitivity analyses for decisions within a game model should be performed for a relevant range of monetary assessments, with the assessments obtained by applying production approaches such as the opportunity cost and preventive cost methods.

Note that suitable methodologies for both cost–benefit analysis and multi-criteria analysis should be applied in the case of complicated contexts: time discounting for cost benefit analysis versus a temporal information system for multi-criteria analysis; spatial discounting for cost–benefit analysis versus a geographical information system for multi-criteria analysis; expected utility or a capital assessment price model for cost–benefit analysis versus expected value or fuzzy sets for multi-criteria analysis; a computable general equilibrium model for cost–benefit analysis versus a social accounting matrix for multi-criteria analysis; and a social welfare function for cost–benefit analysis versus inequality weights for multi-criteria analysis. Moreover, the analytical hierarchy process should be used to estimate relative weights rather than environmental decisions, since spatial, dynamic, uncertainty, linkage, and inequality issues might become relevant in the long-run. Finally, multi-criteria analysis (i.e., in terms of mitigation rather than in terms of adaptation) should be preferred to cost–benefit analysis if decision interactions in the long-run (i.e., a decision interaction and stocks) are relevant, since utility approaches inaccurately measure the welfare of stakeholders.

Tables 5.17 and 5.18 characterise my 27 empirical and theoretical articles from Scopus by using the same categories I used for the 835 articles used in the previous Section. Note that my 25 *empirical* articles cover 60% of the 15 most popular scientific journals and 60% of the 20 most popular scientific journals. Moreover, my article in *International Journal of Environmental Science* (2021) is about participation. Finally, my articles in the *Journal of Environmental Economics and Management* (1998a) and the *Journal of Environmental Management* (2008) are *theoretical* articles on pollution and resources, respectively.

In particular, a detailed analysis of these articles shows that I did not make the specific mistakes that I showed to be statistically significant in Sect. 5.1. Indeed, I have never discussed efficiency in the context of asymmetric information, uncertainty, or imperfect competition (mistake 1); that is, my articles in *Environmental Monitoring and Assessment* (2012), *Environmental Management and Software* (2013), *Natural Hazards* (2013), *Applied Soft Computing* (2016) are about risk. I have also never applied cost–benefit analysis to stock environmental issues (mistake 3); that is, my articles in *Natural Hazards* (2013) and *Water* (2017) are about flows. In contrast:

- In settings where mistake 2 is possible (i.e., stocks or a decision interaction), I applied one or more equity criteria in my articles in *Environment and Development Economics* (1998b), *Water Resources Management* (2010), *Journal of Hydrology* (2014), *Sustainability Science* (2015), *Marine Policy* (2017), *Environment, Development and Sustainability* (2018), *Desalination and Water Treatment* (2020), *Sustainability* (2020), Nature—Palgrave Communications (2020), and *Frontiers in Environmental Science* (2022).
- In settings where mistake 4 is possible, I developed a dynamic model in my articles in *Science of the Total Environment* (2016), *Sustainability* (2019), and *Journal of Environmental Economics and Policy* (2021) by using the three-dimension simplex method for relative weights. In addition, I used many assessment indexes in *Sustainability* (2016).

Moreover, detailed analysis of these articles shows that I applied the appropriate methodologies in more general settings:

- Flows and policy measures: I used efficiency in my articles in *Environmental and Resource Economics* (2017) and *Journal of Environmental Economics and Policy* (2021); equity between countries in my article in *Environment, Development and Sustainability* (2018), and equity between sectors in my article in *Sustainability* (2016).
- Stocks and policy measures: I used efficiency in my articles in *Sustainability* (2016), *Sustainability* (2019), *Journal of Environmental Economics and Policy*

Table 5.17 Topics in the empirical and theoretical research articles listed in my curriculum vitae

Art. no	Authors	Year	Journal	Risk	Resources	Pollution	Flows	Stocks	Renewable resources	Non-renewable resources	Air	Water	Land	Technology support	Decision interaction	International	National	Regional	Local
Empirical articles																			
1	Z	1998	EDE		1			1	1						1	1			
2	Z	2010	WRM		1			1				1			1		1	1	
3	Z	2011	EM		1	1						1							1
4	Z	2012	EMA	1	1		1					1							1
5	Z&R	2013	EMS	1	1		1					1						1	
6	Z	2013	NH	1	1		1					1				1			
7	Z&R	2014	JH		1		1					1			1	1			
8	Z	2015	SS		1			1					1		1	1			
9	Z	2016	SUS		1	1	1	1	1	1	1	1					1		
10	Z	2016	AMM		1		1					1					1		
11	Z	2016	STE		1			1	1									1	
12	Z	2016	ASC	1	1		1					1						1	
13	Z	2017	ERE		1	1	1	1	1	1	1	1	1			1			
14	Z	2017	WAT		1		1					1							1
15	Z	2017	MP		1	1							1		1	1			
16	Z	2018	EDS		1	1							1		1	1			
17	Z	2019	SUS		1			1	1	1									1
18	Z&R	2020	DWT			1	1					1		1	1		1		
19	Z	2020	JEEP		1	1		1	1				1		1	1			
20	Z	2020	SUS		1	1							1		1	1			

(continued)

Table 5.17 (continued)

Art. no	Authors	Year	Journal	Risk	Resources	Pollution	Flows	Stocks	Renewable resources	Non-renewable resources	Air	Water	Land	Technology support	Decision interaction	International	National	Regional	Local
21	Z	2020	NP		1	1							1		1	1			
22	Z	2021	NHSS		1	1	1	1	1				1		1	1			
23	Z	2021	OCM		1	1		1	1			1						1	
24	Z	2022	JCP		1														
25	Z	2022	IJES		1	1							1			1			
Theoretical articles																			
26	Z	1998	JEEM			1	1	1			1				1	1			
27	Z	2008	JEM		1			1	1						1	1			

Abbreviations: EDE = Environment and Development Economics, WRM = Water Resources Management, EM = Environmental Management, EMA = Environmental Monitoring and Assessment, EMS = Environmental Management and Software, NH = Natural Hazards, JH = Journal of Hydrology, SS = Sustainability Science, SUS = Sustainability, AMM = Applied Mathematical Modelling, STE = Science of the Total Environment, ASC = Applied Soft Computing, ERE = Environmental and Resource Economics, WAT = Water, MP = Marine Policy, EDS = Environment, Development and Sustainability, DWT = Desalination and Water Treatment, JEEP = Journal of Environmental Economics and Policy, NP = Nature—Palgrave Communications, NHSS = Nature—Humanities and Social Sciences Communications, OCM = Ocean and Coastal Management, JCP = Journal of Cleaner Production, IJES = International Journal of Environmental Studies, JEEM = Journal of Environmental Economics and Management, JEM = Journal of Environmental Management, Z = Zagonari, Z&R = Zagonari and Rossi. *Notes* None of the papers involved asymmetric information ASY or imperfect competition IMP

Table 5.18 Methodologies in the empirical and theoretical research articles listed in my curriculum vitae

Art. No	Countries	Efficiency	Equity	Cost–benefit analysis	Monetary assessment	Multi-criteria analysis	Relative weight assessment	Time	Space	Uncertainty	Linkages	Inequalities	Participation	Assessment
Empirical articles														
1	DC, LCD		DNS					DYN						
2	BR		N					DYN						
3	DZ					1			SPA					
4	IT					1			SPA					
5	IT					1	AHP			FS				
6	BE, NL			1	CV								1	
7	IQ		K			1	LR							
8	DC, LCD		R											
9	IT	TAX, STA				1	SCE				IO			1
10	IQ					1	LR							
11	IT					1	AHP	DYN/TD	SPA/SD					
12	IT					1	All 4						1	
13	EU	All 7												
14	IQ			1	P*					EV				1
15	BC		N, K, R											
16	DC, LCD		R											
17	IT			1		1	LR	DYN						
18	SA					1	SCE		GIS			IM		
19	DC, LCD	FDI, CBT						DYN						

(continued)

Table 5.18 (continued)

Art. No	Countries	Efficiency	Equity	Cost–benefit analysis	Monetary assessment	Multi-criteria analysis	Relative weight assessment	Time	Space	Uncertainty	Linkages	Inequalities	Participation	Assessment
20	WD		R											
21	WD		H											
22	WD	All 5	1					DYN						
23	IT			1	TC	1	RSP	TD						
24	IT												1	
25	WD		R											
Theoretical articles														
26	DC, LCD		N					DYN						
27	DC, LCD		N					DYN						

Abbreviations: DC = Developed Countries, LCD = Less Developed Countries, BR = Brazil, DZ = Algeria, IT = Italy, BE = Belgium, NL = Netherlands, IQ = Iraq, EU = European Union, BC = Baltic Countries, SA = Saudi Arabia, WD = World, TAX = taxes, STA = standards, All 5 = all main religions discussed in Chap. 2, All 7 = all policy measures discussed in Chap. 4, K = Kalai-Smorodinsky, FDI = foreign direct investment, CBT = cross-border trade, R = Rawls, N = Nash, H = Harshanyi, P* = equilibrium market price, CV = Contingent valuation, TC = travel cost, AHP = analytical hierarchy process, All 4 = all relative weights estimation procedures discussed in Chap. 4, LR = linear regression, RSP = revised Simos' procedure, SCE = Scenario, DYN = dynamic model, TD = time discount, SPA = space discount, FS = fuzzy set, IO = input–output model, IM = inequality measures. *Notes* No study involved population POP and metrics MET; monetary life-cycle analysis and weighted life-cycle analysis are performed in *Sustainability* (2019) and *Ocean and Coastal Management* (2021)

(2021); in contrast, I used equity in my article in *Water Resources Management* (2010), *Sustainability Science* (2015), *Marine Policy* (2017), *Environment, Development and Sustainability* (2018), *Sustainability* (2020), *Nature—Palgrave Communications* (2020), *Nature—Humanities and Social Science Communications* (2021), and *Frontiers in Environmental Science* (2022).

- Flows and investment projects: I used cost–benefit analysis in my articles in *Natural Hazards* (2016) and *Water* (2017); in contrast, I used multi-criteria analysis in my articles in *Environmental Management* (2011), *Environmental Monitoring and Assessment* (2012), *Environmental Management and Software* (2013), the *Journal of Hydrology* (2014), *Applied Mathematical Modelling* (2016), and *Applied Soft Computing* (2016).
- Stocks and investment projects: I used multi-criteria analysis in my articles in *Science of the Total Environment* (2016) and *Desalination and Water Treatment* (2020).

Finally, a detailed analysis of these articles showed that I did not have prejudice against any methodology. Indeed, I adopted both efficiency and equity approaches for policy measures as well as both cost–benefit analysis and multi-criteria analysis for investment projects, depending on the nature of the problem that I studied.

Note that in my article in *Environmental and Resource Economics* (2017), I performed a positive analysis (i.e., the really approved environmental EU Directives, Regulations and Decisions) as opposed to a normative analysis (i.e., the theoretically suggested environmental policies): here, I estimated the relative numbers and dynamics of European Union environmental policies. Moreover, my articles in *Sustainability Science* (2015), *Marine Policy* (2016), *Environment, Development and Sustainability* (2018), *Sustainability* (2020), and *Nature—Palgrave Communications* (2020) relied on the application of a per capita ecological footprint to the current and next generations to transform a dynamic equilibrium into a static equilibrium; that is, I assumed that strong sustainability in terms of inter-generational equity for resources depicts ecological resilience at a global level. Finally, *Ocean and Coastal Management* (2021) used time discounting with a temporal model and travel costs with a range of monetary assessments within cost–benefit analysis for weak sustainability; using these approaches, I compared the obtained decisions with the decisions based on multi-criteria analysis for strong sustainability as the reference methodology.

To summarise this Section, I preferred strong sustainability to weak sustainability in complicated contexts (i.e., asymmetric information, uncertainty, imperfect competition, stocks, or a decision interaction), which explains the larger number of multi-criteria analysis and equity articles compared with cost–benefit analysis and efficiency articles: this supports my previous methodological statements about the suitability of the latter methodologies in a smaller number of settings than the former, although there are no methodologies that should be excluded a priori. In particular, I suggest that researchers avoid:

- disregarding equity in the realistic presence of inefficiency
- disregarding equity when evaluating global issues

- applying marginal and monetary assessment in the presence of social and ecological interdependencies
- using flow assessments when estimating stock issues

Note that I developed many dynamic models when stocks were crucial, both for policy measures (i.e., my articles in *Environment and Development Economics* 1998, *Water Resources Management* 2010, *Journal of Environmental Economics and Policy* 2020, *Nature—Humanities and Social Sciences Communications* 2021) and for investment projects (i.e., my articles in *Science of the Total Environment* 2016 and *Sustainability* 2019).

References

Zagonari, F. (1998a). International pollution problems: Unilateral initiatives by environmental groups in one country. *Journal of Environmental Economics and Management, 3*, 315–326.

Zagonari, F. (1998b). Tropical deforestation: Debt-for-nature vs. debt-for-development swaps. *Environment and Development Economics, 3*, 269–295.

Zagonari, F. (2008). Integrated coastal management: Top-down vs. community-based approaches. *Journal of Environmental Management, 88*, 796–804.

Zagonari, F. (2010). Sustainable, just, equal, and optimal groundwater management strategies to cope with climate change: Insights from Brazil. *Water Resources Management, 24*, 3731–3756.

Zagonari, F. (2011). An optimisation model for integrated urban planning: Development and application to Algeria's Reghaïa and Heraoua municipalities. *Environmental Management, 47*, 937–959.

Zagonari, F. (2012). An optimisation model for Integrated Coastal Management: Development and application to Italy's Comacchio municipality. *Environmental Modelling & Assessment, 18*, 115–133.

Zagonari, F. (2013). Implementing a trans-boundary flood risk management plan: The case of the Scheldt estuary. *Natural Hazards, 6*, 1101–1133.

Zagonari, F. (2015). Technology improvements and value changes for sustainable happiness: A cross-development analytical model. *Sustainability Science, 10*, 687–698.

Zagonari, F. (2016a). A non-probabilistic methodology for reliable sustainability planning: An application to the Iraqi national irrigation system. *Applied Mathematical Modelling, 40*, 10563–10577.

Zagonari, F. (2016b). Choosing among weight-estimation methods for multi-criterion analysis: A case study for the design of multi-purpose offshore platforms. *Applied Soft Computing, 39*, 1–10.

Zagonari, F. (2016c). Four sustainability paradigms for environmental management: a methodological analysis and an empirical study based on 30 Italian industries. *Sustainability, 8*, 504.

Zagonari, F. (2016d). Using ecosystem services in decision-making to support sustainable development: Critiques, model development, a case study, and perspectives. *Science of the Total Environment, 548–549*, 25–32.

Zagonari, F. (2017a). Coherence, Causality, and Effectiveness of the EU environmental policy system: Results of complementary statistical and econometric analyses. *Environmental and Resource Economics, 70*, 1–29.

Zagonari, F. (2017b). Combining econometric, cost-benefit, and financial methodologies in a framework to increase diffusion and to predict the feasibility and sustainability of irrigation schemes: a case study in Kurdistan, Iraq. *Water, 9*, 821.

Zagonari, F. (2017c). Responsibility, inequality, efficiency, and equity in four sustainability paradigms: Policies for a shared sea from a multi-country analytical model. *Marine Policy, 87*, 123–134.

Zagonari, F. (2019a). (Moral) philosophy and (moral) theology can function as (behavioural) science: A methodological framework for interdisciplinary research. *Quality & Quantity, 53*, 3131–3158.

Zagonari, F. (2019b). Multi-criteria, cost-benefit, and life-cycle analyses for decision-making to support responsible. *Sustainable, and Alternative Tourism, Sustainability, 11*, 1038.

Zagonari, F. (2019c). Responsibility, inequality, efficiency, and equity in four sustainability paradigms: Insights for the global environment from a cross-development analytical model. *Environment, Development and Sustainability, 21*, 2733–2772.

Zagonari, F. (2020a). Comparing religious environmental ethics to support efforts to achieve local and global sustainability: empirical insights based on a theoretical framework. *Sustainability, 12*, 2590.

Zagonari, F. (2020b). Environmental sustainability is not worth pursuing unless it is achieved for ethical reasons. *Nature – Palgrave Communications, 6*, 108.

Zagonari, F. (2020c). Foreign direct investment vs. cross-border trade in environmental services with ethical spillovers: A theoretical model based on panel data. *Journal of Environmental Economics and Policy, 10*, 130–154.

Zagonari, F. (2021a). Decommissioning vs. reusing offshore gas platforms within ethical decision-making for sustainable development: theoretical framework with application to the Adriatic Sea. *Ocean and Coastal Management, 199*, 105409.

Zagonari, F. (2021b). Religious and secular ethics offer complementary strategies to achieve environmental sustainability. *Nature – Humanities and Social Sciences Communications, 8*, 124.

Zagonari, F. (2022a). Only religious ethics can help achieve equal burden sharing of global environmental sustainability. *International Journal of Environmental Studies*. https://doi.org/10.1080/00207233.2022.2139559

Zagonari, F. (2022b). Sustainable business models and conflict indicators for sustainable decision making: an application to decommissioning vs. reusing offshore gas platforms. *Journal of Cleaner Production* (under review)

Zagonari, F., & Rossi, C. (2013). A heterogeneous multi-criteria multi-expert decision-support system for scoring combinations of flood mitigation and recovery options. *Environmental Modelling & Software, 49*, 152–165.

Zagonari, F., & Rossi, C. (2014). A Negotiation Support System for disputes between Iraq and Turkey over the Tigris-Euphrates basin. *Journal of Hydrology, 514*, 65–84.

Zagonari, F., & Rossi, C. (2021). A spatial decision support system for optimally locating treatment plants for safe wastewater reuse: An application to Saudi Arabia. *Desalination and Water Treatment, 178*, 1–20.

Chapter 6
Conclusion

This book summarizes the main concepts involved in environmental ethics, sustainability, and the associated decisions, and offers a consistent and transparent sequence for learning about environmental ethics, sustainability, and decisions. In Chap. 2, I presented many environmental ethical rules and ethical reasons, but focused on maximising the average welfare within the teleological (purpose-oriented) approaches (i.e., actions have a goal) and minimising resource inequalities within the deontological (principle-based) approaches (i.e., actions are not based on their consequences). In Chap. 3, I presented many sustainability paradigms, but focused on weak sustainability to maximise welfare and strong sustainability to minimise inequalities. In Chap. 4, I presented two main types of decision (policies and projects), but focused on the decisions based on policies (taxes, standards, subsidies, and permits) and projects (cost–benefit analysis), with the goal being to maximise welfare and to achieve efficiency, and on policies (national laws and regulations, bilateral and multilateral agreements) and projects (multi-criteria analysis) to minimise inequalities and to achieve equity.

My main *assumptions* in this book are that both religious and secular ethics are considered to remain fixed, although they could represent effective strategies to achieve local and global sustainability.[1]

The main (positive) *achievements* of this book from an *education* perspective are:

- It provides summary figures to intuitively depict ethical issues. For example, if a more economic perspective is adopted, then an increasing per-capita consumption must be coupled with decreasing natural capital (Fig. 3.2 in Chap. 3). Conversely, if a more ecological perspective is adopted, then maintaining the current environmental status might not ensure its resilience (Fig. 3.1 in Chap. 3).
- It provides summary tables for the main environmental ethics (Tables 2.1 and 2.2 in Chap. 2), the main environmental sustainability paradigms (Table 3.1 in Chap. 3), the main contexts for policies to achieve efficiency (Sect. 4.1.1 in Chap. 4), and the main equilibria for policies to achieve equity (Sect. 4.1.2 in Chap. 4).

---

[1] Zagonari (2021).

F. Zagonari, *Environmental Ethics, Sustainability and Decisions*,
https://doi.org/10.1007/978-3-031-21182-9_6

- It provides the simplest theoretical frameworks and mathematical models to explain some methodological mistakes. For example, if stock or interaction issues are relevant, then equilibrium market prices are different from both marginal utilities and shadow prices (see the figures and proofs in the Remarks of Chap. 4). Similarly, if asymmetric information or uncertainty or imperfect competition are relevant, then policies to achieve efficiency are not equivalent (see the figures and proofs in Sect. 4.1.1 of Chap. 4).
- It provides simpler versions of complex methodologies to deal with complicated contexts (time, space, uncertainty, linkages, inequalities) both within cost–benefit analysis and within multi-criteria analysis. This will help readers appreciate some of the key methodological concerns. For example, if linkages are relevant, then input–output models should not be applied within cost–benefit analysis, and a computable general equilibrium model should be used instead. Similarly, if space is relevant, then geographical information systems should not be applied within cost–benefit analysis, and a spatial discounting method should be used instead.

Note that, at different education levels (i.e., from elementary pupils to university students) and at different environmental scopes (i.e., from local to global problems), the main alternative approaches to education for environmental responsibility can be summarised in: (i) education about the environment (i.e., here sustainability), (ii) education through the environment (i.e., here exercises on ethics and sustainability), and (iii) education for the environment (i.e., here secular and religious ethics). In particular, (i) refers to theoretical knowledge about human/nature relationships (i.e., both formal concepts or models and informal secular principles/religious precepts or images),[2] (ii) refers to emotional experience from human/nature relationships (i.e., both action and awareness),[3] and (iii) refers to pro-environmental behaviours (i.e., both habits and decisions).[4] However, knowledge about human/nature relationships depends on which science is considered, but (i) is not enough for (iii), since sustainability is an ethical issue. Moreover, there no empirical literature about different impacts on different behaviours to assess whether (i) is more or less effective than (ii) to achieve (iii): a global environmental ethics is impossible, but unnecessary. Finally, decisions should be preferred to habit in (iii), since human/nature relationships change over time.

The main (positive) *achievements* of this book from a *research* perspective are:

- It provides an assessment of important mistakes and concerns in the sustainability literature. In particular, there are few significant concerns, apart from the use of geographical information systems and input–output models within cost–benefit analysis. In contrast, the lack of focus on equity if asymmetric information or uncertainty or imperfect competition are relevant (MIS1) is overall medium (i.e., it is significant in 20% of the cited references) and constant (i.e., it increases with $p > 0.90$). The lack of focus on equity if interaction is relevant (MIS2) is

[2] Payne (2020).

[3] Rousell (2020).

[4] Kopnina (2020).

overall medium (i.e., it is significant in 20% of the cited references) and constant (i.e., it increases with $p > 0.40$). The improper use of utility approaches or some production approaches to the monetary assessment within cost–benefit analysis (MIS3) is overall large (i.e., it is significant in 30% of the cited references) but significantly decreasing (i.e., it decreases with $p < 0.10$). Last, the improper use of subjective methods (SM) for assessment of relative weights within multi-criteria analysis (MIS4) is overall small (i.e., it is significant in 10% of the cited references) but significantly increasing (i.e., it increases with $p < 0.10$).

- All four mistakes that I describe in this book depend on journals. Indeed, panel data probit estimations are better than pooled estimations in the most popular 15 journals. Moreover, there is *no* convergence of journals towards minimizing the types of mistakes that I discuss in this book. Indeed, journals that publish papers with mistakes to a greater extent also show an increasing frequency of mistakes, whereas journals that publish papers with mistakes to a smaller extent also show a decreasing frequency of mistakes. Finally, in terms of mistakes, journals are distinguished into ecological economics (i.e., *Ecological Economics, Environmental Management, Land Use Policy*), environmental economics (i.e., *Environmental and Resource Economics, Climate Policy, Journal of Environmental Management, Environmental Science and Policy, Environment, Development and Sustainability, Journal of Cleaner Production*), and other sustainability journals (i.e., *Water Resources Management, Science of the Total Environment, Environmental Monitoring and Assessment, Sustainability, Water, Environmental Policy*), although *Ecological Economics* and *Environmental and Resource Economics* are similar in terms of incorrectness score.
- It provides an overlook of environmental issues, policies, and methodologies and how they have changed since the late 1980s. In particular, articles focus more on policies than on projects and analyse stocks more than flows. The most analysed environmental issues are climate change and biodiversity loss, with interactions mainly analysed for climate change and trans boundary rivers. Articles focus more on efficiency than on equity, although equity is an increasing concern. Authors apply cost–benefit analysis more than multi-criteria analysis, although the latter is an increasingly popular methodology. Taxes and subsidies within market-based policies to achieve efficiency are increasingly often analysed, whereas the focus on permits is decreasing. The use of standards and harvest rights to achieve efficiency within command-and-control policies is increasingly often analysed, whereas the focus on ecological policies is increasing. International agreements on both pollution and resources are mainly analysed in the years immediately after their introduction. The analytical hierarchy process has become more popular than linear regression among the subjective methods in multi-criteria analysis, whereas contingent valuation and choice experiments are decreasingly and increasingly popular, respectively, among the utility approaches in cost–benefit analysis.
- There are some significant differences among environmental issues in the types of mistakes. In particular, MIS1 is significant for solid waste management, MIS2 for climate change and groundwater management, MIS3 for biodiversity loss, and MIS4 for solid waste management and groundwater management.

Note that, for researchers, I have retained journal names in the text, mainly for seminal research articles, since these will be quoted in many future research articles in the literature. However, for students, I have provided an overall framework, mainly from the perspective of recent review articles, to systematise important articles that have received insufficient attention and the future research articles that should be written for each sustainability topic.

The book's main normative *insights* from an *education* perspective are:

- Students will know that environmental sustainability is an ethical issue, not only because it is not worth pursuing unless it is achieved for ethical reasons, but also because achieving sustainability requires the selection of metrics, objectives, and policies to be implemented or methodologies to be applied (Chap. 4), which are necessarily based on *ethical choices*. For example, in Chap. 2, I compare the maximisation of welfare versus the minimisation of inequalities, with a focus on population. In Chap. 3, I compare efficiency in weak sustainability vs. equity in strong sustainability, with a focus on metrics.
- Although environmental ethics can be instrumental to sustainability, and could be chosen because of their potential to achieve sustainability, the assumptions and their implications will be *transparent* to students. For example, in Chap. 2, I compare the average or total approach vs. the individual approach. In Chap. 3, I discuss different forms of capital in weak sustainability as substitutive rather than complementary, and I compare current vs. past environmental status as a reference in strong sustainability. I consider a lack of intra-generational equity. I discuss absolute decoupling in weak sustainability to achieve ecological continuity, and I focus on capital other than natural capital to achieve social continuity in strong sustainability. In Chap. 4, I compare market-based vs. command-and-control policies and market-based vs. command-and-control equity criteria.
- Whatever the chosen environmental ethic, students will appreciate the *consistency* of the policies to be implemented and assessed as well as the methodologies to be used in evaluating projects. For example, in Chap. 4, I discuss market-based policies (e.g., taxes, subsidies, permits) and agreements (e.g., the Paris Agreement) to achieve efficiency, and command-and-control policies (e.g., standards, nature conservation, exploitation rights) and agreements (e.g., the Kyoto Protocol) to achieve equity. I discuss cost–benefit analysis to achieve efficiency (e.g., time-based discounting, spatial discounting, expected utility or capital asset pricing models, the computable general equilibrium model, social welfare functions) and multi-criteria analysis to achieve equity (e.g., temporal and geographical information systems, expected values, fuzzy sets, social accounting matrices, inequality weights).

Note that I discuss ethical choices most in Chap. 2, with decreasing emphasis in subsequent Chapters, and I deal with transparency issues mainly in Chap. 3, and I discuss consistency issues most in Chap. 4, with decreasing emphasis in preceding Chapters. Moreover, consistency of methodologies is not a hypothetical matter (i.e., it is a real-world concern), since decision-makers who are elected because of their promise to achieve specified environmental goals could allocate public funds based on unreliable assumptions (e.g., to environmental technologies based on absolute decoupling). Finally, transparency of assumptions is not a hypothetical matter (i.e., it is a real-world concern), since the general population could support alternative decision-makers who propose environmental goals based on alternative environmental visions (e.g., a greater emphasis on inter-generational equity than on intra-generational equity).

The book's main normative *insights* from a *research* perspective are:

- Things to avoid in principle: applying marginal and monetary assessment in a context of social and ecological interdependencies; disregarding equity in the case of environmental interactions; disregarding equity in the case of economic inefficiency; and using monetary or non-monetary flow assessments to cope with environmental stock issues.
- Things to avoid in practice: if environmental stock problems exist, then market-based policies are inadequate to maximise welfare within weak sustainability and utility approaches such as contingent valuation and choice experiments are inadequate to estimate monetary assessments in the long-run within cost–benefit analysis.
- Things to do in principle: set priorities, constraints, assumptions, objectives, scales, metrics and indicators, *then* choose policy instruments or decision-making tools to measure and implement sustainability.
- Things to do in practice: If inefficiency or interactions exist, then focus on policies or projects that will improve equity; if environmental stock problems exist, then develop dynamic models to depict both policies and projects; if environmental flow problems exist, then market-based policies should be suggested within weak sustainability (to maximise welfare) and command-and-control policies should be suggested within strong sustainability (to minimise intra-generational inequality); if environmental stock problems exist, then command-and-control policies should be suggested within strong sustainability (to minimise inter-generational inequality), by applying the precautionary principle; if short-run environmental flow problems exist, then the analytical hierarchy process or linear regression should be used to estimate relative weights within multi-criteria analysis or contingent valuation and choice experiments should be used to estimate monetary assessments within cost–benefit analysis by involving stakeholders to increase participation; and if long-run environmental stock problems exist, then a dynamic model within multi-criteria analysis should be combined with the estimation of relative weights of economic, social, and environmental factors by involving stakeholders, with sensitivity analyses for the decisions with respect to relative weights based on the based on a three-dimensional simplex.

Note that the choice of some indicators can simplify the analytical context. For example, the ecological footprint can be used at the long-run equilibrium level, consistently with strong sustainability, to depict a dynamic model with infinite time as a dynamic model with two times (i.e., the current status and the steady status), although ecological continuity is only one aspect of strong sustainability.

Moreover,[5] showed that environmental sustainability is not worth pursuing unless is achieved for ethical reasons (sustainability is a matter of ethical coordination within a deontological approach which could *then* support achievements within a teleological approach). In other words, environmental sustainability is an institutional problem, in which institutions shape the rules of interactions (conventions, norms, and legal rules) and institutions aggregate the values of societies (communication, treatment and representation of complexity[6];). However, referring to the market as the only institution, as is the case in many economic articles, can be misleading in terms of the compromise value provided by the equilibrium market price whenever stocks (long-run values that depend on short-run flow values) and interactions (some humans depend on other humans and vice-versa) are relevant. In other words, even if preferences are fixed and institutions are constraints, markets cannot depict a proper compromise for non-use values[7] and for non-commensurable values,[8] while aggregation and distributional issues are disregarded. Consequently, weak sustainability based on market evaluations of nature and of externalities could lead to the wrong decision, even though this approach is based on high participation. Similarly, strong sustainability, which adopts the reference condition for resilience, could be more reliable, although it is based on low participation, apart from referendums on sustainability issues. In other words, the institutions affecting environmental decisions are crucial components of any sustainability paradigms, *regardless of the labels* attached to given areas of research.[9] For example,[10] represents institutions in a Nash bargaining dynamic equilibrium based on the relative importance of stakeholders within the Brazilian groundwater regional committees; similarly,[11] depicts how religious and secular ethics affect sustainability within an evolutionary model based on alternative representative individuals.

Finally, the introduction of ecological constraints within a weak sustainability paradigm (e.g., tipping points, uncertainties, resilience) is misleading. Indeed, this is an ad hoc modification of a framework, since the results are unpopular; it is therefore a way to hide the ethical choices behind the weak sustainability paradigm.

The benefits of this book (positive results) for students, in the form of simple and intuitive explanations of the main ethical and methodological issues, are less useful for researchers, and the benefits of this book (normative insights) for researchers, in the form of things that should be done and that should be avoided in research, both in principle and in practice, are less useful for students. Nonetheless, the normative insights remain useful for students, particularly in terms of sustainability as an ethical issue, the need for transparency of assumptions, and consistency of policies and methodologies. Similarly, researchers will benefit from the statistical analyses

[5] Zagonari (2020).

[6] Vatn (2009).

[7] Stoeckl et al. (2018).

[8] Groeneveld (2020).

[9] Spash (2021).

[10] Zagonari (2010).

[11] Zagonari (2021).

on the significance of the main mistakes and concerns as well as changes in the main topics and methodologies in the sustainability literature over time and across journals. These benefits for both groups of readers link the book's education and research perspectives. Indeed, by building on specific methodological problems, I show that the two main research groups in the literature are deficient. In particular, the orthodox literature, which represents neo-classical environmental economics, lacks *transparency* to the general population: for example, it fails to declare that the positive social discount rate applied to solve many models means disregarding future generations, and it fails to show that biodiversity could be optimally reduced whenever this is supported by human assessment of ecosystem services or production externalities. Conversely, the heterodox literature represented by ecological economics lacks consistency to sustainability experts outside their research movement: for example, it uncritically uses cost–benefit analysis by pricing externalities; and it adopts mathematical models of optimising human behaviour with humans treated as maximisers of self-interest utility by assuming unrealistic microeconomic axioms.

Next, by referring to the simplest possible models and methodologies, I show that some issues must be *transparently* and *consistently* tackled to properly discuss sustainability issues. For example, under the assumption of commensurability, shadow prices (primarily meant to implement efficiency) are different from marginal utilities whenever stocks (for both pollution and resources) or interactions (between humans) are relevant. Similarly, efficiency might not be achieved where any assumption behind the economic general equilibrium model is not met, while inter- and intra-generational equity is disregarded.[12]

In other words, the main methodological issues explained to students (i.e., unmet assumptions behind the economic general equilibrium model such as information asymmetry, uncertainty, and imperfect competition for policies; the existence of stocks and interactions for both policies and projects; the improper application of methodologies to projects) are enough to *sort out* the sustainability literature in terms of consistency and transparency based on descriptive and estimative statistical analyses. Indeed, by estimating the incorrectness frontier based on a Data Envelopment Analysis, I empirically achieved the same insights suggested in the methodological literature[13]: environmental economics and ecological economics show more similarities than differences.

Note that information asymmetry, uncertainty, and imperfect competition are enough to make policies fail to achieve efficiency, although additional environmental stocks or interactions further complicate this issue. Moreover, environmental interactions require equity, although considering additional stocks makes this issue even harder. Finally, both stocks and interactions can make equilibrium prices more misleading in monetary assessments within cost–benefit analysis. In other words, for the benefit of students, I referred to simple cases to present the main methodological issues.

---

[12] Munda (2014).

[13] Spash and Ryan (2012).

However, both *consistency* based on using methodologies that are logically adequate to a study's stated objectives and *transparency* based on clearly declaring the foundations of the study's results are ethical matters. Thus, sustainability scientists will find an incentive to be transparent and consistent, if the general population can understand the assumptions they rely on and the methodologies they apply. For example, environmental economists must declare their assumptions, whereas ecological economists must declare their true objective or modify the applied methodologies. However, consistency and transparency have costs for both students and researchers: students must spend some time understanding technicalities such as how methodologies can be applied and which assumptions are implicit in those methods, whereas researchers may deal with rejections of their papers if they adopt consistent and transparent procedures, but they use non-conventional methodologies or assume non-standard contexts.

Therefore, it is acceptable that non-humans are neglected, with a teleological approach based on a representative individual perspective. Actually, this book provides many reasons for adopting weak sustainability instead of strong sustainability, with a deontological approach based on a per-capita individual perspective. However, readers should remember the assumptions and procedures that are required to justify this conclusion. For example, within weak sustainability, which is a pure economic model, subject to the assumption of inter-generational equity, the objective is more ambitious: to maximise welfare under the assumption of *individual* (unbounded) rationality embodied in the market as the *only* institution. There is also higher participation in the case of markets (a global approach based on a representative individual), but nature is evaluated by humans and its assessment is inadequate for many environmental issues. In contrast, within strong sustainability, which is a pure ecological model, subject to the assumption of the reference status being resilient, the objective is less ambitious: to minimise impacts under the assumption of *social* (ethical) rationality embodied in the many *fixed* institutions. There is also lower participation (the local approach is based on single individuals), apart from general referendums on environmental issues, but humans account for natural dynamics and their assessment is adequate for many environmental issues.

References

Groeneveld, R. A. (2020). Welfare economics and wicked problems in coastala and marine governance. *Marine Policy, 117*(193945).

Kopnina, H. (2020). Education for the future? Critical evaluation of education for sustainable development goals. *Journal of Environmental Education, 51*, 280–291.

Munda, G. (2014). On the use of shadow prices for sustainable well-being measurement. *Social Indicators Research, 118*(2), 911–918.

Payne, P.G. (2020). Amnesia of the moment in environmental education. *Journal of Environmental Education, 51*, 113–143.

Rousell, D. (2020). Doing little justices: Speculative propositions for an immanent environmental ethics. *Environmental Education Research, 26*, 1391–1405.

Spash, C. L. (2021). A reply to Levrel and Martinet. *Ecological Economics, 179*(106695).

Spash, C. L., & Ryan, A. (2012). Economic schools of thought on the environment: Investigating unity and division. *Cambridge Journal of Economics, 36*, 1091–1121.

Stoeckl, N., et al. (2018). The crowding out of complex social goods. *Ecological Economics, 144*, 65–72.

Vatn, A. (2009). An institutional analysis of methods for environmental appraisal. *Ecological Economics, 68*, 2207–2215.

Zagonari, F. (2020). Environmental sustainability is not worth pursuing unless it is achieved for ethical reasons. *Nature—Palgrave Communications, 6*(108).

Zagonari, F. (2021). Religious and secular ethics offer complementary strategies to achieve environmental sustainability. *Nature—Humanities and Social Sciences Communications, 8*(124).

Zagonari, F. (2010). Sustainable, just, equal, and optimal groundwater management strategies to cope with climate change: Insights from Brazil. *Water Resources Management, 24*, 3731–3756.